T0200739

Paradox and Paraconsistency
Conflict Resolution in the Abstract Sciences

In a world plagued by disagreement and conflict, one might expect that the exact sciences of logic and mathematics would provide a safe harbor. In fact, these disciplines are rife with internal divisions between different, often incompatible systems. Do these disagreements admit of resolution? Can such resolution be achieved without disturbing assumptions that the theorems of logic and mathematics state objective truths about the real world?

In this original and historically rich book, John Woods explores apparently intractable disagreements in logic and the foundations of mathematics and sets out conflict resolution strategies that evade or disarm these stalemates. Among the conflicts to which these strategies are applied are: the disagreement between classical and relevant logicians; Quine's attack on quantified modal logic; disagreement as to whether dialethic logic has an adequate motivation; conflicts about how to understand the paradox of sets and the Liar paradox; the vexed relationship between modern logic and theories of reasoning; and conditions under which logical laws exhibit normative force.

An important subtheme of the book is the extent to which pluralism in logic and the philosophy of mathematics undermines realist assumptions. Woods's response is an account of truth in which realism is an irresistible epiphenomenon.

This book makes an important contribution to such areas of philosophy as logic, philosophy of language, and argumentation theory, but it also will be of interest to mathematicians and computer scientists.

John Woods is Director of the Abductive Systems Group, University of British Columbia.

Paradox and Paraconsistency

Conflict Resolution in the Abstract Sciences

JOHN WOODS

University of British Columbia

CAMBRIDGE
UNIVERSITY PRESS

PUBLISHED BY THE PRESS SYNDICATE OF THE UNIVERSITY OF CAMBRIDGE
The Pitt Building, Trumpington Street, Cambridge, United Kingdom

CAMBRIDGE UNIVERSITY PRESS
The Edinburgh Building, Cambridge CB2 2RU, UK
40 West 20th Street, New York, NY 10011-4211, USA
477 Williamstown Road, Port Melbourne, VIC 3207, Australia
Ruiz de Alarcón 13, 28014 Madrid, Spain
Dock House, The Waterfront, Cape Town 8001, South Africa

http://www.cambridge.org

© John Woods 2003

This book is in copyright. Subject to statutory exception
and to the provisions of relevant collective licensing agreements,
no reproduction of any part may take place without
the written permission of Cambridge University Press.

First published 2003

Printed in the United Kingdom at the University Press, Cambridge

Typeface Times Ten 9.75/12 pt.   *System* LATEX 2_ε [TB]

A catalog record for this book is available from the British Library.

Library of Congress Cataloging in Publication Data
Woods, John.
Paradox and paraconsistency: conflict resolution in the abstract sciences / John Woods.
p. cm.
Includes bibliographical references and index.
ISBN 0-521-81094-9 – ISBN 0-521-00934-0 (pb.)
1. Logic, Symbolic and mathematical.   2. Pluralism.   I. Title.
QA9 .W749   2002
511.3–dc21   2002067744

ISBN 0 521 81094 9 hardback
ISBN 0 521 00934 0 paperback

In Memory of
Richard Sylvan (né Routley), 1936–1996.
Rob Grootendorst, 1944–2000.

Contents

Preface

This work arises from a series of lectures on paraconsistent logic delivered at the University of Groningen in the spring term of 1988. There followed a year later a schedule of lectures on Quine's philosophy of logic. The fruits of these endeavors circulated for awhile as *The Groningen Lectures on Paraconsistent Logic*. My efforts were graced by excellent students and generous colleagues. I am especially grateful to E. M. Barth, Jeanne Peijnenberg, Erik C. W. Krabbe, and David Atkinson for sharp criticism and helpful support. In 1990, a Fellowship at the Netherlands Institute for Advanced Study made it possible for me to join the research group on Fallacies as Violations of Rules for Argumentative Discourse. I worked there on conflict resolution strategies for intractable disagreements in questions of public policy. Only toward the end of my stay in Wassenaar did it occur to me that such strategies might be extended to contentious issues in the philosophy of logic and related fields. I owe much to the stimulation and encouragement of my NIAS colleagues: project leader Frans H. van Eemeren, the late Rob Grootendorst, Sally Jackson, Scott Jacobs, Agnès van Rees, Agnes Verbeist, Douglas Walton, and Charles Willard.

Thus was born a preoccupation with conflict resolution in the abstract sciences, which became the main business of my University of Lethbridge course on Deviant Logic in 1991 and 1992. The distractions of other research projects and heavy administrative responsibilities entailed a postponement of this one until an appointment as Visiting Scholar in the Department of Philosophy at Stanford University in 1994 afforded me the stimulation and leisure to turn my mind again to conflict resolution. In this I was much helped by Michael Bratman, Chair of the Department, Johan van Benthem, Maurice Finocchiaro, David Grover, and Timothy Schroeder. Administrative duties necessitated a further pause, but the project came to life again and was completed thanks to a Visiting Professorship in the Department of Discourse Analysis, Argumentation Theory, and Rhetoric at the University of Amsterdam in the spring term of 1998 and 1999. I am indebted to the Department's Head, Frans H. van Eemeren, and to his (and my) colleagues, the late Rob Grootendorst,

Francisca Snoeck Henkemans, Eveline Feteris, Peter Houtlooser, and Bart
Garssen. For helpful correspondence I am also most grateful to Anil Gupta,
Patrick Suppes, Dov Gabbay, Graham Priest, Kit Fine, and Julius Moravcsik,
and the late Richard Sylvan.

Thanks, too, for the support of the Social Sciences and Humanities Research
Council of Canada, the University of Lethbridge Research Fund, and the Dean
of Arts and Science of the University of Lethbridge, Professor Bhagwan Dua.
Through his efforts, it was possible to appoint as research assistants Jasminn
Berteotti, Dawn Collins, Ethan Toombs, and David Graham, whose talent
and interest were of considerable help. For technical assistance, I am indebted
to Randa Stone and Dawn Collins in Lethbridge, and Willy van der Pol in
Amsterdam. I would also like to thank my editor, Terence Moore, and my
copy editor, Laura Lawrie.

My special thanks are reserved for my students in Deviant Logic over the
years, but especially David Grover, Augustus Butterfield, Kevin Gaudet, James
Hormoth, James Brown, Maurice Lam, James King, Brian Hepburn, and Jack
Kwong.

The Prologue of this book is an expansion of my "Just How Stupid Is
Postmodernism?" in D. M. Gabbay et al. (eds.), *Springer Lecture Notes in
Artificial Intelligence: Quantitative and Qualitative Practical Reasoning*, Berlin:
Springer-Verlag, 1997, 154–8. Small parts of my "Pluralism About Logical
Consequence," in John Woods and Bryson Brown (eds.), *Logical Consequence:
Rival Approaches*, Oxford: Hermes Science Publishers, 2001, show up in Chapters 1, 2, 3, and 4. A small portion of Chapter 1 is adapted from my article "Aristotle" in the File of Fallacies section of *Argumentation*, 13 (1999),
203–20. Chapter 2 absorbs four pages from Chapter 6 of my *Aristotle's Earlier Logic*, Oxford: Hermes Science, 2001. Brief parts of Chapter 3 are taken
from my "Ideals of Rationality in Dialogues," *Argumentation*, 2 (1980), 395–
408 and "The Relevance of Relevant Logic," in J. Norman and R. Sylvan
(eds.), *Directions in Relevant Logics*, Dordrecht: Kluwer Academic Publishers,
1989, 77–86. Chapter 4, "Semantic Intuitions," is an extensive revision of a
paper of the same name in Johan van Benthem, Frans H. van Eemeren, Rob
Grootendorst, and Frank Veltman (eds.), *Logic and Argumentation*, Amsterdam: North Holland, 1996, 177–208. Chapter 6 adapts some material from my
"Fortress Fiction," in C. Mihailescu et al. (eds.), *Fiction Updated: The Theory
of Fictionality and Contemporary Humanities*, Toronto: University of Toronto
Press, 1996, 39–47. I am grateful to all concerned for permission to use this
material.

Prologue

Postmodern Logic

Your discovery of the contradiction caused me the greatest surprise and, I would say, consternation.... It is all the more serious since, with the loss of my Rule V, not only the foundations of my arithmetic, but also the sole possible foundations of arithmetic, seem to vanish.

Gottlob Frege, "Letter to Russell," 1902.

The abstract sciences are those that cannot, and have no need to, negotiate the empirical check. This anyhow is a widely received view of the matter. An abiding question for such theories is this: What sorts of check *can* they negotiate, and does doing so preserve intuitive presumptions of objectivity and realism? There is a particularly vivid context for posing this question and reflecting on how it might be answered. The context is that of *conflict resolution* strategies for rival theories.

In a broadly accepted use, with which I concur, objectivity attaches to things when they exist apart from and antecedently to anyone's thought of them; and objectivity attaches to statements or beliefs when they are true, or false, apart from and antecedently to anyone's conceiving of them as so. Realism in turn is always realism about something – about abstract objects, about universals, about material things, and so on. The realisms that absorb us in this book are those that attribute this twofold objectivity to what I am calling abstract theories when they meet certain properly understood conditions of adequacy.

Two historical developments in the last century suggest a not always tacit acquiescence to the suggestions that objectivity and realism are unrealizable and unrealistic targets for even our most methodologically austere and successful abstract theories. One of these developments is a tolerant and substantial *pluralism* that has taken root and flourished in logical theory. This pluralism relates significantly to the toleration of it. The greater the latter, the more the former does damage to presumptions of objectivity and realism. The greater the latter, the greater the likelihood that theoretical rivalries will be interpreted in such ways that conflict resolution does not matter – or even that it would somehow be a misplaced thing to try to bring off. The other historical development is what could be called the *received view* of the significance of the paradoxes of sets and of truth. The received view concurs widely on the diagnosis of the paradoxes, on estimates of the damage done by them, and on the general character of strategies for set theoretic and semantic recovery. This, too – or so I shall say – puts in a false light objectivity and realism in mathematics and formal semantics.

If we wished to draw dramatic attention to these developments, we could first remind ourselves of the buoyant confidence of 1879, and for a time thereafter, in the imperiousness and canonicity of the new logic, the new mathematics, and what came to be known as "analytic" philosophy. We could then reflect upon their subsequent apparent collapse into the unedifying embrace of *postmodernism*.

Postmodern logic? The very idea! Yes, the very idea; we should not shirk it. As Hintikka sees it, "the main post-Gödelian, not to say *postmodern*, foundation problem is to look for new deductive methods and to analyze them ([Hintikka, 1996, p. 99] emphasis added). And, "[a]mong other features of this [= Hintikka's] concept of negation that have to be *deconstructed*, is the so-called law of excluded middle" ([Hintikka, 1996, p. 161] emphasis added).

The paradoxes play on our reflections in ways that are decidedly queer. Seen in Russell's way, they drive us towards a kind of idealism, a detested thing in twentieth-century English-speaking philosophy; and no wonder inasmuch as it appears to land us in the swamps of postmodernism. I see postmodernism in Eagleton's way; as

a style of thought which is suspicious of classical notions of truth, reason, identity and objectivity, of the idea of universal progress or emancipation or single frameworks, grand narratives or ultimate grounds of explanation. (Eagleton, 1996, p. viii)

Indeed,

[a]gainst these Enlightenment norms, it sees the world as contingent, ungrounded, diverse, unstable, indeterminate, a set of disunified cultures or interpretations which breed a degree of scepticism about the objectivity of truth, history and norms, the givenness of nature and the coherence of identities. (Eagleton, 1996, p. vii)

It will occur to some people that what used to be called the "moral sciences" are in a state that is tailor-made for postmodern summing up, although the very word for it is historically careless (there is not a trait in the postmodern catalog that was not abundantly evidenced in antiquity, with periodic recurrence ever since). Perhaps we should not find it so striking if there was no fact of the matter about Ophelia's acquiescent sexuality, or about what final moral interpretation a given body of data calls for, or about whether or not the Id exists. Postmodern latitude is, if anything else, recognition of the slack that attends our softly scientific judgments. This is postmodernism cheaply bought, and I for one do not think much of it. Should it indeed be the case that the soft sciences have nowhere to go but postmodern, the harder sciences are a harder sell; and it is to them we should turn our attention. It would be a discovery worth making a fuss over if we could show that postmodernism reposes in the very coils of the hard, which is to say not in the history of science (which is soft), but in science's essential methods and settled practice.

Cases in point are the mathematical theory of sets and truth conditional semantics for constative discourse. Each is beset with paradox. With sets, it was the paradox that Russell communicated to Frege in 1902. In truth conditional semantics, it was the Tarski Paradox. Russell, and Frege, too, thought that his paradox destroyed the concept of set (Russell, 1967, pp. 127–8); and Tarski thought that his paradox destroyed the concept of the statement, that is to say, of bivalent sentence (Tarski, 1983, pp. 152–4, 165). However, let me put it on the record early rather than late that Tarski made much too light of these damaging consequences. We shall return to this oddity in Chapters 5 and 7.

On the view that it destroyed the concept of truth or the concept of biva-lent sentence, the Tarski Paradox *is* a devastation. If it destroys the concept of statement, then there is no concept of statement, and *there can be no state-ments*. Even if Tarski were to reconsider his commitment to the nonexistence of the concept of statement, his logical classicism binds him to absolute incon-sistency – every statement of any natural language is true. The first alternative sounds the death-knell of constative discourse; we lack the means even to *try* to say what is the case. The second alternative guarantees an alethic liber-tinism that amounts to nihilism – all that is is precisely what is not, provided that Convention T is true, provided, that is, that truth is disquotational. Even if T did not obtain, statement-making discourse, while not impossible, would be dispossessed of any rationale, since everything anyone ever says is always both true and not.

The first is the greater problem. If the Tarski Paradox demonstrates the impossibility of statements, of constative discourse as such, then it cannot be the case that beliefs have propositional contents. It cannot be that my belief that the cat is on the mat bears any relation at all to anything identifiable as what is stated by the sentence "The cat is on the mat," since neither that sentence nor any other states *anything*. This problem about belief comes to the fore in a rather pressing way when we consider Tarski's method of recovery from the Paradox. If, as I am assuming, Tarski understood his strategy in the same way that Russell understood his own as regards the paradox of sets, then we run into vexations of considerable significance. Russell began his work in the foundations of mathematics as an idealist. It is a commonplace of Russell's way of being an idealist that our ordinary concepts – the concept of space, for example – are inconsistent. The job of the theorist therefore is to repair the concept, to refine the inconsistency out of it. Doing so is subject to what we might call "the principle of consistent similarity," which bids the theorist to make his new concept as similar to the original as consistency allows.

By 1903, with the publication of *Principles of Mathematics*, Russell had abandoned his idealism for something called "analysis," to which he was drawn by the forceful ministrations of G. E. Moore. It is on this analytical perspective that inconsistent concepts do not exist, and, since nonexistent, nothing whatever falls under them. It becomes quickly evident that the idealist strategy for repairing inconsistent concepts cannot be applied when concepts are understood in the analytic way. The idealist strategy requires that the new concept resemble the old as much as consistency allows. But on the analytic approach to concepts, there *is* no original concept. Anything proposed as the successor concept must, on the principle of consistent similarity, resemble – as much as consistency allows – *nothing whatever*. From which we have it either no successor concept satisfies the principle, or that every consistent concept whatever satisfies it, and satisfies it equally.

After much dissembling, Russell did the only thing he could do short of giving up, which is what Frege eventually did. *He stipulated*. Sets were now introduced by nominal definitions, which Russell dressed up as "*mathematical analyses.*" Russell knew as well as anyone ever did that whereas one is free to stipulate as one pleases, no one else is required to bear it any mind. So he imposed a condition governing what would count as acceptable stipulations in mathematics. A stipulation is acceptable to the extent that the right people are disposed to believe it. Thus, someone stipulates that *p*, and perhaps in time the community of *p*-enquirers come to believe it. If so, the stipulation is acceptable.

I need hardly dwell on the postmodernist skeins with which Russell's recovery of set theory is shot through: There are no facts of the matter about sets; sets are a human construct; how sets are is relative to what people are prepared to believe about them; sets are patches of consensus in the mathematical conversation of mankind; and so on.

Can Russell's strategy for recovery be applied to the devastation of the Tarski Paradox? Recall that Russell's strategy is stipulation supported by elite communal belief. If Tarski's Paradox establishes the impossibility of statements, and beliefs are propositional attitudes – psychological states in some kind of apposition to statements – *then there are no beliefs either*, and Russell's strategy fails for sets and statements alike. If it shows anything, Tarski's Paradox establishes that the cost of persisting with the analyst's conception of concepts is the death of discourse, belief, and desire (since it, too, is a propositional attitude).

It is a striking peculiarity of the received view that the utterly radical thrust of the consequences of the Tarski paradox are not much noticed, and certainly not much bothered with. Tarski himself just got on with the business of finding a formalized language suitable for rigging a successor to the demolished concept of truth. In this respect, the received view in semantics resembles the received (pre-Bohmian) view in quantum mechanics with regard to nonlocality. It, too, was not much bothered with by working physicists. If considered at all, it was considered at most as a fit thing for philosophers, who, as is widely believed by scientists, have nothing *scientifically* consequential to say about physics.

Like realism, idealism is always idealism about something. Like any philosophically big notion, it ranges from the commonplace to the theoretically extreme, and is interesting to the extent that it purports to displace a realism antecedently thought secure. Thus, idealism is comparatively uninteresting, but not uncontested, when it is acknowledgment of the mental dependency of ordinary mental events, and it is interesting and important in, for example, Berkeley's celebrated displacement of the external world. In its use here, idealism is a less radical affair. It sees in knowledge something of the knower's creative contribution; it sees truth as comparative and partial; it sees all thinking, except "metaphysical" thinking, as defective and all concepts save "metaphysically" repaired ones as inconsistent; and it sees knowledge as something less than objective.

Even so, I take it without further ado that the death of discourse, belief, and desire is too much to bear even for "the brilliant young zombies who know all about Foucault ... " (Eagleton, 1996, p. 23). What is to be done? One option, obviously enough, is to revert to this "minimal" idealism and fess up about it. It is well to attend to what the reversion buys us. It buys a way of recovering from paradox. Costs are another thing. Human knowledge, whether in politics or in the foundations of mathematics, is now, in part at least, a human artifact; and knowledge is wrought, one way rather than another, for what it is wanted for. Collectively, the cost of the idealist strategy is the abandonment of realism, of the view that how the world is is independent of what we think of it, and that our beliefs are objectively true or objectively false depending on how the world is apart from what we think of it.

Naturalism offers another way of proceeding, and a more attractive one on its face for those who dislike the postmodern cachet of idealism, if the anachronism may be forgiven. Naturalism offers promise of the recovery of realism. For, unlike the old epistemology, naturalism seeks "no firmer basis for science than science itself" (Quine, 1995, p. 16). The naturalist "is free to use the very fruits to science in investigating its roots" (Quine, 1995, p. 16). It is a self-referential process, as is postmodernism itself, but no mind, since it is "a matter, as always in science, of tackling one problem with the help of our answers to others" (Quine, 1995, p. 16). In the case of sets, the naturalist rejigs not to preserve as much as he can of the old concept but, rather, with a view to facilitating the broader aims of mathematics, broadly indispensable in turn, to science. Similarly, our theory of the external world will be a rational reconstruction from modest beginnings – sets of triggered neural receptors at a specious present; and before long bodies will be sets of quadruples of real numbers in arbitrary coordinate systems. Those liking the naturalist option could do no better than to turn to Quine for instruction, for it is he, more than anyone else, who has given the project a commanding and detailed articulation.

But *caveat emptor*; the raw recruit to naturalism may be unprepared for what awaits him there:

Even the notion of a cat, let alone a class or number, is a human artifact, rooted in innate disposition and cultural tradition. The very notion of an object at all, concrete or abstract, is a human contribution, a feature of our inherited apparatus for organizing the amorphous welter of neural input. (Quine, 1992, pp. 687–725)

And,

if we transform the range of objects of our science in *any* one-to-one fashion, by reinterpreting our terms and predicates as applying to new objects instead of the old ones, the entire evidential support of our science will remain undisturbed. (Quine, 1992, p. 8, emphasis added)

Quine, of all our philosophers, is the most French. Consider what he tells us of theories. Theories are pieces of text, sets of sentences having a complex structure, inherently topic neutral, but susceptible to interpretations that are imposed in accordance with what we find it interesting to suppose. They are exercises in our conceptual sovereignty, and stand in complex and convoluted – and dominantly notional – relations to sensory turbulence.[1]

How did the naturalist come to this sorry pass? And why should we not say that the strongest case ever made for the truth of postmodernism in the hard sciences has been made by him? We can say it if we like, but the irony of it all should not be lost on us (more postmodernism still). The naturalist, like the rest of us, begins his scientific account of our access to the world rooted in the *realist stance*. He assumes that the world is objectively there no thanks to us, and that what we come to know of it is objectively so. Once up and running, whether in the precincts of neurological theories of perception, or in theories of the interior of the atom, or in the foundations of transfinite arithmetic, naturalism makes it clear, over and over again, that our best scientific accounts of how beings like us know the world show that we do so in ways that fulfill the canons of idealism. This is our *anomalous realism*. It provides that when we bring to bear the presumptions of realism on our scientific enquiries into how we know the world, it emerges that enquiry itself is idealist. In this, it seems that we cannot help ourselves. The realist stance delivers the goods for idealism every time, but we cannot make ourselves reject the stance. We cannot help *being* idealists while *thinking* that we are realists. This is what Sartre made much of under the heading of *mauvaise fois* – bad faith. Reactivating the realist stance so as to bring it to bear on the persistent and pervasive phenomenon of *mauvaise fois*, there is little to conclude but that it is naturally selected for, that it is needed for survival.

One of the most recalcitrant travails of postmodernism in the arts and letters, and in the soft sciences, is postmodernism's own bad track record with the question, "What now?" What work is there to do in history or in literary studies if postmodernism is true and faithfully concurred with? If there are no

Archimedian points, it is hard to see what the research program could be. Not seeing where the research program should go is like not knowing where you are. It is the kind of lostness that promotes abandonment – for example, the rejection of literature in some departments of English; or it invites intellectual rubbish, exchanged under hostile dialectical conditions of a kind that Aristotle called "babbling." And it invites – it positively begs for – Sokal's hoax.[2]

On this score, naturalism has the edge, not because it evades postmodern commitments but precisely because it abounds in them. The advantage given to naturalism is that it seizes on its own postmodern consequences and lets them shape a coherent research program.

What is the program? It is to employ the best of what naturalism can offer to explain the persistence of the realist stance even in the face of the pervasive endorsement it gives to antirealism. The project, in short, is a naturalistic explanation of the epistemic *mauvaise fois* of the human condition. And that, anyhow, is something.

With theories, says Quine, ideology is everything and ontology hardly counts (Quine, 1983). What matters is what we make the text *say* and that it be made, in the end, to negotiate the empirical checks, however convolutedly. Anomalous realism is the most fruitful way of proceeding. It is constructive make-believe *par excellence*. If we think realistically about what we make theories say, there is a greater chance that we will think up better theories than otherwise. They are better not because they reveal better what is really so, but because they negotiate the empirical checks more smoothly and efficiently, and as structural consequences of how the text itself was contrived. There are two things, then, to be said for the realist stance. It is a tried and true theoretical heuristic, and it is an economical way of paying attention to what happens around us. It discourages our taking the onrushing bus for a phantom. It is an efficient way of staying alive. Such a view is pure *Boul. Mich.*, although with an Ohio accent.

It lies in the nature of our anomalous realism to dislike anomalous realism, indeed to disbelieve it utterly at the level of practice and as a way in which we find it necessary and natural to experience the world. Even if it is our best option, it is not an option we want. It is therefore appropriate to wonder whether there might be a way out of it. I mean to look for a way in the very precincts in which historically it has grown deep roots. So we shall examine in this book the complex dynamics of conflict resolution in the abstract sciences. To this end, I reassume the realist stance, and shall persist with it until and unless our looming reflections knock me from this perch.

In examining the dialectical structure of conflict resolution in the abstract sciences, I have thought it prudent to select as test cases contentions that are comparatively well known, concerning which readers of this book may already have taken a position. One test of the resolution devices developed here is the extent to which they incline readers to alter their positions, or at least to re-tain them more reflectively. These, then, are the test cases: the rivalry between

relevant and classical logic; the rivalry between paraconsistent logic and classical logic; Quine's attack on quantified modal logic; contentions against the received view of the Russell paradox and the Liar paradox; the realist-antirealist
controversy; contentions against the intuitions methodology in philosophy and
other abstract sciences; and contentions against the normative presumptions
of such theories.

1

Conflict in the Abstract Sciences

> How can a philosophical enquiry be conducted without a perpetual *petitio principii*?
>
> Frank Ramsey, *The Foundations of Mathematics*, 1931

CONFLICT RESOLUTION

1905 was an intellectually eventful year. It saw the birth of Russell's "On Denoting" and Einstein's special theory of relativity, to say nothing of the founding of the Bloomsbury Group and the appearance of *The Psychopathology of Everyday Life* and the Binet Test. Relativity theory was attended by conflict right from the beginning, and barely a year passed before disconforming experimental evidence was unearthed.[1] In one of the century's more alluring examples of a theory's resistance of empirical discouragement, relativity hung on until, in 1914–16, it received experimental confirmation strong enough to annul the Kaufmann deviations.[2] While the new physics was awaiting empirical respectability, the foundations of geometry occasioned considerable contention. Frege and Hilbert saw things differently. They clashed over the nature and function of the geometric axioms. Frege saw the axioms as a reflections of conditions necessary for spatial experience, and so as synthetic propositions known a priori. For Hilbert, axioms are the theoretical constructions of the geometer, epistemically secure if consistent. On Hilbert's view, whether a geometric axiom strikes us as a priori true, or, for that matter, as a priori false, is a fact about us, not about geometry intrinsically. Axiom sets are consistent specifications of mathematically possible spaces, whose physical realization, or not, tells neither for nor against the axioms.

We have here two historically important cases of scientific disagreement in the twentieth century. Anyone interested in the dynamics of conflict resolution in the sciences will at once see the two cases as importantly different. The Einstein-Kaufmann conflict was eventually settled. The Frege-Hilbert conflict just went on and on, and ended without resolution, on Frege's death in 1925.

The conflict resolution theorist is bound to make something of this difference and to offer an account of it. On the face of it, he has not far to go for an answer. Relativity theory triumphed in the end on the strength of its *empirical adequacy*.[3] The dispute between Einstein and Kaufmann was settled by Nature. The intractability of the standoff between Frege and Hilbert is similarly explained, but in the opposite direction, so to speak. In this case, empirical adequacy was not an applicable or appropriate resolution device. Theirs was a dispute with regard to which Nature had nothing to offer.

In a rough and ready way, theories divide into those for which the criterion of empirical adequacy is a legitimate standard, if not always a fulfilled one, and those for which the standard is made inappropriate by subject matter and method. This distinction I mean to mark by saying that theories that are properly held to the condition of empirical adequacy are *empirical theories*, whereas those that are not are *abstract theories*.[4] Rough as it is, our present distinction is consequential in a way that I shall try to take the measure of. Empirical theories have inbuilt procedures for conflict resolution – as with the Einstein-Kaufmann dispute – however complex and indirect they may be. Collectively these mechanisms are a theory's empirical check. Abstract theories, such as the epistemology of geometry, lack these mechanisms for conflict resolution, and it is this that makes them methodologically interesting. Among empirical theorists there is a philosophically naive but utterly entrenched inclination to suppose that a theory's empirical check is also a *reality* check for it; that a theory is objectively right in its claims to the extent that it "checks out" empirically. Abstract theories lack an empirical theory's way of negotiating its reality check.[5] On the face of it, this matters. We are left to ask whether abstract theories have reality checks and, if so, what they are and how we come to recognize them. If not, how can the principles and laws of such theories count as true?

Some readers will not much like the putative dualism of the empirical and abstract. Perhaps these skeptics will have been persuaded by Quine's arguments, which for their influence and their artistry demand a certain tarrying over here. Quine is a radicalizer of Duhem's comparatively modest holism about physics. In Quine's hands, the confirmation due to *any* theory applies to it whole and entire rather than sentence by sentence. Confirmation goes global, attaching to individual sentences honorifically, in a mode of attribution that, save for the honorific, would be the ancient fallacy of division.

Mathematics is indispensable to science. Seen in Quine's way, mathematics is essential to a theory's implication of its observation categoricals. Observation categoricals are sentences such as "When it snows, it's cold." They are the "direct expression of inductive expectation," the first intimation of a theory's laws (Quine, 1995, p. 25). This should make us curious about whether its indispensability to theories having empirical checkpoints is sufficient to pass on the status of empirical to mathematics itself, as relativity theory was thought to do for Riemann's geometry. Quine is affirmatively minded.

He asks – rhetorically – whether there is any epistemological advantage in treating the mathematics of a globally confirmed theory differently from what its confirmation requires for the theory itself. Although mathematics lacks empirical *content*,[6] Quine finds no good reason to contrive, for scientifically useful mathematics, a separate epistemology. It is not just that mathematical epistemologies have had a bad track record (as witness, the unhappy careers of synthetic apriority and reductive analyticity); it is also a matter of methodological economics. Why should a scientific theory have two epistemologies – one for the empirical part, the other for its mathematical part – when one could be made to do across the board?

Quine also supposes that the same can be said for a theory's meaning. It is often said that the rejection of verificationism has long been a centerpiece of Quine's philosophy. Thinking so is a serious misapprehension. Quine is an unwavering verificationist. Meaningfulness is conferred by confirmation; not, as we see, sentence by sentence, but on whole theories. Thus, Quine's brand of verification encompasses what is sometimes called "semantic holism," and his complaint against Carnap and other positivists is a complaint not against the verificationism of their semantics but against its atomism; its supposed application to sentences one by one. What, then, of those individual sentences? Do they acquire their meaningfulness from the confirmation conferred on the theories in which they occur? If so, is the achievement of local meaningfulness also honorific, as we supposed in the case of local confirmation? If so, then semantic holism is a dislocater of classical logic. If a theory's sentences are meaningful one by one only in an honorific sense, then they are true or false only honorifically, too – which makes the Bivalence law of classical logic false. Not so for Quine, of course, who rejects any notion of meaning linked to the suggestion that the bivalence of a sentence requires it to have a propositional context. Still, on reflection, we might think better of honorificizing our inferences in *sensu diviso* and plump for more straightforward deductions. In the case of confirmation in isolation, we could say that a sentence is actually, not honorifically, confirmed by its membership in the set of derivations of a confirmed theory. In the logico-semantic case, we could likewise say that a sentence is actually, not honorifically, meaningful by its membership in the set of sentences of a meaningful theory. We would appear to be wrong each time. Equivocation looms. In its application to a theory, "confirmed" means something like "stands in such-and-so relation R to the available evidence," whereas in its application to sentences, "confirmed" means "is derivable in a confirmed theory." Thus it is a non sequitur – the fallacy of division – to infer the confirmation of sentences from the confirmation of the theories in which they are derived. That is to say, any such inference is the fallacy of division *if holism is true*. Whatever the relation R to which a confirmed theory stands to the available evidence, and to which it owes its confirmation, holism insists that it is not *that* relation that any of a confirmed theory's assertions bears to the available evidence. It is the same way with attributions of meaningfulness. When applied to a theory, "meaningful" means "confirmable," that is, "could come to stand in relation R to evidence that becomes available." As applied to sentences, "meaningful" means "is asserted or denied by a theory that might bear R to evidence that becomes available," a relation in which if holism is true sentences cannot stand one by one. This leaves the semantic holist painfully positioned. He can have his truth-valued sentences either honorifically or actually, but at a cost either way. If honorifically, he must – short of Quine's semantic skepticism – reconcile himself to the loss of classical logic. If nonhonorifically, the price is worse; it is the fallacy of division.

Perhaps the dilemma could be slipped if the requisite ambiguities were noted. Then the inferences,

1. T is a confirmed theory
2. Φ is derivable in T
3. Therefore Φ is a confirmed sentence

and

a. T is a meaningful theory
b. Φ is a sentence of T
c. Therefore Φ is a meaningful sentence

would duck the charge of equivocation if the terminal "confirmed" expressed something different from the initial "confirmed," and likewise for "meaningful." But unless we have antecedent knowledge of the sense of these terminals, we shall not know what these inferences convey, never mind whether the conveyance is valid. We could venture that in line (a) "meaningful" means "verifiable," and suppose that in its recurrence in line (c) it means "has a truth value." There is something to be said for this line of thought, since even if "truth-valued" does not appear to follow from "meaningful," it may appear to follow from "verifiable," from which *on the verificationist account* "meaningful" itself follows. The transitivity of following from takes care of the rest. If this is our solution, it is consequential well beyond our interest in the derivation of (c) from (a) and (b). It gives us grounds for thinking that verificationism is not a theory about meaningfulness after all, or, to say the same thing more circumspectly, that it is an account of meaningfulness in a technical and neologistic sense of the term.

If we have found a way to reconcile ourselves to the validity of the derivation of (c) from (a), and (b), I confess that I am at a loss about the move to (3) from (1) and (2). I am unable to contrive an interpretation of "confirmed" in (3) that leaves any chance of the derivation's validity. Perhaps it is just a failure of imagination. I do not, in any case, propose to attempt to bring our discussion of holism to a final solution.

The general specification of R is, of course, not an open-and-shut affair; neither is the tightness of the fit of evidence to confirmations that R affords an easy thing to describe. Proxy functions are part of this problem. As we saw in the Prologue, "a set of sentences can be reinterpreted in any one-to-one way, in respect of the things referred to, without falsifying any of the sentences" (1995, p. 72), and so "if we transform the range of objects of our science in any one-to-one fashion, by reinterpreting our terms and predicates as applying to new objects instead of the old ones, the entire evidential support of our science will remain undisturbed" (1992, p. 8).

Those who do not mind the intended dualism between empirical and abstract theories will welcome the difficulties in which semantic holism finds itself. Perhaps they will go even further, insisting that precisely where the dualism is most sharply edged it does not matter whether semantic holism is true. Its edges are sharpest in the higher reaches of mathematics and pure logic, and it is there that holism – which if plausible at all is plausible for scientifically applicable mathematics – quickly becomes implausible for its attempt to snare inapplicable mathematics as well. Quine himself asks "about the higher reaches of set theory itself and kindred domains which there is no thought or hope of applying in natural science" (1992, pp. 5–6). His answer resembles the stand he takes against a special epistemology for mathematics: It is uneconomical to contrive a special semantics for the higher reaches, whose sentences "are couched in the same vocabulary and grammar as applicable mathematics" (1992, pp. 5–6). Special accommodation would involve "an absurdly awkward gerrymandering of our grammar" (1992, pp. 5–6).

For those who are still not drawn to our dualism, Craig's Theorem beckons attractively (1953). The theorem asserts that for any theory in which a partition exists on empirical and theoretical terms, theorems containing theoretical terms reduce without relevant loss to theorems containing empirical terms only. Thus, in principle, empirical terms are all the terms required for the adequacy of any theory containing theoretical terms as well. There is no effective means of finding a purely empirical reducer for any such mixed theory. Craig's Theorem requires the prior specification of the mixed theory in order that the existence of the pure theory can be proved in the abstract. Therein lies a distinction resembling the one I am seeking to invoke. An abstract theory *modulo* Craig's Theorem is a theory requiring such prior specification.

Ramsey sentences offer the same appearance of relief from dualism. They are Ramsey's way of eliminating reference to theoretical entities in science. Ramsey sentences arise from term-containing sentences by displacement of terms with individual variables and concomitant binding by way of the existential quantifier. Applied to a theory's every theoretical term-containing sentence, Ramsification lays bare the topic-neutral structure of the theory (Ramsey, 1931). Ramsification anticipates Quine on proxy functions, a move that extends a thesis about the reference of theoretical terms to a thesis about the reference of all terms. Dualism is avoided right enough, but it is *term*-dualism (which is what I do not want) rather than *theory*-dualism (which is what I do want).

An abstract science is a discipline that makes its enquiries and reaches its conclusions without the benefit or discipline of empirical checkpoints. It is sometimes contended that the definition is empty, since no science or discipline worthy of the name fails to engage the empirical check, however indirectly. Even the upper reaches of mathematics, it is said, make contact with the empirical by virtue of the indispensability of some branches of mathematics to the hard sciences.

I am unconvinced by this argument, but it does not matter. My conception of abstractness is a practical one. When a set theorist or a topologist or logician announces his axioms, produces his arguments, and draws out his theorems, he rarely, if ever, does so with improvements to physics in mind, and he never allows his conclusions to be judged by their amity toward empirical science, even if in the fullness of time such amity proves to have existed (consider, for example, the surprising applicability of category theory to the methodology of mathematical physics, or the fact that the permanent stoppage of the heart cannot be explained fully without a theorem from topology). This is abstractness at the level of praxis, but it is abstractness enough for the purposes of this book.

The general question is "How do abstract theorists go about their business without the comforts of empirical checkpoints?" The particular question is "How do abstract theories resolve their differences, especially their heartfelt differences about basic things?" The particular question is important in a way that the general question is not, important as it is otherwise. Pressing the particular question of conflict resolution strategies is an efficient way of unmasking bad answers to the first, more general, question.

Our two questions bear on a third. Can the abstract theorist do his business and resolve his quarrels in ways that preserve realist assumptions; that is, in ways that allow him to think that how well he does his business and how well he settles his disputes will be a matter of how close he gets to the objective facts of the matter at hand?

It is easy to see that two methodologies dominate the abstract sciences. One I shall call the *method of intuitions*. The other is the *method of costs and benefits*. On the face of it, the method of intuitions is tailor-made for scientific and philosophical realism. The cost-benefit methodology is more a creature of prudence, an exercise in doxastic economics, so to speak. It delivers the goods for realism, if at all, in a much less obvious and less direct way.

A Medical Analogy

I do not, as I say, intend to pursue the distinction between empirical and abstract theories to its philosophical finality. Imperfectly drawn as it may be, and philosophically questionable as it may also be in the abstract, in practice it is a distinction too attractive not to make use of. In this book, I shall be concerned with disagreements that arise in abstract theories such as logic, set theory, formal semantics, and certain of the normative disciplines. The first task is to specify the dialectical structure of disagreements of the sort that I wish to examine. For this a medical metaphor is an inviting way of proceeding. In medical practice, when an injury or an illness befalls,

* **symptoms** present themselves.

There follows

* a **diagnosis**

and then,

* some **triage**.

Thereupon

* a **treatment** is proposed

in light of which

* a **prognosis** is made.

It is much the same way with conflict in abstract theories. If we take, as an example, the sound and fury that attend the classical theorem known as *ex falso quodlibet* – that if a contradiction is provable then every sentence is provable – our medical figure applies as follows.

Symptoms. In the metatheory of classical propositional logic and in modal systems such as Lewis' S5 *ex falso* is provable.

Diagnosis. A great many theorists are agreed that the derivability of *ex falso* is paradoxical, at least in the sense of being sharply *counterintuitive*.

Triage. Depending on what is made of the verdict of paradox, a number of possibilities present themselves. Triage is a way of answering the question, "How bad is it?" Historically, answers range all the way from "It is not bad at all; *ex falso* is counterintuitive only in a weak sense; it is only a surprise," to "It is very bad. It is counterinuitive in a sense strong enough to convict any theory in which *ex falso* is derivable of the derivation of a falsehood." An even stronger finding is possible: *ex falso quodlibet* violates the very meaning of "is derivable" and "implies," and so is not just false but semantically or conceptually false, hence necessarily so (Anderson and Belnap, 1975, ch. 1).

Treatment. Depending on the results of triage, treatment can range all the way from none to a decision to change one's logic in ways that block *ex falso*. Historically, proponents of systems of strict implication opted for the first treatment-option. For others, such as paraconsistent logicians – relevant logicians being prominent among them – the required treatment is the displacement of the "classical" treatment of implication by some or other deviant variation, such as the relevant system R of Anderson and Belnap (1975, pp. 249–391).

Prognosis. Where treatment is deemed unnecessary there is no cause for prognosis. For those who opt for treatment, there should be some thought as to how to answer the question, "How will the patient now fare?" If one is a relevant logician, there will be a disposition to argue that not only will the patient benefit from the expulsion of a false theorem, but that in its restored state the patient will do a better job in giving a realistic account of rules of deductive inference, for example.

As conceived of by theorists such as Russell at the turn of the twentieth century, set theory threw up an interesting symptom. The symptom was the derivability in intuitive set theory of the Russell Paradox, which demonstrates the existence of a set that is a member of itself if and only if it is not a member of itself. The diagnosis, again, was paradox, and by a broadly accepted triage the paradox was very bad news indeed, since, with the aid of the law of Excluded Middle, it implies an explicit contradiction in which the Russell set both is and is not a member of itself. In the years since 1902, nearly all theorists have agreed on at least the general type of treatment required. The consensus was that intuitive set theory would have to be replaced by a new theory constructed in ways to avert a Russell Paradox. Prognoses varied depending on how close the analyst was to the symptomatic event of 1902. First-generation postparadox theorists took comfort in the presumed consistency of set theories such as ZF (Zermelo-Fraenkel), ZFC (Zermelo-Fraenkel with Choice), and NBG (von Neumann-Bernays-Gödel), but they also were disposed to think of the mechanisms for the exclusion of the Russell set as artificial, *ad hoc*, and counterintuitive. Later generations came to see ZF, or some or other spinoff of the cumulative hierarchy, as capturing the ordinary concept of set – as natural as breathing almost.

Our two problem cases touch on and, so to say, infect one another. What to make of the Russell Paradox hinges in no mean way on what a contradiction implies, hence on whether *ex falso* is true. As a matter of contingent history, opinion has clustered around the position that because *ex falso* is true the Russell Paradox is bad enough to require the replacement of the old set theory with something new and different enough to prevent paradox from reobtruding. Here is a position in which when a theory T collides with classical logic we change T; we do not change logic. It is well to note, however, that in principle the reverse strategy is also available: *Retain* T and *change* logic. Such is the position of paraconsistent logicians, logicians who see *ex falso* as false, and for whom the presumed coincidence between a theory's negation inconsistency and its absolute inconsistency is a mistake. A paraconsistent theory is both inconsistent and not; it is negation-inconsistent and yet absolutely consistent. Beyond these fundamentals, paraconsistentists fan out in two main, and irreconcilable, directions. There are those for whom the negation-inconsistency of a theory T is bad enough, short of implying omniderivability, to call for a successor theory T*. Relevant logicians typify this first sort of paraconsistentist. When faced with paradox in a theory T they are *comprehensive* revisionists, changing *both* logic and T alike. Paraconsistentists of a less meddlesome stripe try to hold the line at a change of logic only. It is more easily said than done, of course. The main idea amounts to a bold new policy for the management of negation-inconsistency, its triage and its treatment. What is proposed is that negation-inconsistency is not so bad after all, certainly not bad enough for surgical removal. Under any such policy, set theory will continue to be done with the old inconsistency left in. But it will not be the old set theory. New or old, what a theory of sets is able to prove depends on what it takes sets to be, and on the implication relation that it embeds. A logic in which negation-inconsistency does not imply absolute inconsistency is a logic different enough from classical logic to produce, in the application of its proof structures even to the old axioms on sets, theorems quite different from those authorized by the old theory, the theory got by applying classical proof procedures to the same axioms.

Paraconsistentists of this second stripe likewise come in two variations, weak and strong. The weak paraconsistentist sees a distinction between inconsistency and contradiction. Say what you like about inconsistency, it is not as bad as contradiction, which is very bad. A theory is inconsistent in the sense presently intended if and only if it is negation-inconsistent, that is, for some sentence Φ both it and its negation $\ulcorner \neg \Phi \urcorner$ are derivable. A theory contains a contradiction if and only if, for some Φ, $\ulcorner \Phi \wedge \neg \Phi \urcorner$ is derivable. Among paraconsistentists of this weak breed there is something to be said for suspension of the Adjunction law, which proves the conjunction of arbitrary pairs of theorems. Those who opt for the cancellation of Adjunction can block outright contradiction, but they tend to vary in their treatment of inconsistency, a matter which I take up in Chapter 3. More radical are paraconsistentists of dialethic stripe. "Dialethic" comes from the Greek words for "two" and "truth." It conveys a tolerance for the truth of contradictory pairs of propositions. Equivalently, it allows in selective cases for concurrent possession of both truth values. Dialethic logic may first have been a gleam in the eye of Heraclitus and – however tacitly and half-bakedly – it has tried to hold the coat of Philosophical idealism in certain of its variations, as witness the *Greater* and *Lesser Logic* of Hegel.

It would be handy to have names for the various ways of being a generic paraconsistentist. I reserve the terms "relevant logician" and "relevantist" for paraconsistantists of the first stripe, that is, for those whose treatment of paradox calls for across-the-board change to the paradoxical theory and its underlying logic alike. Weak paraconsistentists and strong paraconsistentists, or dialethists, agree on a policy for a theory's inconsistency, namely, that it need not destroy the theory even if left in, but they fall out over contradictions, with the dialethist allowing that, on occasion, even they might be true.

Our medical metaphor also can be put to use in the case of a third paradox, the so-called Tarski paradox, but that belongs in truth to Eubulides (thought credited by St. Paul to Epimenides). Consider these statements.

(1) is not true
(2) is a statement (i.e., a bivalent sentence).

(1) is true if and only if it is not. Here, too, the symptoms are the demonstration of something paradoxical. Diagnosis reveals a contradiction, since with the aid of Excluded Middle, the express contradiction "(1) is true *and* (1) is not true" is derivable. As before, most triagists agree that contradiction is a serious problem, certainly serious enough to justify even rather radical steps to evade the paradox.[7] Accordingly most treatments involve – or are represented as involving – the gerrymandering of language in ways that prevent paradoxical recurrence. Prognosticators are hopeful, by and large. Although the Tarski Paradox puts natural language out of business, paradox-free formalized languages are available, either in fact or in principle, to do the serious business of science.

Conflicts in the abstract sciences owe something of their dialectical flavor to our medical metaphor. Theorists can disagree in their diagnoses, in their triagic and treatment judgments, and in their prognoses. The Liar Paradox illustrates diagnostic disagreement. Some theorists think that the Liar *proof* is defective. For them there is no paradox, and if there is trouble anywhere near at hand, they tend to see it in the assertion that the Liar sentence is indeed a statement. Rival reactions to the Russell Paradox and *ex falso* are not typically diagnostic. For the most part, theorists agree that there is something genuinely paradoxical under foot. In the case of *ex falso*, there is substantial disagreement at the level of triage, with judgments ranging from "not at all bad" to "not all that bad" to "horrible." Beyond that, contentions ramify noticeably. Strictists (so called after Lewis's systems of strict implication) require no more by way of treatment than the reassurance of a supplementary proof, revealing that *ex falso's* triagic worst is "not all that bad." Among those of harsher triagic judgment, contentions and alarums cluster around treatment options, and to a lesser extent around prognostication. With set theory we see a different contention space: Broad symptomic agreement (there is a paradox here); broad diagnostic agreement (the paradox proves a contradiction); a solid if not perfect consensus about triage (a bad problem); and a flourishing dissensus about treatment, both as regards *what* should be treated, and by what *means*.

The historical record reveals, for both the Liar and the Russell Paradoxes, diagnoses and triages more dire than those we have examined so far. Concerning the latter, Frege and Russell saw the paradox as a proof of the inconsistency of the concept of set. Tarski thought that the Liar established the inconsistency of the concept of truth;[8] or in greater strictness, as I have suggested, that it showed the inconsistency of the concept of statement, that is bivalent sentence.[9] In each case, the proof of the "concepts" inconsistency destroyed the concept. There is no concept of set and there is no concept of statement. Suppose we dub Frege's reaction to the Russell Paradox *Frege's Sorrow*. It may strike us as an extreme response, a trifle on the hysterical side. Frege opined that arithmetic was toppled by the paradox, that it lacked secure foundations. So harsh a triagic judgment places great weight on treatment options, needless to say. I shall reserve discussion of these options (one of which will surely be the *null* option: there is nothing to be done) until Chapter 5. For now it suffices to see something of the structure of *Frege's Sorrow*. It may be understood as the following argument, generalized to any concept K

(1) The putative concept K is inconsistent   [diagnosis]
(2) Therefore, there is no concept K   [triage]
(3) That is, there is nothing to the very idea of K   [restatement of (2)]
(4) Since there is no concept K, K has no extension. Alternatively, K has the null extension.   [from the analysis of concepts]
(5) Therefore, there are no K-things.

Some readers will see a non sequitur in the move from (1) to (2), from (2) to (3), (3) to (4), and (4) to (5). I count myself as one of them. For present purposes, the passage from (4) to (5) stands out. The derivation is valid only if the existence of K-things requires or guarantees the existence of the concept, that is something like the class of those very K-things. Whether this is so has been a philosophical vexation throughout Western Philosophy, and it afflicted Cantor's and Dedekind's and Zermelo's efforts to get a usable concept of set up and running for service in transfinite arithmetic. The question is affirmatively answered in two Philosophical traditions – platonism and idealism. In the first instance, there can be no K-things unless there is a Form of K, and, in the second, there can be no K-things if there is no idea of K-things, that is unless K-thingness is more or less successfully *conceived*.

It is true that I have characterized the reactions of Frege, Russell, and Tarski somewhat starkly. To the extent that their reactions have become something approaching the received wisdom among Philosophers, it may even be supposed that I have misrepresented their positions. It is customary among Philosophers to say that what the paradoxes cost us (or them) is the *intuitive* idea of set and the *intuitive* idea of truth (or the *intuitive* idea of statement). So understood, *Frege's Sorrow* proclaims the nonexistence not of sets, but of sets in the intuitive sense; in application to the Liar Paradox it proclaims the nonexistence not of truth, but of truth in the intuitive sense, and of statements in their natural language sense. I shall say in Chapter 7 why I think this softer reading is wrong, that is, a misreading of the original texts. But even if I am wrong in resisting this softer reading, it is not all that soft. It lands the set theorist and the semanticist alike in the thicket of having to think up set theory and the theory of truth without the aid of intuitions about sets and truth.

Tough questions are triggered. How do they know how to proceed? How do disputants know when they have got it right? And thereupon: How are rival ways of proceeding, yielding rival theoretical outcomes, to be adjudicated? Here, too, the narrow fact is comparatively clear. The Russell set and the Liar statement lead to contradictions. Everyone who has ever granted those facts and reflected on them is ready to admit consequences more or less wide. Everyone, in other words, who has granted these facts and reflected on them is prepared also to grant consequences more or less momentous.

The modern history of *ex falso* is one in which it follows from the strictist's *definition* of implication. Thus

Def: Φ (strictly) implies ψ iff it is not possible that Φ and $\neg\psi$.

If Φ is some contradiction, say, $\ulcorner\chi \wedge \neg\chi\urcorner$, there is no possibility that it is true, hence no possibility both that it is true and something else is false. So $\ulcorner\chi \wedge \neg\chi\urcorner$ implies anything whatever. The reaction of the strictist was, first, that the only thing at all wrong with *ex falso* was its counterintuitiveness – and an especially benign sort of counterintuitiveness at that. Lewis and Langford thought their theorem merely surprising. This did not stop them from offering reassurance in the form of a conditional proof, which may have originated with Alexander Nekham as early as the year 1200 (Lewis and Langford, 1932, p. 252; Nekham, 1863, ch. 173, pp. 288–9):[10]

(1) $\Phi \wedge \neg\Phi$		Hypothesis
(2) Φ	1,	Simplification
(3) $\Phi \vee \psi$	2,	Addition
(4) $\neg\Phi$	1,	Simplification
(5) ψ	3,4	Disjunctive Syllogism.

Hence, by the Conditionalization Rule, $\ulcorner\Phi \wedge \neg\Phi\urcorner$ implies ψ. The proof is offered in the spirit of reassurance precisely because its forwarders thought, or should have, that it avoids a dialectical problem, which the provability of *ex falso* from the definition of strict implication attracts. To the strictist the worst that can be said against *ex falso* is that it is *weakly counterintuitive*, that is, true though initially implausible. On the other hand, to the critic of *ex falso*, the problem is *strong counterinuitiveness*, strong enough to establish its falsehood. Antagonists who disagree over the strength of *ex falso's* counterintuitiveness and who plight their cases on nothing more than how the counterintuitiveness strikes them are guaranteed to beg one another's questions. To the credit of those who advanced it, the Lewis-Langford proof was a strategically adroit move. It attempted a resolution by deriving *ex falso*, not from a definition that was now in doubt, but from elementary principles of logic that were *not* in doubt. Here was a perfect example of what Locke called *argumentum ad hominem*, not the fallacy of later traditions but the wholly legitimate "pressing a man with consequences drawn from his own principles, or concessions"

(1975, p. 686). Locke saw to that an *ad hominem* was not an argument *ad judicium*, that is, an argument that purports to advance us in the truth of things. He saw instead that a person confronted with consequences he was not happy to accept was bound on pain of inconsistency either to swallow the unwanted consequence or give up something from which it followed. In the general case, the *ad hominem* argument would not pick out the falsehood among the refutee's inconsistent concessions. It would establish only that at least one falsehood was present there. Arguments *ad judicium* are arguments that seek to advance our learning by way of the foundations of knowledge and probability. In the case of the Lewis-Langford proof, the advancers of it believed that it invoked no principle of derivation not acceptable to a critic of *ex falso*. Precisely because *ex falso* is a narrow issue and the validity of the set {Simplification, Addition, Disjunctive Syllogism, Conditionalization} is a wider issue, the makers of the proof might have expected that resistance to *ex falso* would not also be accompanied by resistance to a goodly chunk of the proof apparatus of ordinary logic. What is more, since to each of the principles in question there attaches not an iota of counterintuitiveness – indeed, it is the other way round; they are so intuitive as to be obvious – the advancers of the proof had good reason to suppose that they were affecting a non-question-begging resolution to the dispute at hand.

Whatever the forwarders of the proof may have expected of their opponents, the disposition of disputes to widen is again evident. A decision to disclaim Disjunctive Syllogism, for example, may seem a narrow affair, but in fact is nothing less than a decision for a nontrivial rejigging of logic.

The astonishing thing is that the Lewis-Langford proof failed in its mission. As things turned out, the presumption of the proof (that it was pressing into service consequences of principles antecedently accepted by the generic paraconsistentist) was false. A good many critics came to see that they disliked Disjunctive Syllogism. I will not review until the next chapter the scope and variations of the attack on Disjunctive Syllogism except to cite the judgment of Anderson and Belnap that the proof is "self-evidently preposterous" and that "it is immediately obvious where the fallacious step occurs" (1975, pp. 164–5).[11] It is the step licensed by Disjunctive Syllogism, which commits "a fallacy of relevance" (1975, pp. 164–5). With that remark, the dispute descended to its prior dialectical level, though more deeply so. It now has a narrow and semantic feel about it. It strikes us as a dispute about the analysis of the concept of alternation, or of negation, or both. The disagreement is also a disagreement not about how strongly counterintuitive a theorem is, but rather a dispute about whether a general principle of logic is counterintuitive *at all*. Frank Ramsey once asked, "How can philosophical equiry be conducted without a perpetual *petitio principii*?" (1931, p. 2). Let us call this *Ramsey's Question*. At this juncture, not only has the gap between the strictist and the paraconsistentist widened alarmingly, they seem doomed to provide for *Ramsey's Question* a negative answer. If we lose Disjunctive Syllogism, we lose either negation or disjunction, or both. If we lose either, we lose the other familiar connectives, owing to their interdefinability. Indeed, thanks to the Functional Completeness Metatheorem, we lose every other truth functional connective, and this has the effect of intensionalizing the new logic, as we will see in greater detail in due course.

The Lewis-Langford proof is thus a serviceable introduction to a large problem for the conflict resolution strategist. It is a problem large enough and vexing enough to deserve a name.

Philosophy's Most Difficult Problem

Let $A = \langle\{P_1, \ldots P_n\}, C\rangle$ be a valid argument, a sequence in which C is a logical consequence of preceding steps. *Philosophy's Most Difficult Problem* is that of adjudicating in a principled way the conflict between supposing that A is a sound demonstration of a counterintuitive truth, as opposed to seeing it as a counterexample of its premises.

Philosophy's Most Difficult Problem extends well beyond impacted disagreement in logic and other areas of technical philosophy. Its provenance is huge, and its presence is ubiquitous. Consider, for example, the classic argument for determinism.

Determinism

1. All human actions are (macro-) natural events
2. All (macro-) natural events have a cause[12]
3. If there are any free actions, they are uncaused
4. Therefore, there are no free actions.

It is easy to see that we can react to this argument in one of two ways. We could hold that the argument is sound and that, notwithstanding its extreme counterintuitiveness, its conclusion is true. It is, so to speak, a *surprising truth*. On the other hand, we could see the argument as valid, but as a *reductio ad absurdum* of the premises that imply it. On this view, the conclusion, far from being a surprising truth, is an utter and transparent falsehood. Their disagreement is not about whether (4) is a logical consequence of the preceding lines, but rather about what the consequence of (4)'s being a consequence of those premises is. People who see the argument in the first way are *determinists*. Those who see it in the second way are antideterminists; and if they select premise (2) as that which the *reductio* argument discredits, are *libertarians*.

Determinists and antideterminists thus find themselves landed in *Philosophy's Most Difficult Problem*.

The essence of determinism is the argument:

Det: Since the law of causality is universally true of natural events, since all human actions are natural events, and since causality contradicts freedom, no human action is free.

The essence of antideterminism is the argument:

AntiDet: Since at least some human actions are free, and since causality contradicts freedom, then either the law of causality fails for certain natural events, or not all human actions are natural events.

It takes little reflection to see that determinism and antideterminism are almost, but not quite, the total opposites of each other. The significance of this opposition is that neither can succeed as a critique of the other. If we try to refute *Det* by forwarding *AntiDet*, we *beg the question* against *Det*. Similarly, if we try to refute *AntiDet* by forwarding *Det*, *we beg the question* against *AntiDet*. Something interesting follows from this. Although *Det* makes a case *for* determinism, it does not make a case *against* antideterminism; and although *AntiDet* makes a case *for* antideterminism, it does not make a case *against* determinism. When any two arguments find themselves in this position, we may say that a *stalemate* exists with respect to some disputed issue.

What is the structure of stalemates? In schematic form, *Det* is:

Schema Det: P and Q and R; therefore not-S.

On the other hand, the schematic form of *AntiDet* is

Schema AntiDet: S; therefore either not-P, or not-Q, or not-R.

It is notable that *Schema Det* and *Schema AntiDet* are equivalent arguments-schemata in elementary logic. They are the (argumental) *contrapositives* of each other.

We now see why *Det* cannot be a case against *AntiDet*, nor *AntiDet* against *Det*. If *Det* is valid, so is *AntiDet*; and if *AntiDet* is valid, so is *Det*. There is a sense, then, in which *Det* and *AntiDet* are the same argument. But if this is so, how can it possibly be the case that in forwarding *Det* as a refutation of *AntiDet*, or *AntiDet* as a refutation of *Det*, we would be begging the question each time?

The answer is that, as we have seen, *Schema Det* and *Schema AntiDet* constitute a stalemate. They do so because they cannot be coforwarded in any contention space without begging the question. And they beg the question because each has a premise that is the negation of the other's conclusion.

The argument for *ex falso quodlibet* is also an example of *Philosophy's Most Difficult Problem*. Here, too, there are two different ways in which logicians have seen this argument. In one of these ways, the argument is seen as a valid demonstration of the proposition that if a contradiction were true, so would every thing else be; that is to say, as a proof of the equivalence of negation-inconsistency and absolute inconsistency. On the second way of seeing it, the conditional conclusion (that a contradiction implies everything), is absurd or utterly and transparently untrue, and therefore at least one of the proof rules employed by the proof is defective. Seen this way, the argument is a *reductio ad absurdum* of at least one of its proof-rules. Which, then, is it?

How do we answer such a question? How do we tell the difference between a valid conditional proof and a *reductio ad absurdum* of an embedded proof-rules? In the actual history of the dispute over *ex falso* – for example, in the dispute between classical logicians and relevant logicians such as Anderson and Belnap – the following dynamic reveals itself.

Ex falso is the claim that a contradiction implies everything whatever. Virtually everyone agrees that this is a *counterintuitive* thing to say. Classical logicians are of the view that it is counterintuitive but true. Relevant logicians see it as a counterexample. In 1932, C. I. Lewis and C. H. Langford produced their proof.[13] They made a point of saying that the proof rests on logical principles that were nowhere in doubt; on principles therefore that both sides would see as *highly intuitive*. Lewis and Langford issued a challenge to their would-be critics. Which of these principles – Simplification, Addition, or Disjunctive Syllogism – would they be prepared to give up? It was, of course, a rhetorical challenge. It never occurred to them that anyone would be disposed to abandon any of these elementary principles of logic.

They were wrong. In 1959, in a paper on truth functions, Anderson and Belnap – on having had it pointed out by the journal's referee that Disjunctive Syllogism failed on their treatment – decided to make a virtue out of this situation. So they asserted that *DS should* fail (Anderson and Belnap, 1959, p. 302).

It is not surprising that in later writings Anderson and Belnap responded to the challenge of Lewis and Langford by rejecting their proof's use of *DS*. It is necessary to emphasize that in 1959 Anderson and Belnap did *not* think that *DS* was counterintuitive or suspicious in any way. They were part of a quite general consensus that regards Simplification, Addition, and Disjunctive Syllogism as highly intuitive and indisputably valid proof-rules. Prior to the intervention of *The Journal of Symbolic Logic's* referee, Anderson and Belnap would have found the rejection of *DS* to be heftily counterintuitive.

We can now begin to see the basic structure of their thinking in regard to *ex falso*.

   I. Because it is highly counterintuitive, *ex falso* is false.
  II. It follows that the Lewis-Langford proof is invalid. At least one of the set {Simplification, Addition, Disjunctive Syllogism} is defective.
 III. Never mind that it is a highly counterintuitive thing to do, we must pin the blame on *DS*.

In other words, we must *preserve* our intuition that *ex falso* is false by *violating* our intuition that *DS* is valid! This is incoherent unless Anderson and Belnap are able to show that the intuition that *ex falso* is false constitutes a counterexample of it, whereas the intuition that *DS* was valid is such that the counterintuitiveness involved in rejecting it reflects only a surprising but correct decision. (I shall return to this point.)

Cost-Benefit Considerations

I shall here sketch a schematic account of what seems to me to be one of only two methods of conflict resolution in the abstract sciences with any chance of being effective, that is, of generating affirmative answers to *Ramsey's Question*. Against this I shall attempt to give due weight – in several chapters to follow – to a way of proceeding that has enjoyed a long run among analytic philosophers. It is what we might call *the Method of analytic intuitions*. In my cost-benefit approach, there is an apparent asymmetry between how *disputes* are *resolved* and how *undisputed results* are *established*. In the other approach – the intuitions approach – there is an apparent symmetry between how conflicts are removed and how uncontested results are obtained, as we shall see. We must not suppose that this cost-benefit rationality is a maximizer of expected utility, in the manner, say, of neoclassical economics. In its dialectical setting it requires shared awareness and joint behavior. It presupposes an interaction of argument and counterargument between parties. The interactive structure is less than game-theoretic, since in the theory of games, players are utility maximizers. The cost-benefit approach we are proposing leaves room for suboptimal choice. Game theory assumes cardinal utilities, whereas our cost-benefit players employ ordinal preferences. Even so, there are significant similarities with cooperative game theory. There are at least two parties; it is assumed by them both that each is a rational player; the parties are aware of the interdependence of their moves; the parties are in a state of conflict; and both parties are pledged to the idea of the best outcome overall. Beyond this, it is not plausible to model the flux of conflict resolution on a matrix game. Our conflict resolution routines do not constitute a zero-sum game, and more closely resemble coordination games.

A better point of comparison for the resolution strategies we have in mind is social choice theory. The theory has two conceptual forebears. In one approach it is an extension of utilitarianism or, in a variation, of welfare economics (Sen, 1986). In the other, the background theories are mathematical modelings of elections and committee decisions (Arrow, 1951). Social choice theories specify ways in which information states can be aggregated, conditions under which positions can be considered defeated, and how conclusions can be grounded in aggregated information. Such theories also presuppose stable and uncontroversial consequence and inference relations. However, since the conflicted issues we examine in this book are either explicitly disputes about consequence, or inference, or both, or carry fairly direct consequences for how such relations are to be treated, it is more difficult, though not impossible, to presuppose for our purposes a stable consensus on consequence and inference.

I shall not take the time to indicate in detail how the various moves of our evolving game of conflict resolution show up in the formalism of social choice theory. A prior thing needs doing. It is to get as clear as we can about such procedures conceptually and, so to speak, informally. Formal embedment I leave for another occasion.

At the heart of the cost-benefit approach to conflict resolution is Locke's *argumentum ad hominem*.[14] It is a procedure in which, from concessions one's opponent makes or is prepared to make, and by deductions he concedes to be correct, consequences are derived that, as he agrees, cannot consistently be added to his original thesis – the matter in dispute – and that he is less disposed to reject than he is to persist with the original thesis. As I have described it, the cost-benefit strategy is playable both critically and cooperatively. In its critical variation, the drawer of his opponent's consequences need not share the concessions from which they were drawn and need not himself accept the principles of deduction that begot them. The method is cooperatively employed when the deriver of consequences shares the concessions from which they were derived, and accepts as valid the means of their derivation. There is something to be said for each variation. A critical application of the strategy is useful for deducing, by one's opponent's own lights, an inconsistency in the set composed of his original thesis, his other concessions, and the rules employed in their deduction. Thus, a classical logician might seek to convince a paraconsistent logician that he is mistaken about something on his own, not the classicist's, principles.

In its critical application, there is an asymmetry at the point of final resolution. The party whose thesis is on trial has no option but to give up his original thesis if his attachment to his further concessions and the applicable rules of deduction that were applied to them is strong enough. What the interrogation has shown is that the answering party has produced an inconsistent defense of his own thesis. There is nothing in this to prevent the *questioning* party from believing the thesis to be true. Cooperative applications restore symmetry to the process. If both participants accept the thesis-holder's additional concessions and the rules of deduction, and do so with requisite conviction, then they must agree on a final resolution, namely, that the original thesis must go.

Except for vigorously streamlined situations under idealized conditions, there are no algorithms for the construction of successful cost-benefit arguments. On reflection, it is a matter of some wonderment that human beings are as adept as they are in making such arguments. It is also surprising that so much of what counts as argument and counterargument in the abstract sciences has nothing of this cost-benefit character. This can be explained. The cost-benefit methodology of *ad hominem* argumentation carries no requirement of objective correctness. There are no facts of the matter whose demonstration is a condition of its success. With the possible exception of *reductio ad absurdum* arguments and arguments *per impossibile*, every successful *ad hominem* resolution could rest on a tissue of objective falsehoods, if there were any. This disturbs the idea that logic and mathematics and the other abstract sciences are paradigms of objectivity, or bedrocks of certainty (correspondingly, it is precisely this presumption that keeps the method of intuitions in business, as we will see). It is on this point that the asymmetry between methods of conflict resolution and means of deriving uncontested consequences – ordinary, business-as-usual-theorem proving – seems to show itself. It is not, in general, true that the party who takes the cost-benefit approach to conflict resolution in mathematics, say, is someone who is also ready to think of his own nonconflictual work as deriving from principles whose significance is that he *concedes* they are all right, or consequences of statements whose significance is that he also *concedes* they are all right. What drives the asymmetry-stake deep into the ground is a supplementary parity-argument. We said that a conflict is resolved when it responds to cooperative cost-benefit argument; and we said that when this is the case the facts of the matter may be different from those conceded in the argument. By parity of reasoning, mathematical proofs in nonconflictual contexts would be good even when they reported no mathematical facts. There are lots of people who see no parity here. For them the asymmetry is established by the absurdity of nonobjective mathematics. Cost-benefit conflict resolution is inherently dialectical; ordinary, uncontested theorem-proving is inherently nondialectical. There is no mathematics by agreement. It hardly warrants saying that from inside his head the ordinary theorem-prover, the workaday transactor of the business of an abstract science, is not pressing himself with *ad hominem* arguments. But we do no have it from this that he is not tacitly attentive to consequences of what he already thinks; still less does it follow that in his solo efforts he is not a participant in a doxastic economy driven by estimates of what can profitably be given up or retained. *From the inside*, the phenomenology of the affair can suggest the objective engagement of independent facts, but the suggestion is not self-verifying (recall, we are beings who take the realist stance; it would appear that we are rigged for realist presumptions).

The asymmetry is striking enough to make us wonder why conflictual reasoning about mathematics should be so different from *non*conflictual reasoning in the same domain. The difference is stark enough to want labeling. Why is it, then, that cost-benefit arguments about mathematical propositions are allowed to satisfy *antirealist* presumptions, whereas ordinary proofs are held to realist presumptions? If this indeed is the case, why do not we just bite the bullet and admit that conflicts in the abstract sciences are (realistically) irresolvable? This, indeed, we might have to concede if realism in mathematics were true. Suppose it *is* true. Then something else becomes inexplicable. How, apart from sleepiness, drunkenness, brain damage, or inattention is disagreement in the abstract sciences about basic matters *possible*?[15] There is an entrenched answer to this question: It is not possible. Mathematical error always rests on a misunderstanding of some objective aspect of the issue at hand, since understanding a mathematical statement (or any other necessary truth) is also knowing its truth value. Chapter 4 picks up an aspect of this theme by examining the a prioristic pretensions of the method of intuitions.

Having invoked the name of Locke, it would be well to expose in some detail how he conceived of *ad hominem* arguments, and how in turn they are to a considerable extent an adaptation of the style of argument that Aristotle called *refutations*.

Locke writes:

Before we quit this subject [i.e., "Of Reasoning"], it may be worth our while a little to reflect on *four sorts of arguments* that men, in their reasonings with others, do ordinarily make use of to prevail on their assent, or at least so to awe them as to silence their opposition. (1975, p. 685)

Locke then introduces in order arguments *ad verecundiam, ad ignorantiam, ad hominem* and *ad judicium*. Concerning the *argumentum ad hominem*, we have it that

Thirdly, a third way is to press a man with consequences drawn from his own principles, or concessions. This is already known under the name of *argumentum ad hominem*.

Then comes the *ad judicium*.

Fourthly, The fourth is the using of proofs drawn from any of the foundations of knowledge or probability. This I call *argumentum ad judicium*. This alone of all the four brings true instruction with it and advances us in our way to knowledge. For: (1) It argues not [as does the *ad verecundiam*] another man's opinion to be right because I, out of respect or any other consideration but that of conviction, will not contradict him. (2) It proves not [as does the *ad ignoratiam*] another man to be in the right way, or that I ought to take the same with him, because I know not a better. (3) Nor does it follow [as with the *ad hominem*] that another man is in the right way because he has shown me that I am in the wrong. I may be modest and therefore not oppose another man's persuasion; I may be ignorant and not be able to produce a better; I may be in error and another may show me that I am so. This may dispose me, perhaps, for the reception of truth but helps me not to it; that must come from proofs and arguments and light arising from the nature of things themselves, and not from my shamefacedness, ignorance, or error. (Locke, 1975, p. 686)

It is sometimes not noticed by commentators that although, as Locke insists, the *ad judicium* stands apart from the other three on grounds of probity, nevertheless all *four* are dialectical arguments. Each is involved in "reasonings with others" in ways designed "to prevail on their assent, or at least so to awe them, as to silence their opposition" (Locke, 1975, p. 686). An *ad judicium* argument is a proof, or a probative argument *used* in a certain way in a certain dialectical setting. In that use, the proof enters into reasonings with others intent on prevailing on their assent or encouraging their retreat into silence. The proof *itself* is none of these things. It is not a form of reasoning with others, and its object is not to prevail upon someone else's assent or to shut him up. The proof itself is nondialectical, and resembles well enough what paragraphs ago we were calling "ordinary, uncontested theorem-proving."

Our interest here is in conflict resolution, an essentially dialectical enterprise, on which we may think the *ad judicium* will have little bearing. I will say in a moment why I think that this is a mistake. It is agreed, in any case, that center stage is reserved for the *ad hominem*. It is a matter of some interest that Locke does not think that *ad verecundiam*, *ad ignorantiam*, and *ad hominem* arguments are fallacies. As I say, Locke's *ad hominem* is little more than Aristotle's refutation. An Aristotelian refutation is a syllogism produced in the following way. There are two parties to a discussion, a questioner, Q, and an answerer, A. The answerer announces a thesis T. Q's job is to refute T. This he does by putting to A question that can be answered Yes or No. Then out of the answers to these questions Q tries to pick premises that syllogistically imply the contradictory of A's thesis T. Thus, T's refutation comes from A's own mouth, since Q has pressed A with consequences of his own principles or concessions. Consider a simplified example:

P_1

P_2

———

$\neg T$

in which P_1 and P_2 are answers given by A to Q's questions and $\ulcorner \neg T \urcorner$ is contradictory of A's thesis. We may ask, "What is the import of an Aristotelian refutation? What does such a refutation show?" What it does *not* show, of course, is the falsity of A's thesis T. In the absence of further information all it shows is that A has given an inconsistent defense of his thesis and so, in Locke's words, has committed an "error"; that he is "in the wrong." In strictness, a refutation of T is never a falsification of it. At times, of course, we may have independent knowledge of the truth of a refutation's premises. In those cases, and only they, we could say with a certain looseness that the refutation is a falsifying one. The looseness is explicable as follows. The falsification of T comes not from the argument that constitutes its refutation, but from the *supplementary* argument:

(1) There exists a refutation R of T
(2) R's premises are true
(3) Therefore, T is false.

I shall say that a refutation is a falsifier of what it refutes if and only if there exists such a supplementary argument and the argument is sound.

Aristotle discusses refutations in the *Topics* and *Sophistical Refutations*. In a number of passages he appears to think that refutations are proofs, but in a looser sense of "proof" than *reductio* arguments.[16] In other passages, refutations are not proofs at all.[17] For

I mean "proving by way of refutation" to differ from "proving" in that, in proving, one might seem to beg the question, but where someone else is responsible for this, there will be a refutation, not a proof. (*Metaphysics* 1006^a 15–18)[18]

In such matters there is no proof simply, but against a particular person, there is. (*Metaphysics*, K5 1062[a] 2–3)[19]

In Barnes's translation, we have it that

About such matters there is no proof in the full sense, though there is proof *ad hominem*.[20]

Aristotle is of two minds about *ad hominem* argument. He is inclined to think of them both as proofs of no kind, and as proofs of some kind. Arguments "against the man" he expressly contrasts with arguments against a man's position, and the former he regards as substandard in *some* way.[21] At times,[22] Aristotle tries to mark a distinction between refutations that turn on factors *ad hominem* and those that do not. The former he roundly condemns outright as sophistical refutations.[23]

Thus we have two contrasts to keep clear about, whose confusion precludes a satisfactory understanding of Aristotle's position. There is the contrast, first, between a proof "simply" and a refutation, or argument *ad hominem*.[24] About this contrast, two things can be said: (1) *Ad hominem* arguments are not sophistical or fallacious; and (2) since *ad hominem* arguments are refutations, they are not proofs "simply." What is it to fail to be a proof "simply"? Aristotle discusses two possibilities. One such way is to be a *nonfalsifying* refutation. This is because nonfalsifying refutations are in *no* sense proofs against the propositions they refute. The other way of not being a proof simply is to be a *falsifying* refutation. Falsifying refutations are proofs in some sense, but they are not proofs in every sense. They are not *demonstrations*, for example, since in demonstrations there exists an epistemic priority rating on the propositions of the embedded syllogism, in which preceding lines are more certain than those that come after. There is no such requirement for falsifying refutations. Then, too, as we have seen, falsifying refutations are not, *strictly speaking*, proofs of the falsity of what they refute. Falsification demands a supplementary argument in which the existence of the refutation is forwarded as one of its premises.

This takes care of the first contrast. The second is another matter. It is a contrast between *ad hominems* in the first sense and those cases of them that are *ad hominems* in some *second* sense. In the first sense, an argument is *ad hominem* if and only if it is a refutation. An argument is an *ad hominem* in the second sense just in case it appears to be an *ad hominem* in the first sense, but is not in fact. Arguments that are *ad hominem* in the second sense are defective would-be *ad hominems* in the first sense, and can reasonably be taken as fallacious.

Aristotle does not include the *ad hominem* in his catalogue of thirteen fallacies in the *Sophistical Refutations*. For him, the central notion of *ad hominem* argument is Lockean (if the anachronism can be forgiven). It is a concept of *ad hominem* that is nicely, if not perfectly, captured by the structure of refutation, concerning which there is no question of fallaciousness as such. Aristotle's particular problem is how closely refutations resemble proofs.

Whereas it is accurate to say that any Aristotelian refutation is a Lockean *ad hominem*, not every correct Lockean *ad hominem* is an Aristotelian refutation. The core of Locke's *ad hominem* "is to press a Man with Consequences from his own Principles or Concessions" (1975, p. 686). Lockean *ad hominems* include what Aristotle and others called "instruction arguments," arguments of a kind used with such dramatic effect in Socrates' introduction of the slave-boy to Pythagoras' Theorem (Plato, ca. 350/1980, pp. 353–4). Here there is no question of refuting a position the boy already holds, but rather of deriving the right consequence of what he already holds, or is now prepared to admit, or both. Still, the Theorem comes from Socrates' pressing the boy with consequences drawn from his own principles or concessions: it is a Lockean maneuver. I said a while ago that we may think the *ad judicium* will have little bearing on the essentially dialectical enterprise of conflict resolution, and I said that thinking so is a mistake. It is indeed a mistake. *Ad judicium* arguments are every bit as dialectical as Locke's other three. All four are arguments we deploy in "Reasonings with others…to prevail on their Assent or at least awe them as to silence their Opposition" (Locke, 1975, pp. 685–6). This being so, a certain pressure is put on Locke's insistence that *ad judicium* arguments differ from the other three. The purported difference is between arguments that are probative and those that are not. But consider: In what ways could one employ a probative argument to serve Locke's dialectical ends without, so to say, reconfiguring the argument as a cooperative cost-benefit strategy or, as we may now also say, a cooperative *ad hominem*? Unless reconfigured in such a way as these, the transmission of the proof that is the core of the *ad judicium* would fall on deaf ears and would bring on itself charges of question-begging.

For obvious reasons, I want my conception of the *ad hominem* methodology to leave room, though not exclusively so, for refutations in something like Aristotle's sense. It is not necessary to think of these exactly as described by Aristotle, since it is an invariant requirement of Aristotle's refutations that they be *syllogisms*. Syllogisms are restrictions on valid arguments. They are valid arguments whose conclusions repeat no premise, and that contain no idle premises. This last condition makes syllogisms interesting to relevant logicians and logicians working in the AI tradition. The condition requires that if an argument is a syllogism then it has no valid proper subarguments. Thus, all premises must be *used* in a syllogism (a kind of relevance condition), and no new premises can be added (a very strong nonmonotonicity requirement). For the sake of generality, I do not want to restrict our conception of conflict-resolution strategies to syllogisms, but I note in passing that syllogisms are extremely interesting structures. When we add as a further condition that syllogisms not have multiple-conclusions, we have that the first logical theory is an intuitionistic, relevant, nonmonotonic and strongly antidialethic logic (Woods, 2001, Ch. 6).

BEGGING THE QUESTION

We have had occasion to suggest that an important feature of impacted disagreements in the abstract sciences is their propensity for *begging the question*, for providing negative answers to *Ramsey's Question*. In its conversational uses of the present day, begging the question can be many things that I want it not to be for the purposes of our enquiry. In one such use, I beg the question of P whenever I say something that occasions enquiry about, or reflection on, P. Such we have if in saying, for example, that Jones's application is in default of governmental policy, someone is moved to enquire whether that policy is sound (or available in French, or well publicized, or whatever else). He will say "Of course, that begs all sorts of questions about the policy of the government. Is it sound; is it in French; is it well-publicized, and so on?" In other words, he begs a question in saying so-and-so when his saying so-and-so *raises* that question.

The first theoretical pronouncements on begging the question were Aristotle's. We would do well to pause a while for some elementary instruction from the founder of logic. We may then see more clearly how begging the question dogs the heels of conflicted disagreement of the sort we are enquiring into. Aristotle provides several different treatments. In the *Sophistical Refutations* it is a direct contravention of the definition of "syllogism" (hence of "refutation"). If what is to be shown is also assumed as a premise, that premise is repeated as the conclusion, and the argument in question fails to be a syllogism, hence cannot be a refutation. In the *Posterior Analytics (86^a21)*, begging the question is a *demonstration* error. Demonstrations are deductions from first principles. First principles are themselves indemonstrable, and as we have said in any demonstration every preceding step is more certain than succeeding steps. If one inserts the proposition to be demonstrated among the premises, it cannot be the case that all premises are more certain than the conclusion, since no proposition is more certain than itself.

In *Topics* 162^b 34ff, Aristotle cites five ways in which a question can be begged:

People appear to beg their original question in five ways: the first and most obvious being if anyone begs the actual point requiring to be shown: this is easily detected when put in so many words; but it is more apt to escape detection in the case of different terms, or a term and an expression, that mean the same thing. A second way occurs whenever any one begs universally something which he has to demonstrate in a particular case.

The third way is to beg a particular case of what should be shown universally; the fourth is begging "a conjunctive conclusion piecemeal" (Hamblin, 1970, p. 74); and the fifth is begging a proposition from a proposition equivalent to it.

Readers are likely to find the expression "begging the question" rather obscure (and oddly theatrical). There is no mendicancy to begging the question apart from the questioner's soliciting a premise for his evolving syllogism by pressing to his opponent with a Yes-No question. Aristotle actually ignores the ordinary distinction between the question that produces the answer and the premise that answer is eligible to be. What Aristotle calls a question *is* in this context the premise produced by the answer to it. Begging such a question, then, is using it as a premise in a would-be syllogism. The "original question" of the two parties is the answerer's thesis, T. A questioner begs the question in the first way if he begs the proposition to be shown, that is, $\ulcorner \neg T \urcorner$. If $\ulcorner \neg T \urcorner$ were in fact a premise of any deduction whose conclusion were also $\ulcorner \neg T \urcorner$, the deduction would be circular; hence would be neither a syllogism nor a refutation. Aristotle is aware that such cases are not likely to deceive or trip up actual reasoners, but he reminds us that if a synonym of $\ulcorner \neg T \urcorner$ were used as a premise, it would look different from the conclusion, and the circularity might go undetected. This also appears to be what Aristotle has against the fifth way, that is, deriving a proposition from one equivalent to it.

In the second way of begging the question, Aristotle is thinking of a certain form of what in a later development was to be called "immediate inference." It is typified by the subalternation argument

All A are B,
Hence, some A are B.

Apart from the fact that the single-premise arguments seem not to qualify as Aristotelian syllogisms, it is difficult to discern a logical difficulty with them. Bearing in mind that the fault, whatever it details, is the questioner's (i.e., premise-selector's) fault, not the answerer's, one might wonder what is wrong with a questioner's asking a question that if answered affirmatively would give him a desired conclusion in just one step. According to Ross, Aristotle thinks that a refutation is worth having only if every premise (individually) is consistent with the answerer's thesis:

A syllogism is distinguished from *petitio principii* in this, that while in the former both premises together imply the conclusion in the latter one premise alone does so. (Ross, 1953, p. 38)

So in a good refutation, the thesis is refuted, even though it is consistent with each answer given (Woods, 2001, Ch. 9).

The third way of begging is illustrated by the invalid form of argument

Some A are B
Hence, all A are B

As the examples make clear, begging the question is a mistake of *premise-selection*. If the questioner's job is to elicit premises that syllogistically imply the contradictory, "All A are B," of the answerer's thesis. In the present case, the questioner has begged the *wrong* question, that is, he has selected the wrong premise. It is a premise that does the answerer's thesis no damage; and in any extension of this evolving argument in which real damage were to be done, this premise "Some A are B" would be idle.

Genuinely perplexing is the fourth way of begging. By the requirement that syllogisms be constructed from propositions, no genuine syllogism could conclude with a conjunctive statement. The reason is that these are not propositions in Aristotle's technical sense, which among other things, makes it impossible for compound statements to be propositions. But syllogisms must at every line contain propositions and nothing else (Woods, 2001, Ch. 1). Then consider the argument

P

Q

_____

Therefore, P and Q

It is clearly valid. Aristotle appears to think that it is useless for the questioner first to beg for P, then for Q, if his intention is to conclude "P and Q." The obvious validity of the argument might fool someone into thinking he had produced a syllogism. But the mistake lies less with his premise-selection than with his choice of target conclusion.

Not all of Aristotle's five ways resonate in the modern ear. For example, there is not likely to be much use for a conception of the *petitio* that, as with the fourth way, turns entirely on a technical peculiarity of syllogisms. Closer to home are the views of Whately, who held that the *petitio* is the fallacy

in which the premise either appears manifestly to be the same as the conclusion or is actually proved from the conclusion, or is such as would naturally and properly be so proved. (1853, bk. 3, sect. 13)

Mill says of this passage

By this last clause I presume is meant that it is not susceptible of any other proof, for otherwise there would be no fallacy. (1974, p. 820)

Whately and Mill are endorsing a twofold conception of the *petitio*, which Douglas Walton and I have dubbed the "equivalence" and "dependency" conceptions.[25] Aristotle's case of the argument whose conclusion repeats a premise is a case of Whately's premise being "manifestly the same as the conclusion," and a case, too, of the equivalence conception of Woods and Walton. Aristotle's demonstration error, apart from also being a conclusion-as-premise error, does not show up in the Whately or Woods-Walton taxonomies. Aristotle's third way of proving, for example, that Italians are charming by showing that this particular Italian is charming, is not to modern ears the *petitio*, but a non sequitur, or perhaps the hasty generalization fallacy. The fourth way is strictly a matter for syllogisms and, in any event, seems not to be the *petitio* in either Whately's or Woods and Walton's conceptions. On the other hand,

Aristotle's fifth way is clearly the Woods-Walton equivalency conception and Whately's conclusion-as-premise conception.

How are we to interpret the claim that intractable conflicts in the abstract sciences have a propensity for question-begging? Consider again the Lewis-Langford proof of *ex falso*, which, just as it stands, generates the mutually question-begging interpretations of *Philosophy's Most Difficult Question*. Anderson and Belnap found the proof to be "self-evidently preposterous." Perhaps they just mean "preposterous," since otherwise they would lack the occasion to demonstrate their charge in finding the application of Disjunctive Syllogism to commit a fallacy of relevance. In any event, there is something dialectically untenable about the "self-evidently preposterous" complaint. Schematically we have it thus, for disputants P and A ("protagonist" and "antagonist"):

P: Here is a good proof of *ex falso*.
A: The proof is self-evidently preposterous.

We could, of course, further represent A's move as

Φ
The proof is preposterous

where Φ is the empty set of premises.

What is so *maladroit* about the Anderson-Belnap complaint is that it is what we might call a "conversation-buster." In responding to P's "Here is a good proof of *ex falso*" with "The proof is self-evidently preposterous," Anderson and Belnap would be hard-pressed to think of any response with a lower likelihood of being acceded to by P. Without supplementary explanation and support, the conversation is over. The conflict cannot be resolved by *it*.

The further complaint by these same authors (that the proof is defeated by its employment of an invalid rule) falls into the same category of conversation-buster. If some protagonist P has just delivered a proof in which it is clear that he is committed to the validity of Disjunctive Syllogism, then any antagonist A will end the conversation, and preempt its potential for conflict resolution, if all he has to say in response is "Disjunctive Syllogism is invalid." Of course, in real life we are used to "changing the subject" or shifting the direction of our argument, especially if doing so carries some promise of bearing on the original issue, however indirectly. As such, we must slightly amend the assertion that these responses are conversation-busters. They are conversation-busters unless the party to whom they are directed is prepared to have his accuser act on his newly acquired burden of proof to demonstrate his charge. His failure to do *that* is an unqualified conversation-buster. The question is: Is there any theoretical point in calling these dialectical indiscretions *question-begging*? There is a contemporary sense of "begging the question" that is reflected in the legal notion of "assuming 'facts' not in evidence." It is also apparent in what modern theorists (but not Aristotle) call the fallacy of Many Questions:

"Have you stopped snorting cocaine (or beating your dog, or reading Foucault)?" It is easy to see that our two cases – the self-evident preposterousness of the proof, and the invalidity of Disjunctive Syllogism – are representable as instances of the " 'fact' not in evidence" type. Also representable in this way is the further thing that Anderson and Belnap allege against the proof. At the step resulting from the application of Disjunctive Syllogism, the proof was guilty of a "fallacy of relevance." Of course, this is not something a protagonist would accede to *just so* (hence is a "fact" not in evidence); and when it later became clear in the actual controversy between Anderson and Belnap that the irrelevance charged in the complaint was transparently *not* irrelevance in either of the two conceptions of it acknowledged and developed in their logic of relevance, the "fact" remained not-in-evidence.

If we think in a certain way of the logical structure of the "fact"-not-in-evidence-maneuver, it does indeed become apparent that it is a worthy claimant for the name of question-begging. Schematically, it is this:

P: T
A:...
P:...
⋮
A: Since S, not-T

where S has not been conceded – or even entertained – by P. It is not the smuggling in of A's desired conclusion as a premise of the evolving argument, but it is the next closest thing. It is the smuggling in of a nonpremise (since P never conceded it), which either directly entails the conclusion A desires, or does so in conjunction with what P has conceded. It is a close enough thing to be called question-begging (and it covers the "self-evidently preposterous" -case as well).

Begging the question by way of illegitimate attribution of virtual premises is also one half of the modern fallacy of *sophisticus elenchus*, sometimes also called "the straw man fallacy." It is the fallacy of refuting or otherwise discrediting a proposition one's opponent does not in fact hold, or has not conceded.

If anything is obvious in the theory of conflict-resolution it is that resolution strategies should not be question-begging. It is a thing more easily said than done, as we shall have ample occasion to see as we proceed.

Even so, without a liberal reconception of it, it cannot be right that the *ad hominem* is the heart and soul of conflict resolution in the abstract sciences. The reason is logistical. A great deal of conflict resolution plays out in highly attenuated variations of dialectical interaction. Even in a series of articles – replies, rejoinders, and so on – most of the statements made by either party in the advancement of his case will fail outright any anti-question-begging rule in the form "Use no statement that does damage to an opponent's case unless the opponent has conceded it." Any such rule would also seriously compromise real-life, face-to-face encounters as they are actually transacted by human reasoners. We may take as unacceptable any conception of begging the question on which actual disputants beg the question against one another with significant frequency and, when their contentions are transacted in print, with dominating frequency.

A natural extension of our (bad) rule suggests itself: Use only those propositions that are conceded by the opponent or are acknowledged by him to follow from what he has conceded, or which answer to the following standards:

1) The opponent would concede it because everyone knows it. (The *common knowledge condition*.)
2) The opponent would concede it because he is a K and the K-community believes it. (The *stereotypical condition*.)
3) The opponent would concede it because he supposes that it would help his case. (The *self-help condition*.)

I have no idea whether I have come close to pinning down the rule the violation of which gives the intended conception of begging the question. There are two reasons for this. One is that it is not clear whether the rule is close to completely stated (or even sketched). The other is that it is rarely self-announcing when the rule has been violated. This places dialecticians-at-a-distance at a particular disadvantage. It is inefficient to ask my opponent whether he thinks that my forwarding P would satisfy the rule; but if I forward it, my opponent may reject it and I will have begged the question against him unless the fact of his rejection can be squared with my having complied with the rule. There is little doubt that in practice beings like us are fairly adept at making such determinations and getting others to agree with them. Let us say for generality that we are fairly adept in attributing to others their *virtual* premises. But knowing this is not knowing much about the details of such dynamics.

Let us not press the matter. It is enough to say that if there is to be any hope for the *ad hominem* as the dominant mode of conflict resolution in the abstract sciences, we must weaken the conception of begging the question in ways that take note of two facts: (1) *Ad hominem* arguments should not have so high a propensity to beg questions or to compromise their general utility. (2) But we should also be realistic. *Ad hominem* maneuvers, especially at a distance, do have a propensity to beg questions. Negative answers to *Ramsey's Question* can be expected as a contingent liability of *ad hominem* maneuvers on the hoof, that is, transacted under condition of real-life argumentation.

It is also desirable that we emphasize the dialectically attenuated character of some of the most important and most notorious examples of conflict in the abstract sciences. Many such contentions attain their finished status in the form of a published work. Sometimes an author will have the views of a particular person in mind and want to criticize them in ways that that person himself sees as telling. If I am the author and Barry my critical target, I will attribute to Barry views, $P_1, \ldots, P_n$, which help me to damage Barry's views.

Since Barry is not present to comment on the accuracy of the P_i attributed to him, is there an attribution rule by which I should feel myself to be governed? Should I attribute to Barry only those P_i whose (contrafactual) refusal by Barry would, without independent support, convict him of what I shall call, *converse* begging the question (*CBQ*), the illegitimate refusal of a virtual premiss? Hard as it is to determine, in a nonquestion-begging way, whether the conditions of such a rule have been satisfied in particular cases, there are further complications. A work's readership, even if small and select, is diffuse and mainly anonymous. This matters for the author's objectives. He may have it in mind to "do a job" on Barry by showing that Barry's position has consequences incompatible with his concessions or virtual concessions. By and large the work's reader, unless he is Barry himself, does not confine his interest to whether Barry has a consistent defense, or virtual defense, of Barry's views. The reader also wants to know whether Barry's views are right or wrong; that is, he will want to see how Barry's views fare in reaction to views of the reader himself. In this way readers are drawn into the author-Barry dialectic. What is it appropriate for the author to do to accommodate his reader without surrendering his *ad hominem* obligations to Barry? There is both a narrow involvement and a wider one for the author to consider for his readers. In the narrower involvement, the author attributes to Barry P_i, which he believes fair-minded readers would say Barry ought to accept, that is, are virtual concessions of Barry. In the broader involvement with his readers, the conditions on the narrower involvement remain in force, and are supplemented by the qualification that the P_i in question are forwarded in the hope that the reader also accepts them as true.

There is a natural tension between narrow and wide involvements. It seems to be a natural human characteristic to prefer the wide for the narrow. We have a fondness for speaking to the crowd, rather than exclusively to our principal target, Barry. If we did not have this preference we would publish only our person-to-person correspondence without commentary. Books and articles afford us a chance to make headway against an opponent in ways that are faithful to our dialectical obligations to him and yet in ways that engage a readership, even a readership initially indifferent to the conflict between the author and his principal target. For all its attractions to human advocates, the publication-readership dynamic harbors a particular way of begging the question against the principal target in ways that cause no general disapproval. It consists of attributing to Barry the views of the readership.

ASYMPTOMATIC ARGUMENTS

I want now to turn to a line of attack that gives our medical metaphor a new twist. Up to now, we have been supposing that disputants always agree that there is something "there" that requires, or at least admits of, *diagnosis*; thus they agree on the metaphorical presence of medically relevant symptoms. It is not always this way with disagreements. In what we have been saying so far, disputes of the sort we are interested in here have always been disputes at some stage of the symptoms-prognosis hierarchy higher than that of symptom. But, as a little attention to the actual histories of theoretical conflict makes clear, often enough a critic will make what is, in effect, an *asymptomatic* diagnosis and in so doing will assert the existence of a problem unsuspected and undetected hitherto. Quine's disenchantment with modal logic is a case in point. Modal logic has been with us since Aristotle's founding efforts. It flourished at various times thereafter, notably in the work of Stoic and Medieval theorists, and it enjoyed a vigorous renaissance in the present century. Everybody knew that the modals were difficult things to bring to satisfactory theoretical account, and – as if in confirmation of this point – that there was perhaps more sheer *pluralism* in modal logic than was good for it. But nobody suspected that modal notions were the disaster that the concept of set is if *Frege's Sorrow* is true. Nobody thought this, that is to say, until Quine launched a barrage on the very coherence of modal notions. "Asymptomatic diagnosis" is a bit of a mouthful, and not very euphonious. I propose adoption of a synonym by stipulation. Henceforth, these are *ambush arguments*, hijackers of the presumption that all is well.

In the life of theories, induction and autoepistemic reasoning coexist in an interesting way. Judging from the historical track record, induction tells a global story of the pervasive vulnerabilities of our theories, in which nearly always the theorist has got things wrong in ways that matter. More locally, from an inside perspective, an autoepistemic argument is at work. It resembles Popper's fix on the significance of failed attempts to falsify a theory. In its autoepistemic variation, it is argued that if there were difficulties with a theory T, we would know of them by now. Since difficulties have not announced themselves, we may presume with sober-sided provisionality that T is all right. What makes for the coexistence of inductive pessimism globally, and autoepistemic hopefulness locally, is their reciprocal logical inertia. Nothing in the one contradicts anything in the other. If someone wants to complain about a particular T, it is not enough that theories do badly in general. To complain of *this* T, it is necessary to see signs of trouble particular to it. As our autoepistemic arguments make plain, failure of such signs to appear is a happenstance we are prepared to make something of. This makes ambush arguments especially valuable. They are upsetters of autoepistemic complacency. They do so by exploiting the connection between a theory that lacks troubling symptoms and a theory protected by autoepistemic calm. Ambushes that work effect changes in the diagnostic spaces of theories. The judgment that an ambush is an asymptomatic diagnosis of trouble is a judgment before the fact. After the fact, if the ambush succeeds, asymptomativity is lost. What the ambush formerly complained of is now seen as part of a revised symptom-diagnosis matrix. I shall say something further about ambush arguments in the chapter to follow.

DIALOGUE LOGIC

Much of what we have been saying about *ad hominem* procedures for conflict resolution is shaped by recent work in *dialogue logic* (which is my "dialectic," nearly enough). Dialogue logic responds to broader interests than those at large in this book. It is a logic of conversation, in which conversation is seen as a dialogue in which assertion, question-asking, reason-giving, challenging, and retraction occur. Dialogue logic is a system of rules for the construction and termination of conversation, and thus subsumes the matter of conflict resolution.[26]

It is natural to wonder whether systems of dialogue logic have the technical wherewithal to advance us in our understanding of conflict resolution in the abstract sciences. Could they, for example, be deployed with profit in the contention space of *ex falso*? Let us see. There are a goodly number of dialogue logics, differing from one another often in nontrivial ways. Our purposes here require a representative example. I shall sketch Roderic Girle's **DL3**, which is rich enough to be representative while keeping technical complexity fairly in check (Girle, 1996). **DL3** is a logic for two parties, X and Y, and two roles, speaker (S) and hearer (H). **DL3**'s permissible speech-acts are: *statements* (of three sorts), *declarations, withdrawals, tf-questions, wh-questions, challenges,* and *resolution demands*.

Categorical statements are statements, such as Φ, $\ulcorner\neg\Phi\urcorner$, $\ulcorner\Phi$ and $\psi\urcorner$, $\ulcorner\Phi \vee \psi\urcorner$, $\ulcorner$ If Φ then $\psi\urcorner$, and statements of ignorance, $\ulcorner\iota\Phi\urcorner$.
Reactive statements are grounds (e.g. "Because Φ, " symbolized as $\ulcorner\because\Phi\urcorner$).
Logical statements are immediate consequence conditionals, such as $\ulcorner$ If Φ and if Φ then ψ, then $\psi\urcorner$.
A term declaration is the utterance of some term τ.
A withdrawal of Φ is an utterance in the form $\ulcorner$I withdraw $\Phi\urcorner$, $\ulcorner$I do not accept $\Phi\urcorner$, $\ulcorner\neg\Phi\urcorner$ or $\ulcorner$I no longer know whether $\Phi\urcorner$. The first two of these are symbolized as $\ulcorner\_\Phi\urcorner$.
The tf-questions are of the form $\ulcorner$Is it the case that Φ?$\urcorner$, symbolized by $\ulcorner?\Phi\urcorner$.
The wh-questions are $\ulcorner$What (when, where, who, what, which) is an (the)Π?$\urcorner$ It is assumed that for each such question it is true that $\exists\chi(\Pi\alpha)$.
A challenge is $\ulcorner$Why is it supposed to be that Φ?$\urcorner$, symbolized by $\ulcorner$WhyΦ?$\urcorner$.
The resolutions demands are the utterances $\ulcorner$Resolve$\Phi\urcorner$.

Locutions in **DL3** are ordered triples of a number, a party, and the party's locution. The number numbers the steps in the dialogue sequence. **DL3** also recognizes "justification sequences," which are ordered quadruples in the form, $\langle\ulcorner$If Φ then $\psi\urcorner$, ψ, $\ulcorner$Why ψ?$\urcorner\rangle$; thus, they are, in order, the antecedent of a conditional, the conditional, its consequent, and a *challenge* of the consequent.

Parties to a dialogue possess **commitment stores (CS)**, for which there are rules.

(C1) **Statements.** After a locution ⟨n, S, Φ⟩, where Φ is a statement, Φ is added to the commitment stories of both S and H, unless the preceding locution was a challenge.

C1 is logic's answer to negative-billing. It is assumed that anything "offered" is "purchased" unless it is explicitly rejected.

(C2) **Defenses.** After the event ⟨n, S, Φ⟩, when: ⌜Why ψ?⌝ and ψ are in S's **CS**, the justification sequence ⟨Φ, ⌜If Φ then ψ⌝, ⌜Why ψ?⌝ enters the **CS** of both S and H. The challenge ⌜Why ψ?⌝ is deleted from the **CS** of S and H.

(C3) **Withdrawals.** When Φ is in S's **CS**, then after any of these three locutions,

⟨n, S,—Φ⟩

⟨n, S, ¬Φ⟩

⟨n, S, ιΦ⟩

(a) Φ is deleted from S's **CS**, and
(b) if the withdrawal was of the form ⌜¬Φ⌝ or ⌜ιΦ⌝ then the withdrawal enters the **CS** of both S and H, and
(c) if Φ is a statement association with a wh-question, for example, ⌜(Q)Tα⌝, where **Q** is the interrogative quantifier, then Φ is deleted from the **CS**, and ⌜(Q)Πα⌝ is deleted from H's **CS** and S's, too, and
(d) if the withdrawal was preceded by the locution ⟨n − 1, H, ⌜Why Φ?⌝⟩, then ⌜Why Φ?⌝ is deleted from the **CS** of both H and S, and
(e) if ⟨Φ, ⌜If Φ then ψ⌝, ψ, ⌜Why ψ?⌝⟩ is in S's **CS**, it is deleted; and if ψ is in the **CS** of either S or H, it is deleted.

C3 is a rather relaxed – and realistic – withdrawal rule. It allows for the inconsistency of commitment stores, and it provides that they are not deductively closed.

(C4) **Challenges.** After the locution ⟨n, S, ⌜Why Φ ?⌝⟩ the challenge, ⌜Why Φ?⌝ is added to the **CS** of both S and H. If Φ is not in the **CS** of H, then Φ is added to it. If Φ is in S's **CS**, it is deleted. If Φ is in the speaker's **CS** as a constituent of a justification sequences, the justification sequence is deleted.

(C5) **Information.** After the locution $\langle$n, S $\ulcorner(Q\alpha)\Pi\alpha\urcorner$, the associated state-ment $\ulcorner(\exists\alpha)\Pi\alpha\urcorner$ is added to the **CS** of both S and H, and the wh-question $\ulcorner(Q\alpha)\Pi\alpha\urcorner$ is added to the **CS** of H.

(C6) **Reply.** After the declaring of a term in a locution, for example, $\langle$n, S, $\tau\rangle$, when the previous locution was a wh-question, for example, $\ulcorner(Q\alpha)\Pi\alpha\urcorner$, then $\ulcorner\Pi(\tau)\urcorner$, which is the wh-answer to $\ulcorner(Q\alpha)\Pi\alpha\urcorner$, is added to the **CS** of both S and H.

(C7) **True/False.** After the locution $\langle$n, S, $\ulcorner?\Phi\urcorner\rangle$, if the statement Φ is in S's **CS**, it is deleted.

Next come the **Interactive Rules**.

 i. **Repstat**. No statement may be made if in the **CS** of both S and H.
 ii. **Imcon**. Logically true conditionals must not be deleted.
 iii. **LogChall**. Immediate consequence conditionals (i.e., modus ponens), must not be deleted.
 iv. **TF-Quest** After $\langle$n, S, $\ulcorner$Is it the case that Φ?$\urcorner\rangle$ the next locution must be $\langle$n + 1, H, $\psi\rangle$, where ψ is either
   a. the statement that Φ, or
   b. the statement that $\neg\Phi$, or
   c. the withdrawal of Φ, or
   d. a statement of ignorance $\ulcorner\iota\Phi\urcorner$.
 v. **Chall**. After $\langle$n, S, $\ulcorner$Why Φ?$\urcorner\rangle$, the next locution must be $\langle$n + 1, H, $\psi\rangle$, where ψ is either
   a. $\ulcorner\_\_\Phi\urcorner$, or $\ulcorner\neg\Phi\urcorner$, or
   b. the resolution demands an immediate consequence conditional whose consequence is Φ and whose antecedent is a conjunction of statements to which the challenger is committed, or
   c. a statement of grounds acceptable to the challenger.

A statement of grounds $\ulcorner$Because $\Phi\urcorner$ is *acceptable* to S if and only if Φ is not under challenge by S, or if Φ is under challenge by S, then there is a set of statements to each of which S is both committed and not committed to challenge, and Φ is an immediate consequence of that set by *modus ponens*.

 vi. **Resolve**. The resolution demand in the locution $\langle$n, S, $\ulcorner$Resolve whether $\Phi\urcorner\rangle$ can be made only if either
   a. Φ is a statement or a conjunction thereof which is immediately in-consistent and to which H is committed, or
   b. Φ is $\ulcorner$If ψ then $\chi\urcorner$ and ψ is a conjunction of statements to each of which H is committed and χ is an immediate consequence of ψ, and the preceding locution was either $\langle$n − 1, H, $\ulcorner$I withdrew $\Phi\urcorner\rangle$ or $\langle$n − 1, H, $\ulcorner$Why χ?$\urcorner\rangle$.

vii. **Resolution**. After the locution ⟨n, S, ⌜Resolve whether Φ⌝⟩ the next locution must be ⟨n + 1, H, ψ⟩, where ψ is either

a. the withdrawal of a conjunct of Φ, or

b. the withdrawal of a conjunct of Φ's antecedent, or

c. a statement of Φ's consequent.

viii. **Enlightenment**. After the locution ⟨n, S ⌜(Qα)Παᐧ⌝⟩ the next locution must be ⟨n + 1, H, ψ⟩ where ψ is either

a. the declaration of a term τ, or

b. the withdrawal or the denial of the associated statement ⌜(∃α)Παᐧ⌝, or

c. a statement of ignorance ⌜¬(∃α)Παᐧ⌝.

Moreover, when H declares a term, the poser of the wh-question must, at the earliest subsequent locution, state the *wh-answer* to the wh-question, or its denial. The "earliest subsequent locution" is the locution separated from the term declaration by no locutions except those in which the asker asks additional wh-questions, which in turn can be wh-answered, and in which the wh-question contain no predicates or terms not contained in the wh-answers immediately preceding them.

DL3 is a system that models the sort of ordinary, practical reasoning that flows through everyday conversations; conversations on the hoof, so to speak. Perhaps classicists will think that it strains the concept of logic to say so, but something can be said for seeing **DL3** as a *logic of practical reasoning*. Its practicality is one thing; its ordinariness is another. Both are important. **DL3** gives us occasion to reflect on an interesting question: Does the ordinary practical reasoner possess the wherewithal for conflict resolution in the abstract sciences? If he does, we might expect some headway on filtering conflicted issues in our abstract sciences through something like the apparatus of **DL3** . If not, this would be reason to suppose that the canons of conflict resolution in the abstract sciences – if they exist at all – rest on principles of reasoning that are somehow out of the ordinary. But out of the ordinary *how*?

Let us see. In a standard "dialectification" of the Lewis-Langford proof, S is the classical supporter of *ex falso* and H is the other side. Though **DL3** does not offer specific guidance about how to make and share assumptions, we may take it that S and H agree to assume step 1, and that in so doing ⌜Φ ∧ ¬Φ⌝ enters the commitment store *CS*, of both parties, with its provisional status suitably flagged. Subsequent steps come by way of **DL3**'s standard rules. Thus, line 2 enters the *CS* of both S and H by way of the rule of simplification together with **DL3**'s rule, *LogChall*. Similarly for lines 3 and 4. It is the same way with line 5, or so it would appear. *Imcon* and *LogChall* "prevent the withdrawal or challenge of logical principles." The prevention embeds an ambiguity. Does it, in the case before us, disenjoin the challenge or abandonment of DS *because* it is a principle of logic? Bearing in mind that S is our classicist and H our generic paraconsistentist, and that they disagree about *whether* DS is a principle of logic, we see that *Imcon* and *LogChall* fail to be usable rules precisely where a usable rule is needed. The Lewis-Langford problem stops cold *any* prospect of **DL3**-resolution. We might think that the dispute now moves to the question of whether *DS* is a principle of logic. No. This is not an askable challenge until it is established that *DS* is *not* a rule of logic. Our rules oblige us not to challenge *DS* unless it is invalid. But its invalidity is precisely what S and H are deadlocked over. The present result easily generalizes. **DL3** is unable to resolve any disagreement about any "logical principle."

   DL3 was designed to capture the dynamics of ordinary conversational give-and-take. It was not engineered for the resolution of conflicts about basic logical principles. In one way, it is not at all surprising that **DL3** fails as a decision tool for such conflicts. On the other hand, **DL3** is typical of the best of the current crop of realistic systems of dialogue logic. It makes a strong enough claim on representational plausibility to support suggestions that might explain the terrible track record of the conflict resolution history of problems such as *ex falso*, or of DS itself. One possibility is that principles as basic as DS suffer the same fate as befell the law of Noncontradiction in Book Γ of Aristotle's *Metaphysics*. In a word, there is no arguing for them. If this were the case here, then it might start to dawn on us that arguments *pro* taken from the realist stance are bound to be bad – even disgracefully bad. It would be no leap to the suggestion that from the realist stance settlement of basic logical disagreements is not possible. Not that this would discredit the realist stance necessarily; it could be a reflection on logic itself, on its distinctive resistance to resolution under realist presumptions. Itself a realistic model of dialogical give-and-take, the importance of **DL3** could be seen as indicating this very thing – and the concomitant need to consider resolution measures other than arguments *pro* and *contra* about *what is the case*.

   Beings like us adopt the realist stance as naturally as they breathe. In taking the realist stance, the question about *ex falso* or DS is the question of their truth or validity. It is the question as to whether things are such that *ex falso* or *DS* are objectively so. Such beings have to *learn* how to ask alternative questions. Such a possibility is sometimes lost on disputants even when they play *ad hominem* strategies. When played cooperatively, it is natural to think of them as played from the realist stance. Even when played critically, the protagonist – or in **DL3** the speaker – is given no discouragement from taking the realist stance.

   Cost-benefit arguments, when *directed to matters that do not initially strike us as having a cost-benefit character*, are more easily seen as transacted from a perch other than the realist stance. Since costs and benefits can be conceived of as consequences of a person's principles or concessions, cost-benefit arguments are reconcilable to the *ad hominem* model in ways that make it especially clear that *ad hominem* arguments need not be taken from the realist stance.

   A second possibility is that there is nothing wrong in adjudicating conflicts about *ex falso* or DS from the realist stance, and that what the impotence of

DL3 suggests is that their adjudication is by way of case-making strategies that the competent ordinary practical reasoner does not possess. If this were true, there are two ways in which it could be so. In one way, there would be a principled and deep distinction between practical reasoning and theoretical reasoning whose significance here is that the adjudication of our present conflicts is a matter for theoretical, not practical reasoning. The other possibility is that conflict resolution in the abstract sciences needs no deep distinction between the practical and the theoretical, and that the significance of **DL3** 's impotence is that *ordinary* thinking will not do for the matters before us.

These are possibilities worth considering. I admit to thinking that the second of them, together with its attendant subpossibilities, is too speculative, too light on content and detail, to give much guidance about how to proceed with actual cases. At the same time, I admit to a certain fondness for the first possibility, although I have no satisfactory idea how it might be demonstrated. Beings like us reason and argue from the realist stance. To the extent that **DL3** is a realistic model of how beings like us reason and argue, it will carry that same presumption, even if **DL3** makes no contribution to the resolution of the conflicts about DS or any other basic principle of logic. I conjecture, therefore, that the moral of this tale is that satisfactory resolution strategies for the kinds of conflict here under review are those for which presumptions of the realist stance are unnecessary, if not simply inappropriate.

But what of the fate in **DL3** of the received diagnostic and triagic reactions to the Russell and Tarski paradoxes? Using the apparatus of **DL3**, let us mimic the (epistolary) conversation between Russell and Frege in 1902. Russell is R and Frege is F.

1. R: If **R** – the set of all non-self-membered sets – exists, then $\mathbf{R} \in \mathbf{R}$ and $\mathbf{R} \notin \mathbf{R}$.
2. F: Yes.
3. R: By axioms we both accept, **R** exists.
4. F: Agreed.
5. R: So we're in trouble.
6. F: Indeed.

At this juncture, we break away from the actual, historical conversation between Russell and Frege. I want to emphasize that in the opinion of the historical Russell and Frege the paradox of sets at a *minimum* destroyed the intuitive concept of set.

This view, *Frege's Sorrow*, would *not*, however, have been the conclusion had this conversation been regulated by Girle-rules. To see how this comes to pass we restart the conversation at step 7.

7. R: Since our resolution rules tell us to drop a conjunct if a statement in our *CS* is an immediate contradiction, let's drop "$\mathbf{R} \in \mathbf{R}$."

8. F: All right. But there is also a rule about honoring immediate consequences of what's left, i.e., in this case, "**R** ∈ **R**." The trouble is that
"**R** ∉ **R**" immediately restores "**R** ∈ **R**"; and we're right back where we
started.

9. R: Yes, the rules seem to drive us into an endless cycle of resolution and
paradox-rebirth.

10. F: Of course, there is no prospect of wriggling out of Excluded Middle,
is there?

11. R: No; it's a principle of logic.

12. F: But look, R. You've shown that if **R** exists then **R** ∈ **R** and **R** ∉ **R**.

13. R: Unfortunately.

14. F: Now the consequent of that conditional is a logical falsehood, is it
not?

15. R: Yes; and of course its negation is a logical truth.

16. F: Right. *And* we can't give up that logical truth and we can't give up
your fateful conditional.

17. R: Nor can we give up *modus tollens*, another principle of logic.

18. F: Which, together with the conditional and our logical truth produces
as an immediate consequence the negation of its antecedent.

19. R: You mean, that R doesn't exist, after all?

20. F: Right.

21. R: So arithmetic isn't toppling?

22. F: No, it isn't.

What is so striking about the Girle apparatus is that it imposes, perhaps inad-
vertently, the Barber solution on the paradox of sets.[27] Whether it is right to
do so is not here an issue, although I shall try to make it one in Chapter 7. The
point rather is that it takes a radically deflationary stand against paradoxical
hysteria. This is not only for the Russell paradox, but also for the paradox of
the Liar, putting "If L is a statement, then L is true and L is not true" in place
of "If R is a set, the R ∈ R and R ∉ R" and repeating the Girle conversation.

DL3 is of no consequence for disputes about what are and are not principles
of logic. But it is tailor-made for paradoxes, and should be welcomed by them.
We have been allowing ourselves the assumption that systems such as **DL3**
do a plausible job of modeling ordinary practical reasoning. **DL3** 's inability
to engage the resolution issue with regard to *DS* we took as suggesting that
such issues are not effectively dealt with from the perspective of the realist
stance. If this is right then the opposite subject applies to the paradoxes. That
DL3 diagnoses them with ease, and that **DL3** – a system of ordinary practical
reasoning – takes the realist stance may incline us to think that there is a fact
of the matter about what the paradoxes show. Also important is **DL3** 's utter
resistance to the symptoms of *Frege's Sorrow*.[28] Jointly we would have it that it
is a fact that the Russell set does not exist, and a fact that the Tarski statement
does not exist, and a further fact that the intuitive *concepts* of set and of truth

are left standing. This being so, the whole rationale for a mode of recovery that instantiates the idealism sketched in the Prologue – and that catapults mathematics and semantics into postmodernist embrace – now lapses. It is too much to suppose that **DL3** proves these things that I say it suggests. I am content with the suggestions. They give us something to think about.

Pages ago we met with the idea of ambush arguments. We said that Quine's attacks on modal logic typify such arguments. This is a matter that will occupy us in the chapter to follow.

2

Modalities

[T]he bold bridgeheads seized by intuition must be secured, by thorough scouring for hostile bands that might surround . . . and destroy them.
    Morris Kline, *Mathematics: The Loss of Certainty*, 1980

As widely understood, a modal logic is one whose logical particles include sentential operators that in their well-formed use inhibit the generality of substitution procedures. Among these we find the alethic modals, "necessarily" and "possibly," the deontic modals, "it is permissible that," and "it is obligatory that," the epistemic modals, "it is known that," and "it is believed that," the causal modals "because," and "it is causally sufficient that," and the temporal modals, "at t," "before t"; and so on. Modal systems are intensional. It is largely a matter of etymological advertence that a full-blown equivalence between the two has not taken root in our taxonomic practice, no doubt a reflection of the fact that not every category of syntactic element that makes a system intensional is, in the schoolboy's grammar of such things, a modal term. Still, there is nothing against the equation for those who find themselves drawn to it. In this more broadly conceived sense, a modal logic is one that sanctions *any* constraint that precludes full-bore extensionality. Seen in this broader way, modal logics include those made intensional by the English modals they embed, those made intensional by constraints required for relevance, and those made intensional by their recognition of semantic relations among elementary sentences. So the reach of the more expansive conception stretches from modal logics more narrowly conceived, to relevant logics, and onward to quantum logic. Each of the three has been the center of considerable and unabated controversy. Any account purporting to say something illuminating about the dynamics of conflict resolution in the abstract sciences would do well to give its attention to controversies of this kind. The first two will occupy us in the present chapter; the third is reserved for separate treatment in work in progress (Woods and Peacock, forthcoming).

QUANTIFIED MODAL LOGIC

S5 is one of the cluster of systems of modal propositional logic that Lewis started studying as early as 1912. In the basic systems, Φ, ψ, χ range over sentences, "M" is for possibility, "N" is for "$\neg M \neg$.", $\ulcorner \Phi \!-\!_3 \psi \urcorner$ is for $\ulcorner N(\Phi \supset \psi) \urcorner$ and $\ulcorner \Phi \equiv \psi \urcorner$ is for $\ulcorner \Phi \!-\!_3 \psi \wedge \psi \!-\!_3 \Phi \urcorner$. Here, for those who have not seen them before, is a brief sketch of the Lewis systems.

Rules
Substitution: If χ' is like χ except for containing ψ at some place(s) where
   χ contains Φ, then $\vdash \ulcorner \Phi \equiv \psi \cdot \supset \cdot \chi \equiv \chi' \urcorner$
Adjunction: If $\vdash \Phi$ and $\vdash \psi$ then $\vdash \ulcorner \Phi \wedge \psi \urcorner$
Detachment: If $\vdash \Phi$ and $\vdash \ulcorner \Phi \!-\!_3 \psi \urcorner$ then $\vdash \psi$

Axioms
For S1[1]:
B1: $\Phi \wedge \psi \!-\!_3 \psi \wedge \Phi$
B2: $\Phi \wedge \psi \!-\!_3 \Phi$
B3: $\Phi \!-\!_3 \Phi \wedge \Phi$
B4: $(\Phi \wedge \psi) \wedge \chi \!-\!_3 \Phi \wedge (\psi \wedge \chi)$
B5: $\Phi \!-\!_3 \Phi$
B6: $(\Phi \!-\!_3 \psi \wedge \psi \!-\!_3 \chi) \!-\!_3 (\Phi \!-\!_3 \chi)$
B7: $(\Phi \wedge \Phi \!-\!_3 \psi) \!-\!_3 \psi$

For S2:
B1 − B7 + B8: $M(\Phi \wedge \psi) \!-\!_3 M\Phi$

For S3:
B1 − B7 + B8: $(\Phi \!-\!_3 \psi) \!-\!_3 (\neg M\psi \!-\!_3 \neg M\Phi)$

For S4:
B1 − B7 + C10: $\neg M \neg \Phi \!-\!_3 \neg M \neg \neg M \neg \Phi$

For S5:
B1 − B7 + C11: $M\Phi \!-\!_3 \neg M \neg M\Phi$

For S6:
S2 + C13: $MM\Phi$

For S7:
S3 + C13: $MM\ \Phi$

For S8:
S3 + $\neg M \neg MM\Phi$.

S5 is one of the propositional systems that extends in a natural way to a quantification theory, and this is one reason for my selecting it here. S5, like the other systems, gives implication by definition:

$$\Phi \!-\!_3 \psi = df \neg M(\Phi \wedge \neg \psi).$$

It is also true that what appears to be the same concept is captured in the metatheory of the propositional fragment of classical first order logic (**CFL**): Φ implies ψ iff $\ulcorner\Phi \supset \psi\urcorner$ is a valid sentence, that is, true in every model.

The fate of modal logic bears on our question about *ex falso*. Modal logic cashes the philosophical commonplace that implication is a modal relation, that is, implication is logical *necessitation*. Whether it is or not, settling the matter does not decide *ex falso* for us. One does not have to be a strictist to have an account of implication in which *ex falso* is true. This is made apparent by the Lewis-Langford proof, which holds in nonmodal proof theory. A simple way of seeing the significance of this is to bear in mind that in the proof each derived line is taken to be implied (really implied, as opposed to materially) by the cited preceding line or lines, and implication (real implication, not material) is taken to be transitive. Decisions as to whether the rules used in the proof establish implications at each new line and whether implication is transitive do not require a decision on whether implication is a modal relation. A deeper and more interesting discouragement of the idea that implication needs to be understood modally is supplied by Richard Cartwright's elegant construction of strict implication (nearly enough) out of a nonmodal language (Cartwright, 1987a). Cartwright's approach will repay a short detour. So consider the classical language S of the propositional calculus. S satisfies these three conditions: (i) every member of S is either true or false; (ii) for every $\Phi \in S$, $\ulcorner\neg\Phi\urcorner$ is true iff Φ is false; and for every Φ, $\psi \in S$, $\ulcorner\Phi \supset \psi\urcorner$ is true iff Φ is false or ψ is true. Jointly these conditions are equivalent to a fourth:

(iv) The correct valuation on S is in B(S), where B(S) is a set of Boolean valuations on S that induces an implication on S.

We have it straightaway that Boolean implications for classical propositional languages are truth-preserving. But we do not have it that they are strict. Perhaps we can get strictness from a stronger result, concerning which a few preliminaries are needed.

A *homomorphism* of a propositional language $\langle S, P, C, N\rangle$ into a propositional language $\langle S', P', C', N'\rangle$ is a function h from S into S' that preserves the sentence-forming operations; that is, $h(N\Phi) = N'h(\Phi)$ and $h(C\Phi\psi) = C'h(\Phi), h\,\psi)$, for every Φ and ψ in S.

Where $h''\chi$ is homomorphic image of χ, that is, the set of values of h for members of χ as arguments, we obtain

(1) If h is a *homomorphism* of $\langle S, P, C, N\rangle$ into $\langle S', P', C', N'\rangle$ then $h''\chi \vdash_{B(S)} h(\Phi)$ whenever $\chi \vdash_{B(S)} (\Phi)$

And from (1) it is easy to get

(2) If $\langle S, P, C, N\rangle$ is classical and h is a homomorphism of $\langle S, P, C, N\rangle$ into a classical propositional language $\langle S', P', C', N'\rangle$ then $h(\Phi)$ is true whenever every member of $h''\chi$ is true and $\chi \vdash_{B(S)} (\Phi)$.

What does (2) tell us? Let us put it that an argument in a propositional language L is an ordered pair $\langle\chi,\Phi\rangle$ such that χ is a set of sentences of L (the premises of the argument) and Φ is a sentence of L (the conclusion of the argument). Let us also say that a counterargument to an argument ⟨χ,Φ⟩ in L is an argument ⟨χ′,Φ′⟩ in a propositional language L′ that has true premises and a false conclusion but that is nevertheless a homomorphic image of ⟨χ,Φ⟩ in the sense that $h\chi = \chi'$; and $h(\Phi) = \Phi'$ for some homomorphism h of L into L.

We can now see that (2) comes to this: if an argument in a classical propositional language belongs to the Boolean implications for the set of sentences of the language, then no counterexample to it exists in any classical propositional language.

Can one conclude from this that Boolean implications for sets of sentences of classical propositional languages are strict? No. Even if we assume that all possible classical propositional languages are actual, we can conclude, together with (2), that if an argument in a classical propositional language belongs to the Boolean implication for the set of sentences of that language then no counterexample to the argument exists in any possible classical propositional language. But nothing excludes the possibility that the argument should have had true premises and a false conclusion.

Of course, for those who distrust the modalities, (2) has a certain charm. It is a clear invitation to make do with the idea of the nonexistence of a counterexample, never mind that it does not quite give us strictness. It is good enough and modal-free, and implications that satisfy (2) reflect the converse of following from or, as we say, logical consequence.

It is easy to see that this rough concurrence between S5 and Cartwright's system could well persuade a conservatively minded logician to the view that if getting implication right were the sole object of S5 then there is no need of S5, since implication is got right in a simpler system. In fact, of course, Lewis had further designs. He thought that the concepts of necessity and possibility were worthy of logical analysis in their own right.

Contra the Modalities. In truth, then, it seems that Lewis need not be troubled by this conservatism about implication. But another objection – indeed an ambush – lies in wait. The objector is Quine. Here is how I read the basic structure of his ambush:

(1) A system of modal propositional logic is a *bona fide* logic only if it has a quantificational extension.

(2) S5, for example, has a quantificational extension, but there is something disastrously wrong with quantified modal logic (**QML**).

(3) The difficulties of **QML** are of such a character as to show – or strongly indicate – that they derive from the modal peculiarities of the embedded propositional system.

(4) So not only is **QML** a failure, so, too, is any propositional system that extends its hospitality to the modalities.

The heart of the ambush of modal logic is premise (2) of Quine's complaint. If (2) fails, his argument collapses. Quine pursued the theme of modal apocalypse from the earliest days of his philosophical work. These arguments constitute Quine's ambush of *the very idea* of necessity and possibility. He insists on their incoherence, never mind that there were no prior symptoms of it. Quine forwards six arguments against **QML**.

1. ("Notes on existence and necessity," 1943). The statement (1) $(\exists x)N$ $(x > 7)$ is curious. " ... would 9, that is, the number of planets be necessarily greater than 7?"
2. ("The problem of interpreting modal logic," 1947).

If we accept the substitutional account of the quantifiers then we have the problem. Let **c** be the congruence relation on morning star (MS), evening star (ES), and Venus, such that

(1) MScES $\wedge$ N(MScMS)
(2) EscES $\wedge$ ¬N(EScMS)

Thus, we get

(3) $(\exists x)$ (xcES $\wedge$ N(xcMS)   from 1)
(4) $(\exists x)$ (xcES $\wedge$ ¬N(EScMS)   from 2)

But right-hand 3 and right-hand 4 are one another's contraries and cannot be satisfied by the same object, Venus. In fact, there must be at least three different objects occurring in our congruence relation. Astronomy is overturned.

3. (Quine, 1953a)
   One and the same number x is uniquely determined by the condition

   (1) $x = \sqrt{x} + \sqrt{x} + x \neq \sqrt{x}$

   and

   (2) There are exactly x planets.

   But (1) has "$x > 7$" is a necessary consequence and (2) does not.
4. (Quine, 1966)
   Suppose those we try to get around this difficulty as follows. We restrict our universe of discourse to objects x such that any two conditions uniquely determining x are analytically equivalent, that is, we put it that

   (i) $(\forall y)(Fy \equiv y = x) \supset N (\forall y)(Fy \equiv y = x)$

   But introducing "①$= z$", for "F①" in (i) and then simplifying and closing, we get

   $$(\forall x)(\forall z)(x = z) \supset N(x = z).$$

5. ("Three grades of modal involvement," 1953.)
Quine gives an argument to the same effect. He also claims that essentialism, which is wholly untenable, is unavoidable in any system of **QML**. For if we have

(1) $NFx \land Gx \land \neg NGx$

then we have

(2) $(\forall x)(NFx \land Gx \land Gx)$

putting "x = y" for "Fx" and "x = x.p" for "Gx," where p is any contingent truth.

6. (Quine, 1960)
If we accept

(1) $(\forall y)(Fy \equiv y = x) \land (\forall y)(Gy \equiv y = x) \supset N (\forall y)(Fy \equiv Gy)$

and if

(2) p

is any true sentence, and

(3) $w = x$

then we have

(4) $(\forall y)(p \land y \equiv w \equiv y = x)$

and

(5) $(\forall y)(y = w \equiv y = x)$

Then, putting "p . ① = w" for "F①" and "① = w" for "G①" in the first formula above we get

(6) $(\forall y)(p \land y = w \equiv y = x) \land (\forall y)(y = w \equiv y = x)$
$\supset N(\forall y)(p \land y = w \equiv y = w)$

But (6), (5), (4) jointly imply

(7) (7) $N(\forall y)(p \land y = w \equiv y = w)$

and (7) implies

(8) $p \land w = w \equiv w = w$

which implies

(9) p.

But since (7) arises from a necessary truth, p must be necessary. Hence

(10) Np.
Modal distinctions collapse.

Quine's objections to **QML** all turn on one or more of the following matters: (a) essentialism, (b) definite descriptions, and (c) the identity relation. If we are properly to take the measure of his complaints, it is important to be clear about these things.

ESSENTIALISM: Quine regards the doctrine as unintelligible. He says that there is no defensible reason to say of a bicycling mathematician that he is necessarily rational and only contingently two-legged.

Virtually all essentialists have wanted to mark the following distinction:

(i) Necessarily Socrates is rational [N(Fa)]

and

(ii) Socrates is necessarily rational [F^Na]

In (i) we have a *de dicto* modality, in which necessity is represented as a trait of sentences, in such a way that (i) is true if and only if the sentence "Socrates is rational" is true in all possible worlds. On the other hand, (ii) imputes a *de re* modality, in which the property of being necessarily rational is attributed to the object Socrates, true of him in every possible world in which Socrates exists.

The principal difference between (i) and (ii) is that (i) imputes to Socrates *necessary existence*, whereas (ii) does not. On its most natural reading, (ii) says not that Socrates exists in every possible world but, rather, that there is no possible world in which he exists but fails to be rational.

Accordingly, we may propose a definition of essentialism by way of necessity *de re*:

Def: a is F essentially iff a F^Na iff (a is F) $\wedge$ N($\neg$(a is F) $\supset$ $\neg\exists$x(a = x)).

Thus, any property a thing possesses essentially is one whose loss would extinguish that thing's self-identity.

DEFINITE DESCRIPTIONS: Definite descriptions are expressions in the form "the so-and-so" (e.g., "the present Queen of England," the "only even prime," the "sum of 2 and 7," the "the wife of Jean Chrétien," etc.) Definite descriptions form a subset of singular terms. They are noun phrases in referring position in singular sentences, or in embedded singular clauses. As such, definite descriptions are subject to what might be called "the standard policy" on singular referring terms.

I. *The Substitutivity of Identicals*: Let us say that if α and β are terms for which $\ulcorner \alpha = \beta \urcorner$ is true, then α and β are *coreferential* terms. Then for any pair of coreferential terms v_1 and v_2, v_1 may be substituted for any occurrence of v_2 in any extensional sentential context, *salve veritate*.

II. *The Principle of Semantic Constancy*: If a sentence ψ arises from a sentence Φ by substitution at one or more places of coreferential terms α and β, then where ψ differs from Φ (vis., at α and β), α and β may differ from one another only syntactically; that is, both α and β must perform *the same semantic function* in both Φ and ψ.

Consider the terms "Cicero" and "Tully" made coreferential by virtue of the truth of

(1) Cicero = Tully.

It is known that

(2) Cicero was a Roman orator.

Then, by the Substitutivity of Identicals, we may infer

(3) Tully was a Roman orator

but only if the semantic roles of "Cicero" in (2) and of "Tully" in (3) are precisely the same (for otherwise, (1) could not be true). In particular, Semantic Constancy precludes that in (2) "Cicero" refers to a city in Illinois, near Chicago; and it likewise precludes "Tully" in (3) referring to a recent Dean of Arts at the University of British Columbia.

IDENTITY: Essentialists and nonessentialists alike want to recognize a distinction exemplified by

(1*) Cicero = Cicero

and

(1) Cicero = Tully

On the face of it, this is a problematic distinction. For one thing,

(1**) $\forall x(x = x)$

can be taken as a necessary truth *de dicto* (for even in the empty world, (1**) is vacuously true; that is, it will still be true that $\neg \exists y(y \neq y)$). And although (1*) looks like a straightforward universal instantiation of (1**) it cannot be both true and necessary *de dicto*, since it is not a necessary truth that Cicero exists. Therefore, we say that (1*) is necessary *de re*. On the other hand, (1) is purely contingent. It is necessary neither in the *de dicto* sense nor the *de re* sense.

Consider now the following claim:

(4) The only way in which it could be false that Cicero is identical to Cicero is if Cicero did not exist.

If we apply to (4) the Substitutivity of Identicals, and if Semantic Constancy is also honored, we obtain

(5) The only way in which it could be false that Cicero is identical to Tully is if Cicero did not exist.

But this is wrong. Or on its most natural interpretation it is wrong. For here we want to say that *one* way in which it could be false that Cicero is identical to Tully is if the existent Tully were named "Charlie" rather than "Cicero."

Hence, if (5) is false, Semantic Constancy is breached. So there must be a pair of contexts in which it is clear that the semantic roles of "Cicero" and "Tully" differ from context to context. Two such contexts are (1*) and (1). For if Semantic Constancy is honored at (1*) and (1), then *both* are necessary *de re*, or *both* are purely contingent. But this is not what they are on their most natural readings. Hence "=" does not occur in (1) as the sign for numerical identity. That is, (1) is equivalent to

(5) $\exists x$("Cicero" names $x \land$ "Tully" names x).[2]

When a sentence such as (1) stands so to a sentence such as (5), logicians say that there is a term in (1) that is *contextually eliminated* in this process. Russell called an expression that is contextually eliminable from a context an *incomplete symbol* in that context. In the present example, we see that "=" is an incomplete symbol in (1) but not in (1*). Hence, if we say that (1*) is a statement of numerical identity, (1) is not a statement of numerical identity.

We now have the wherewithal to appraise Quine's six objections to **QML**.

Objection One (Quine, 1953a)

(1) $\exists x N(x > 7)$.

is accessible to the Existential Instantiation rule. Here are two applications of it:

(2) $N(9 > 7)$

and

(3) N (The number of planets >7)

Although it may be the case that (2) is all right, (3) is absurd. Hence, there must be something wrong with (1). Here we meet with the basic structure of Quine's objection. Existential Instantiation is a valid rule. In its application to (1), it produces (2) and (3), something apparently unproblematic in the first instance and highly problematic in the second. Contrary to initial appearances, there must be something wrong with (1). *Diagnosis* is required. Certainly there is nothing wrong with (1) *minus* the modal operator. It is a sentence from which the instantiations (2) and (3) follow unproblematically. By simple elimination, "N" is the culprit. The least damaging thing to say against (1) is that "N" occurs in the wrong place, viz., in the scope of the quantifier "$\exists x$." But a system of logic is not a **QML** unless it tolerates such placement of its modal expressions. Hence, there is something wrong with **QML** as such.

Reply: There is no particular reason to give "N" in (1) a reading *de dicto*. It is already problematic whether even (2) is true (where "N" is *de dicto*), short of ascribing necessary existence to the integers. So the commonplace claim that there are some things that are necessarily greater than 7 should be analyzed as

(1*) $\exists x \, (x > {}^N 7)$.

As for (3), if the Principle of Semantic Constancy is honored in the move from (1) to (3), then it is essential that in (3) the definite description "the number of planets" names the number 9. This is because (3) comes from (2) by the Substitutivity of Identicals on

(4) The number of planets = nine.

If "=" is the identity relation in (4), and if (4) is true, it must be flanked by coreferential terms. Hence, if (2) were all right, so, too, would (3) be. But (3) does not *look* right. On its most natural reading it is synonymous with

(5) Necessarily, there are more than seven planets.

Now (5) is wrong for both *de dicto* and *de re* reading of "N." But it does not matter. We cannot get (5) from (1) except by breaching Semantic Constancy.

Objection Two (Quine, 1943, p. 123). It is essentially the same objection as before. I give it essentially the same reply.

Objection Three (Quine, 1947, p. 47). Let c be the congruence relation on the set {The Morning Star, The Evening Star, Venus}. Then we have both

(6) (MS c ES) $\wedge$ N(MS c MS)

and

(7) (ES c ES) $\wedge$ ¬N(ES c MS).

But from (6) we have

(8) $\exists x(x$ c ES$) \wedge$ N$(x$ c MS$)$

and from (7) it follows that

(9) $\exists x(x$ c ES$) \wedge$ ¬N$(x$ c MS$)$.

Since the rightmost clauses of (8) and (9) are contraries, no one object can satisfy these clauses. Thus, our original set gives us three *different* objects, which is absurd. Here, too, the Quinean paradigm is evident. Existential Generalization is a valid rule applied to sentences that appear to be unproblematic. The appearance is deceiving, since (8) and (9) jointly falsify astronomy. It follows that here is something wrong with at least one of (7) and (8) after all. In their unmodalized version all this trouble disappears. The diagnosis is unmistakable. The culprit is the modal "N."

Reply: If we want to say that the Morning Star is necessarily self-congruent, the necessity must be read as *de re* (otherwise the MS would exist necessarily). Proposition (6) is a completely unacceptable reading of this innocent claim. Moreover, if in (7) Semantic Constancy is honored, then its rightmost clause is as acceptable or not as the rightmost clause of (6). Of course

(10) MS = ES.

But if this is to be a purely *contingent* claim, it is equivalent to

(11) $\exists x$("MS" names $x \land$ "ES" names x).

As we see, we get (8) and (9) only if we violate Semantic Constancy, which we must not do if (6) is a genuine statement of numerical identity. Of course, on its most natural reading it is not. That is, "=" is contextually eliminable in (10) in favor of (11). In which case "MS c ES" is not even necessary *de re*.

Objection Four (Quine, 1953a, p. 149). One and the same object is uniquely determined by the conditions

(12) $x = \sqrt{x} + \sqrt{x} + \sqrt{x} \neq \sqrt{x}$

and

(13) There are exactly x planets.

But (12) has a necessary consequence.

(14) $x > 7$

whereas (13) does not.

Reply: What Quine is trying to say is that 9, which satisfies (12), is necessarily greater than 7; and that 9, which satisfies (13), is not necessarily greater than 7. Quine is confused. Being necessarily (*de re*) greater than 7 is a trait of 9 each time. What we do *not* have is

(15) There are necessarily more than seven planets

that is,

(16) There are of necessity more planets than seven.

Now (16) is certainly not true. On the other hand, (16) is not derivable from anything that goes before.

Objection Five (Quine, 1966/1976, p. 173). Suppose we grant that

(17) $(\forall y)(Fy \equiv y = x \land (\forall y)(Gy \equiv y = x) \supset N(\forall y)(Fy \equiv Gy))$

Then we obtain

(18) $(\forall y)(Fy \equiv y = x) \supset N(\forall y)(Fy \equiv y = x)$

which gives

(19) $(\forall x)(\forall z)(x = z \supset N(x = z))$.

Hence, there are no contingent truths of identity, which is absurd.

Here the Quinean strategy changes. Let us, he says, imagine a domain of objects in which (17) might be true. Certainly no one other than a neo-Leibnizean would see it as true in the general case. Perhaps it is true in the natural numbers. Perhaps it is correct to say, for example, that the two conditions on the integer four, "sum of 1 and 3" and "sum of 2 and 2" are necessarily equivalent. But (17) turns out badly; it gives us the collapse of the modal distinction on the identity relation. The trouble disappears when (17) is demodalized, and with it (18) and (19) too. By simple elimination, the diagnosis must be that the culprit is "N," notwithstanding that in (17) it does not occur in the scope of a quantifier.

Reply: The conclusion that there are no contingent truths of identity is true, not false *if* Semantic Constancy is honored and *if* necessity is construed as *de re*. For consider

(20) $\text{Cicero} = {}^N\text{Cicero}$

which is true, and

(21) $\text{Cicero} = \text{Tully}$

which is also true. Then by the Principle of Substitutivity of Identicals, *together* with Semantic Constancy, we substitute "Tully" for the second occurrence of "Cicero" in (20) and obtain

(22) $\text{Cicero} = {}^N\text{Tully}$

which is true. Of course it is *not* true on the most natural reading of (21), viz.,

(23) $\exists x(\text{"Cicero" names } x \wedge \text{"Tully" names } x)$.

For on that reading, (22) would be construed as

(24) $\exists x(\text{"Cicero" names}^N x \wedge \text{"Tully" names}^N x$

which is indeed absurd. But note that (24) does not come from (20) by Substitution. This is because (21) is not a statement of numerical identity. The sign "=" is contextually eliminable in (21), in the manner of (23).

There is a final thing to say against this objection. We saw that (17) was ventured only for domains in which it might plausibly be taken for true. One such domain is the set of natural numbers. But according to the received wisdom, the natural numbers are distinctive: No theorem is contingent. No non-Leibnizean modalist will allow that every true identity whatever is a necessary truth, but this is precisely what he will allow – indeed, insist on – in the domain of natural numbers. Assumption (17) is plausible precisely for those domains in which modal collapse is already a desired consequence. The derivation of

(19) therefore fails to stick the modalist with anything he would regard as absurd.

Objection Six (Quine, 1960, pp. 197–8). Here, too, we begin with assertion (17). Quine goes on to show that modal distinctions collapse under that assumption, that is, that for all Φ

(25) Φ iff N Φ

Reply: The same point applies to (17). The objection produces no absurdity, but rather the wholly welcome consequence that, for example, arithmetic truths are necessary. Apart from that, at one place in Quine's proof, it is asserted that

(26) $(\forall y)(p \wedge y = w \equiv y = x) \wedge (\forall y)(y = w \equiv y = x) \wedge N(\forall y)p \wedge y = x \equiv y = w).$

Note that "N" is in *de dicto* position. It has no right to be. It belongs in *de re* position only. But thus positioned, the proof collapses; we cannot validly derive (25). It is well to remember that Quine's acceptance of (17) is both tactical and highly restricted. It is intended for domains such as the integers, in which unique specification is necessarily unique specification. In such a domain that the number 4 is uniquely determined by the condition "2 + 2" and by the condition "3 + 1" gives us all we need to judge the hard equivalence of "is the sum of 2 and 2" and "is the sum of 3 and 1." But, again, there is nothing whatever surprising that in the theory of such a domain, viz., *arithmetic*, modal distinctions collapse. That is what every modalist already thinks, that is, that arithmetic truths are not contingent. This is an ironic twist for Quine; his embrace of an assumption that he does not like serves to deliver a consequence to which the modalist is already pledged, and in which he sees no reason for concern.

All six of Quine's objections turn on wildly implausible readings of commonplace truths. These objections are so obviously defective and their shortcomings so apparent that it can only be wondered whether Quine has not been trifling with us. If so, it becomes necessary to ask what his real objections are. It would seem that they are two (and they are linked).

I. **QML** carries commitments to essentialism.
II. Modal contexts make it difficult (and sometimes impossible) to individuate the ontology of our best theories.

Essentialism is a trait of well-individuated individuals that retain their sharp identities over time and under change of circumstance and attribute. Essentialism holds for "Aristotelian substances." When he made his objections, Quine was a physicalist. To the extent that ontology counts for anything in science, all that there is are "classes of quadruples of numbers according to an arbitrarily adopted system of coordinates" (Quine, 1981a, p. 17). But sets have extremely fragile identity conditions. *Any* change in the membership of a set destroys it. Sets are not Aristotelian substances, and since essentialism is a *consequential* metaphysical doctrine only as applied to Aristotelian substances, it is not a credible or useful doctrine otherwise. Since "otherwise" is precisely what obtains under Quine's physicalism, essentialism is a useless vestige of a naive and discredited metaphysics.

As for the individuation problem Quine sees in modal logic, consider again **QML** as a quantificational extension of a basic propositional system, S5, let us say, supplemented by identity and accompanied by a policy for definite descriptions. If in the general case a formula

(1) NA

is paraphrased as

(2) "A" is necessary

then since (2) is a singular assertion made up of a name (of a sentence) and a linguistic predicate, Quine is prepared to allow (1) as a paraphrase of (2), since (2) is an extensional sentence.

But consider the sentence

(3) There is someone who could meet the train.

What will its linguistic paraphrase be? What about

(4) $\exists$x ("*x* meets the train" is possible)?

No. Quotation is an opaque context. The quantifier of (4) fails to bind any variable in the open sentence.

Most current systems of quantified modal logic treats the modalities by way of possible worlds. A sentence is necessary if and only if it is true in all possible worlds, possible if and only if true in some, and impossible if and only if true in none. If we wish to say, for example, that there is something, viz. the number two, that is necessarily prime and even, that is,

(5) $\exists x$N (*x* is prime $\wedge$ *x* is even),

this will cash out in the possible world semantics only if the bound variable takes the same value in all possible worlds. Similarly for singular sentences such as

(6) The Countess of Snowden is necessarily rational.

Here, too, it is a condition on the truth of (6), assuming for now we take it to be true, that in every possible world "The Countess of Snowden" denotes the same person as it denotes here in the domestic world. At the heart of the possible worlds project is the necessity to preserve the transworld identity of individuals, this is, in David Kaplan's memorable phrase, "the problem of transworld heirlines." Quine believes that the problem has no nonarbitrary solution.

The best way to understand Quine's skepticism is to consider first the question of *transtemporal* identity in a *given* world, say this one.

We have it that

(7) The person who is the Countess of Snowden used to be
Princess Margaret Rose.

Is there one person persisting through time or not? We cannot strictly say. The reason, as we have mentioned, is that Quine has foresworn the ontology of spatio-temporal particulars, that is, of discrete substances or bodies. In the case before us, since the Countess of Snowden is an embodied person, the ordinary idea of person is also foresworn. One speaks instead of "the material content of any portion of space-time, however irregular and discontinuous and heterogenous" (Quine, 1981a, p. 10). This is what physical objects have come to. To ask whether Margaret Rose is whom the Countess of Snowden used to be is to ask whether there are objects in this new sense containing all the events occurring in the space-time of (loosely) the Countess of Snowden and whether they include all of those events occurring in the space-time of (loosely) Margaret Rose. The answer is that some do. Others contain some but not all, and others contain none. Which of these gives precisely the physical history of that British Royal? It is perhaps not quite clear that there is an answer, although it also is true that we are not free to say just what we please. Transtemporal evolutions are imposed on us; they arrive already structured, to some extent, as cross-sections of our domestic world that do indeed seem uniquely vouchsafed (Quine, 1981a, p. 12). If Quine is right, then even though we cannot account for the transtemporal identity of (7) via old-fashioned bodies we can account for it somehow.

Modal logic demands a further liberalization of the notion of body, and that precisely is the trouble with modal discourse:

Physical objects would be simply the sums of physical objects of the various worlds, combining denizens of different worlds *differently*. (Quine, 1981a, p. 126, emphasis added)

The modal theorist demurs. No, he says, they surely would be the objects most resembling themselves as they are in a given world. Quine thinks that any such talk of transworld intimacy is misguided. For whereas transtemporal evolutions are already structured and imposed on us, "all manner of paths of continuous gradation from one possible world to another *are free for the thinking up*" (Quine, 1981a, pp. 126–7, emphasis added).

The only way to defeat the liberality of such free creations is essentialism. (Quine, 1981a, p. 118; cf. Quine, 1960, p. 199). But essentialism is "surely indefensible" (Quine, 1960, p. 200). Modal theorists are not convinced. Adam cannot be Noah, they say, because they differ essentially – Adam, in not having any parents, and Noah, in arising from the fusion of just that ovum and just that sperm. So there are essences, and the transworld identity of things is underwritten and secured by them.

It is interesting that Quine concedes that essences make sense in context:

Relative to a particular inquiry some predicates may play a more basic role than others, or may apply more fixedly; and these may be treated as essential. (Quine, 1981a, pp. 120–1).

But talk of essences belongs to talk of old-fashioned substances, and substances are long gone; they are no "proper annexe to austere scientific language." They are not needed for describing the facts because they are not needed for physics.

In the end, Quine's basic worry about **QML** is not that it is not logic properly speaking. His reservations run deeper: *There is nothing there to accommodate.* Talk of quantified modalities requires talk of essences, and essences make sense, if at all, only as traits of substances. But there are no substances, anyhow no scientific need to think so.[3]

Quine's affection for physics is philosophically momentous. It induces him to withhold official recognition from, among other things, any philosophical theory that tries to make sense of our old fashioned world, the world as we experience it. Modal logicians, for all their technical and abstract enthusiasms, are old-fashioned theorists in just this sense. They take the world more or less as it comes; but it is precisely this world that is not needed for physics.

THE DIALECTICS OF AMBUSHES

Ambush arguments have their own dialectical peculiarities. As we have proposed to understand them here, ambushes purport to demonstrate hitherto unsuspected trouble. They are proofs that demonstrate the heretofore unrecognized necessity of diagnosis, triage, treatment, and prognosis. Often an ambush is all four things at once. In the example of Quine's attempted demolition of modal logic, one begins in innocence: Everyone knows that modal notions are difficult to analyze, but who would so much as suspect that they are *incoherent*? But before he is finished with them, all three things are in place: the proof of unsuspected trouble; then a diagnosis (the modals cannot be made to behave in contexts of identity and hard equivalence); next a triage (the modals are incoherent); and finally a treatment (give up the modalities). There is also a prognosis of sorts: Epistemology will fare better freed from the modal presumptions attaching to analyticity and syntheticity.

It remains true and distinctive of the ambush that it sets out to prove unsuspected trouble. In a dialectical sense, the "other party" is anyone previously inclined to see no trouble, the modal theorist in the present case. It falls to him to yield or fight back by attacking the proof of unsuspected trouble. There is a standard protocol for this. The respondent will either attack the validity of the proof or he will try, often at the point of triage, to slip the punch it delivers. In my own replies to Quine's proofs, I have tried both methods. All the arguments – so I said – are either non sequiturs, or turn on aberrant interpretations of commonplace truths; and some deliver consequences, or purport to, to which the modal theorist is already pledged and that cause him no trouble.

Quine is a famous debunker. He also ambushed the analytic-synthetic distinction. This has produced a huge literature of counterattack and rejoinder and it has won legions of philosophical converts. The ambush of modal logic has produced, in comparison, little more than nothing. Historically, it has not been the provocation of a big fight. Although there has been some discussion in the literature, it has not been the dominant reaction to Quine's proofs. The dominant reaction has been no reaction. Modal logicians go about their business without so much as a murmur about the ambush, as we see in the standard textbooks of Hughes and Cresswell, and of Chellas (Hughes and Cresswell, 1968; Chellas, 1980). On the face of things, this is a remarkable development. It is possible, of course, that modal logicians have simply shirked their intellectual obligations. It is also possible that their indifference reflects a certain view of conflict in the abstract sciences which we have yet to bring to the surface (doing so will be part of the business of Chapter 4).

Perhaps modal notions are fully coherent. Even so, if modal logic does imply essentialism, it is open to Quine to press into service a principle that resembles his principle M^3, the **maxim of minimum mutilation**. *We* might call it M^3p: Do not make concessions that injure the commitment to *physicalism*. Physicalism ordains physics in a hegemonic role for all of science. Hence, if essentialism is true anywhere, it must be true in physics. Quine's view is that it cannot be true in physics, since the objects of physics are mathematical, hence set-theoretic. I shall not here quarrel with this second claim, but I reject the first, physicalism. Quine himself has recently and reluctantly done the same (1990/1992, p. 36). The point remains that *ex falso* is true in classical logic never mind what we say about the modals. What is more, even if the modals are coherent notions, Quine could still be right in supposing that **QML** is not a logic. Perhaps the study of modals more properly belongs to the philosophy of language or some such thing (Harman, 1972, pp. 75–84). Logic or not, the propositional fragment of **QML** gives *ex falso*, which in any event is also given classically in the metatheory of the standard logic of propositions. If our interest is in settling the furor that attaches to *ex falso*, we need not concern ourselves with the status and prospects of modal logic. This was the moral of Cartwright's essay (1987a). In one way, the status of *ex falso* is a more urgent question than the status of **QML**. The triage and treatment of paradoxes such as Russell's and Tarski's depend substantially on what contradictions imply. Even so, the mere fact, or appearance of it, of logical pluralism is not something to be ignored. Pluralism is problematic in two ways. It forces us to reflect on methods for the adjudication of contested claims made by rival systems. But, as has emerged from the discussion of **QML**, it faces us with the challenge of answering the following question. Let T be an abstract theory that, so far as we can tell, makes no mistakes and, in particular, produces correct analyses of its target concepts. Among examiners of T there is a cleavage of opinion among those admirers who hold that T *is a logic*, which is what the others deny. If we assume that the disagreement is neither casual nor uninformed, that it is a disagreement among people wholly at their ease with the methods and results of T, we expose ourselves to further questions. One is, "Is there a fact of the matter here?" Another is, "Would the existence of a fact of the matter *matter*?" What would it matter *for*? In practice, the question of when a theory is a logic does not attract much attention from logicians. Among working logicians there is a comfortable latitudinarianism. Logic is not a contested concept, much less an "essentially contested" one. It is something for which the idea of family resemblance seems tailor-made. Historically, such interest as there is in the question of when a theory is a logic comes straight out of a *philosophical* preoccupation. The best-known modern example of this is Frege's epistemological program in the foundations of mathematics. Frege wanted to prove the analyticity of arithmetic. This he set out to do by reducing arithmetic to logic. Here is a case in which our question seems to take on some real importance. For if what Frege called logic is not logic after all, then, even if the reduction came off, Frege would appear not have succeeded in reducing arithmetic to logic, and his program would have failed. I doubt it. It was not at all central to Frege's purposes that what he would reduce arithmetic to would be logic. What was fundamental was that it be *analytic*. Frege's system was an extensional second-order quantification theory. By some lights, second-order logic is not logic. But if Frege's system were analytic, and had Frege succeeded in reducing arithmetic to it, nobody disbelieving that second-order quantification theory is logic would have the slightest reason to question the analyticity of arithmetic.

Except when driven by an external agenda, our question has the look of a curiosity. It is a question that somehow we seem to think need not be much pressed. Perhaps we should be shocked that this is so, but it *is* so, and we should seek some instruction from it. It may bear on the matter that there should be so much untroubled ecumenism in what passes for logic – a good deal more than is to be found in topology or category theory, for example. We see in what passes for logic high levels of agreement to disagree. Among logicians who are also philosophers, Quine is perhaps preeminent in pressing our question. It is well to remember that the question is not whether this or that account of something that has caught our interest is correct, but rather, even if it were, is it logic? This serves to remind us that there is room to like *ex falso* even while rejecting any theory of implication in which *ex falso* is derivable but in which implication is a modal relation. Quine's resistance to the modals is a case in point. In their most extreme form, modalities are incoherent; so there is nothing in them for a theory or a logic to be a theory or a logic of. In a more subdued variation, even if modal notions had legs enough to make possible a theory of them, no such theory could count as logic. This leaves Quine and others who think like him oddly positioned. A modal theory might give an account of implication in which implication is inherently a modal relation. What is more, the theory might, in principle, be correct. Yet, on this subdued version of modal skepticism, the theory could not be a logic. One way of saying this with a certain verve is that, on the present view, even a correct theory of *entailment*, were such a thing possible, could not be a *logic of implication*.

PARACONSISTENT OBJECTIONS

For all his skepticism about modal theories of entailment, there is something that Quine and the strictist agree on: namely, that a condition on something counting as logic is that *ex falso* be derivable in it and that the Lewis-Langford proof be recognized as good. The strategy of the Lewis-Langford proof is a variation on Lockean *ad hominem* arguments. Instead of pressing an opponent with agreed consequences of what the opponent has conceded and, so to speak, put on the record, this proof derives the negation of the opponent's thesis from nothing he concedes, but only what he is asked to assume, by derivation principles whose validity he is *presumed* to accept. The proof is cooperative in that the principles are part of "common knowledge," for in the history of the dispute about *ex falso* up to the presentation of this proof there is no record of disagreement between the two parties over these principles. Here again is the proof.

(1) $\Phi \wedge \neg\Phi$ Hypothesis
(2) Φ 1, Simplification
(3) $\Phi \wedge \psi$ 2, Addition
(4) $\neg\Phi$ 1, Simplification
(5) ψ 3, 4, Disjunctive Syllogism

Hence, by Conditionalization, an arbitrary ψ follows from $\ulcorner\Phi \wedge \neg\Phi\urcorner$. Since for each of these valid principles there is a corresponding true entailment statement, for entailment is the converse of the relation of following from, and since entailment is transitive, we have it that $\ulcorner\Phi \wedge \neg\Phi\urcorner$ entails ψ. The effect of the proof is to present the opponent with a choice. He must either accept *ex falso* or he must assume the burden of showing that at least one of the deduction principles is invalid. The proof is made sincerely when its makers can justifiably assume that opponents would not not be prepared to accept this onus. As events turned, this assumption proved to be wrong, and, because it did, the dispute widened. Not only does *ex falso* stay unresolved, but it now extends to the validity of one or more of Simplification, Addition, Disjunctive Syllogism, and Conditionalization. Here is another situation tailor-made for ambush. We must look for attacks on principles or assumptions never before in doubt. Although this is the general strategic pattern of the proof, there is at least one historical example of a rebuttal of it that steps out of the strategy's ambit, but that still constitutes an ambush. It is the objection of the *School of Cologne*.

For the Lewis-Langford proof to work, we must allow that for all valuations v

$$v(\Phi \wedge \neg\Phi) = F$$
$$v(\Phi) = T \text{ iff } v(\ulcorner\neg\Phi\urcorner) = F$$
$$v(\Phi) = F \text{ iff } v(\ulcorner\neg\Phi\urcorner) = T$$
$$v(\psi \wedge \chi) = T \text{ iff } v(\psi) = T \text{ and } v(\chi) = T$$
$$v(\psi \wedge \chi) = F \text{ iff } v(\psi) = F \text{ or } v(\chi) = F$$

Moreover, with $\ulcorner\Phi \wedge \neg\Phi\urcorner$ taken as the first step of the proof, it is necessary that the assumption be made without prejudice to these facts.

The Cologne objection is that it is incoherent to think that we can assume $\ulcorner\Phi \wedge \neg\Phi\urcorner$ without prejudice to these facts. For in assuming $\ulcorner\Phi \wedge \neg\Phi\urcorner$, we assume that $v(\Phi \wedge \neg\Phi) = T$. But if $v(\Phi \wedge \neg\Phi) = T$, and if Simplification is a good rule (as the proof itself assumes), then the occurrence of $\ulcorner\neg\Phi\urcorner$ at line 4 cannot displace the occurrence of Φ in $\ulcorner\Phi \vee \psi\urcorner$ at line 3.

If the objection were sound, nothing but logical truths would follow from a contradiction. For let ψ be any contradiction and let χ be any proposition. Now consider any valid Σ:

1. ψ assumption

　　·

　　·

　　·

n − 1.

χ

Since ψ is assumed at line 1, then in this argument $v(\psi) = T$. It is nowhere in dispute that if Σ is valid then it is not possible for $\psi, \dots,$ to be true and χ false. Holding ψ true, as one must in this proof, there is precisely one circumstance in which the possibility that ψ is true and χ is false is blocked, namely, when χ is true. For this case to obtain, χ must be a logical truth; for if it were not, it would be possible that ψ is true and χ is false.

How does χ get to be true in Σ? Certainly not in the way that ψ does. ψ gets to be true in Σ *by assumption*. Σ aspires to show what propositions follow from ψ, thus assumed, without any constraints on χ except those required by the entailment of χ by ψ. So, again, when does χ guarantee the impossibility of ψ being true and χ false when it is guaranteed (by assumption) that ψ is true? The answer is, when χ is a logical truth. This is how χ gets to be true in Σ; it is true everywhere.

It is a *reductio*, of course. In particular, it requires that Simplification fail with respect to the conjunction of any nonlogical truth and its negation, provided that some contradictions entail some propositions (apart from themselves). Thus the following set of claims is inconsistent:

{The Cologne objection to the proof, the proposition that some contradictions entail at least some propositions, the law of Simplification (and also Reflexivity)}.

The Cologne objection, and the criticism of it too, make free with the idea of "truth *in the proof*." We should look at this again. The classical logician will insist that if the refutation of the proof is to succeed, it is necessary to give to its connectives the classical interpretation. Otherwise, the refutation is not a refutation of this proof, but of some other in an already nonclassical logic. Against this, the Cologne School argues that it is impossible to write line (1) of the proof without prejudice to these facts. Line (1) comes by assumption, and in assuming that $\ulcorner \Phi \wedge \neg \Phi \urcorner$ one assumes that $\ulcorner \Phi \wedge \neg \Phi \urcorner$ is *true*. However, if $\ulcorner \Phi \wedge \neg \Phi \urcorner$ is true, and if Simplification is a good rule (which the objectors readily grant), then the occurrence of $\ulcorner \neg \Phi \urcorner$ at line (4) cannot contradict Φ as it occurs in $\ulcorner \Phi \vee \psi \urcorner$ in line (3). Thus Disjunctive Syllogism fails in just this kind of this case. Why is this? The objection supposes that Simplification is not only truth preserving but also *assumed-truth* preserving. At line (2), Φ is assumed true and at line (4) $\ulcorner \neg \Phi \urcorner$ is assumed true. With Φ in $\ulcorner \Phi \vee \psi \urcorner$ of line (3) also assumed true, the objection supposes that propositions assumed true cannot have different truth values, hence cannot in fact contradict one another. This can only be the case if assuming a proposition true makes it true (or guarantees that it is true).

It is too much to bear, of course. The objection poleaxes all *reductio* proofs. The objection pivots on the assumptions (i) that assuming the truth of Φ makes Φ true (or, more carefully, that assuming Φ true in a proof makes Φ true for the purposes of the proof) and (ii) that the undisputed rules of deduction are not only truth-preserving but also assumed-truth-preserving. Consider, then, any valid argument Δ ending in a contradiction, in which initial lines are assumed true and subsequent lines are deduced from previous lines by rules not here in doubt.

(1) Φ_1
(2) Φ_2

.

.

.

(n − 1) Φ_{n-1}
(n) $\psi \wedge \neg \psi$

Under the assumptions at hand, $\ulcorner \psi \wedge \neg \psi \urcorner$ also an assumed-truth, hence for the purpose of argument Δ is true. Thus, Δ cannot be a *reductio* argument or a *per impossibile* argument; an interesting outcome to derive from a proof of the impossibility of *reductio* proofs that is itself a *reductio* proof.

If indeed they are logics at all, relevant logics make relevance a kind of modality. This being so, we might expect the very idea of relevance to have run into the same heavy weather that logicians such as Quine have tried to rain down on necessity and possibility. Quine has no time for *dialethic* relevant logicians. In weakly paraconsistent systems, all that is required of a relevant logic is the blockage of *ex falso* (together with compensating adjustments). In such systems – the basic systems of Anderson and Belnap, for example – two concepts of relevance are on offer.[4] In what we might call the "content-overlap" sense, relevance falls within the ambit of the following two principles.

Ent : Φ entails ψ only if Φ is relevant to ψ.

In systems of Anderson and Belnap, and many others, a further condition is imposed:

Rel∗ : If Φ is relevant to ψ than Φ and ψ share a propositional letter.

A second idea we could call the "full-use" sense. A derivation from hypothesis, $H_1, \ldots, H_n$ is relevantly valid when it is classically valid and makes use of all the H_i. Full-use relevance is a descendent of one of Aristotle's conditions on syllogisms. Syllogisms are valid arguments that satisfy additional constraints. One is that conclusions may not repeat premises. Another is that there be no idle premises, that is, that there be no valid proper subarguments. This is also what full-use relevance comes down to. Both relevant derivability and syllogisticity are strongly nonmonotonic.

No Quinean I know of has queried the *very idea* of full-use relevance or the very idea of content-overlap relevance. Ultraconservative logicians press two complaints. One is that relevant logic is not logic. The other is that *ex falso* is true and that the Lewis-Langford proof is correct. Against this, relevant critics hold, in effect, that irrespective of whether relevant logic is logic, *ex falso* is false and its proof defective.

The target of such criticisms is Disjunctive Syllogism. In one ambush, advanced by Stephen Read, DS fails to be a universally correct principle because there are instantiations of it that commit a fallacy of equivocation. In a second ambush, fashioned by Anderson and Belnap, DS is also the target, and Anderson and Belnap declare its employment in the proof to commit a fallacy of relevance.

Here is Read's argument: (1) To conclude Q from $\ulcorner\neg P\urcorner$, one must know at a minimum that if ¬P then Q. (2) Assuming that DS is universally correct, then if Q is legitimately inferable from the set $\{\ulcorner P \vee Q\urcorner, \ulcorner\neg P\urcorner\}$, then $\ulcorner P \vee Q\urcorner$ must imply $\ulcorner$If ¬P then Q$\urcorner$. (3) It appears that the implication of (2) is true. (4) It also appears that $\ulcorner P \vee Q\urcorner$ is legitimately inferable from P. (5) However, it is "not plausible" to suppose that $\ulcorner$If ¬P then Q$\urcorner$ can correctly be inferred from P. (6) In *one* sense ⌜P ∨ Q⌝ follows from P alone and does not imply ⌜If ¬P then Q⌝. (7) In *another* sense ⌜P ∨ Q⌝ does imply ⌜If ¬P then Q⌝ but fails to follow from P alone. (8) In the move from P to ⌜P ∨ Q⌝, "∨" has another sense. (9) The Lewis-Langford proof fallaciously equivocates on "∨" (Read, 1988, pp. 22 ff).

Read's argument pivots on statement (3), which concedes that if Q is correctly inferred from {⌜P ∨ Q⌝, ⌜¬P⌝} then it follows that

I If P ∨ Q then if ¬P then Q.

But we also have it from I by the logic of truth functions that

II If ¬P ⊃ Q then if ¬P then Q.

By the rule known as *Impl*,

III If ¬P then Q then ¬P ⊃ Q.

Accordingly, Read has pledged himself unawares to the equivalence of implication and material "implication," which is something he is particularly intent on not doing.

We may take it, then, that I is not a condition on the inference for Q from {⌜¬P ⊃ Q⌝, ⌜¬P⌝}, or from ⌜¬(P ∨ Q) ∧ ¬ P⌝, and that there is nothing in *DS* that licenses I in this role.

Now, it is open to Read to try for the high road. "What I have shown," he might say, "is that if it is true that if P ∨ Q, then if ¬P then Q, then material "implication" and implication are one and the same. I have done nothing at all to disturb the claim that *DS* is such that if Q is correctly inferred from {⌜¬P ∨ Q⌝, ⌜¬P⌝} then indeed if P ∨ Q, then if ¬P then Q. By simple *modus tollens*, it cannot be the case that Q is correctly inferred from {⌜¬P ∨ Q⌝, ⌜¬P⌝}. Thus *DS* collapses."

It will not work. What the *DS* rule expressly endorses is this: if Q is correctly inferred from {⌜¬P ∨ Q⌝, ⌜¬P⌝} then it follows that if P ∨ Q *and* ¬P, then Q, from which the fatal "If P ∨ Q then if ¬P then Q" is not derivable. And a good thing, too. It establishes that there is *no* admissible interpretation from the connective "∨" for which it is the case that if Q is correctly inferred from {⌜¬P ∨ Q⌝, ⌜¬P⌝} then it is true that if P ∨ Q then if ¬P then Q. This strikes at the heart of the equivocation-argument that Read and others direct against the Lewis-Langford proof. In its most general form, the equivocation argument is this:

(i) There is a sense of "∨" in which the rule of Addition is universally valid.
(ii) There is a sense of "∨" in which the rule of Disjunctive Syllogism is universally valid.
(iii) They are not the same sense of "∨."

It is not in contention that the "∨" of proposition (i) is the extensional or truth functional "∨." Those who press the equivocation complaint against the proof of *ex falso* are united in thinking that the "∨" of proposition (ii) is an intensional, non-truth functional "or." It is only fair to ask for a characterization of this "or." For Read, it is any "or" such that if Q is correctly inferred from {⌐¬P or Q⌐, ⌐¬P⌐} then if P or Q, then if ¬P then Q. But, as we now see, there is no such "or."

In the argument advanced by Anderson and Belnap, the Lewis-Langford proof is "self-evidently preposterous." Indeed, "it is immediately obvious where the fallacious step occurs, namely in passing from [(3)] and [(4)] to [(5)]" (Anderson and Belnap, 1975, p. 165). This is a move sanctioned by DS. What is the nature of the fallacy that Anderson and Belnap see in it? It is, they say, a fallacy of relevance, which is an argument the archetype of which would "enable us to infer that the Van Allen belt is doughnut-shaped" or indeed anything else (Anderson and Belnap, 1975, p. 165).
Consider the principle

Rel: Φ is relevant to ψ if Φ and ψ share a propositional letter.

Anderson and Belnap are careful to say that **Rel** is not their own condition on relevance. "A formal condition for 'common meaning contents' becomes almost obvious once we note that commonality of meaning in propositional logic is carried by commonality of propositional variables. So we propose as a *necessary*, but by no means sufficient, condition for the relevance of A to B in the pure calculus of entailment, that A and B must share a variable" (Anderson and Belnap, 1975, pp. 32–3). Thus, Anderson's and Belnap's relevance is not given by condition **Rel** but rather, as we said, by condition

Rel*: If Φ is relevant to ψ then Φ and ψ share a propositional letter.

It is clear at once that **Rel*** is unconvincingly weak. It penalizes any would-be entailment ⌐Φ⊩ψ⌐ when Φ and ψ fail to share a letter, but it refuses to "reward" would-be entailments when they do. On a well-entrenched conception of such things, letter-sharing mimics "common meaning contents," which ought to be sufficient for relevance. Not allowing it to be overwhelms a powerful intuition with strategic cynicism. **Rel*** does not preserve Anderson's and Belnap's own claim that "commonality of meaning in propositional logic *is carried by* commonality of propositional variables" (1975). And if **Rel** were allowed to stand, that is, if letter-sharing were sufficient for relevance, then it could be said that each step of Lewis-Langford defense of *ex falso* fulfills it – an embarrassment for the relevantist. The strategic allure of **Rel*** over **Rel** is, of course, that the employment of **Rel*** to some extent abates that embarrassment. For it we persist with the necessity-only of a relevance condition, then we can no longer say that the steps of Lewis-Langford's proof fulfill the condition, but only that they do not violate it. Even so, the complaint that DS involves a "fallacy of relevance" is not explained by **Rel***. And this, no doubt, is part of what lies behind the opinion of those who find that relevance as construed in **Rel*** is of virtually no use in helping to determine which rules are valid (Diaz, 1981, p. 17).

It is insufficient to explain the necessity-only of relevance by noticing that content-overlap cannot be allowed to be sufficient for entailment. It cannot, but the required constraint is already supplied by a further principle to which relevant logicians are drawn. It is our principle,

Ent: Φ entails ψ only if Φ is relevant to ψ.

So, then, the decision to fetter the relevance condition by **Rel*** rather than by **Rel** is problematic in three ways.

    a.   It disconforms to the strong conviction that commonality of content suffices for relevance on at least one established conception of that relation. In any case, it is at odds with Anderson's and Belnap's own conviction that commonality of content is "carried by" variable sharing.

    b.   The preference for **Rel*** over **Rel**, having little or no plausible basis, is wholly strategic; it helps avert the discouragement of the Lewis-Langford proof.

    c.   The idea of relevance as formulated by **Rel*** is of no use in deciding which deductive principles are valid, that is, genuinely entailment-generating. In particular, the fallacy of relevance that DS is supposed to be afflicted cannot be characterized in terms of **Rel***.

Readers will have noticed that in recent pages I have been making rather free with complaints of counterintuitiveness and *ad hoc*ness, and of the cynicism of the relevantist's strategic maneuver. Surely this offends my inclination in this chapter to play up the importance of strategy and play down the importance of intuitions. The point is well taken; but it can be answered. I am making these complaints against certain relevant logicians precisely because they are in the same currency – the language of analytic intuitions – in which their complaints are advanced against classical positions. In short, I am doing what my cost-benefit strategy requires. I am making *ad hominem* arguments against those relevant logicians.

Here now is a second objection from Read.

    (1) We grant that the Deduction Metatheorem is true. In particular it is conceded that $\Phi \wedge \psi \Vdash \chi$ iff $\psi \Vdash \Phi \rightarrow \chi$

for an appropriate interpretation of "$\rightarrow$," which symbolizes ordinary "if then"

    (2) Is the truth of *ex falso*, if indeed it is true, compatible with the Deduction Metatheorem?

    (3) It is compatible only if implication is material implication.

    (4) But implication is not material implication [an irony!].

    (5) So {*ex falso*, Deduction Metatheorem} is an inconsistent set.

(6) Something must go. Since the Deduction Metatheorem is central to any logic of implication, *ex falso* should be abandoned.

Proof of the main point of the objection:

1. $\Phi \wedge \neg\Phi \Vdash \psi$ classical account
2. $\neg\Phi \Vdash \Phi \to \psi$ 1, Deduction Metatheorem
3. $\Phi \wedge \psi \Vdash \psi$ classical account
4. $\psi \Vdash \Phi \to \psi$ 3, Deduction Metatheorem
5. $\neg\Phi \wedge \psi \Vdash \Phi \to \psi$ 2,4, classical account
6. $\Phi \wedge \neg\psi \Vdash \neg (\Phi \to \psi)$ classical account
7. $\Phi \to \psi \Vdash \Phi \supset \psi$ def. implication
8. $\Phi \supset \psi \Vdash \Phi \to \psi$ 5, def. $\supset$
9. Hence $\to$ is $\supset$ 7,8.

Assessment: Consider, for arbitrary Φ and ψ, the conjunction $\ulcorner \Phi \wedge \psi \urcorner$. By Simplification, which is not here in doubt,

$$\Phi \wedge \psi \Vdash \Phi.$$

And by the Deduction Metatheorem

$$\psi \Vdash \Phi \to \Phi.$$

Hence, by Contraposition,

$$\neg(\Phi \to \Phi) \Vdash \neg\psi.$$

Since ψ is arbitrary, the present argument holds for the negation of any χ substituted for ψ in $\ulcorner \neg\psi \urcorner$. Hence, it holds for arbitrary χ, by Double Negation. But this is *ex falso* if $\to$ is $\supset$. Since it is precisely this that Read is at pains to deny, our result, while bad, is not *ex falso*. Still, it is a bad result. It establishes the inconsistency of the set {Contraposition, Deduction Metatheorem (for$\to$), Double Negation, Simplification}.

Suppose we say that the Deduction Metatheorem does not hold for $\to$ (as opposed to $\supset$). If this is to be our solution, then Read's own proof fails at line 2, viz., at the claim that $\Phi \wedge \neg\Phi \Vdash \psi$ only if $\neg\Phi \Vdash \Phi \to \psi$. On the other hand, if the Deduction Metatheorem does hold for $\to$, then the set {Contraposition, Double Negation, Simplification} is inconsistent. Here, too, my rejoinder to Read is a pure *ad hominem*. I am pressing him with consequences of his own concessions or principles.

Central to the relevantist's standard objection is the insistence that DS is invalid for the truth functional "or." Is there a way of providing more or less direct immunization against the assertion? Suppose that we pointed out that if *DS* were an invalid rule for the truth functional "or," then there would be an admissible interpretation of I on which $\ulcorner P \vee Q \urcorner$ and $\ulcorner \neg P \urcorner$ are true and Q false; but there is no such interpretation that, as required, honors the classical extensionality of the connectives. Would this provide the immunization the classicist wishes to have? Not according to the relevantist. Very well, he will concede, the immunization would be perfect if it were the case that a deduction rule is valid if and only if there is no admissible interpretation on which its antecedent is true and its consequent false. But relevantists do not grant this biconditional; they grant only one-half of it, that if a rule is valid then there is no such interpretation as I. In particular, then, the relevantist's position is that DS is a valid rule if and only if (1) there is no such interpretation as I and (2) some further condition K is satisfied. For present purposes, it hardly matters what the details of K are. It suffices to know that, on the truth functional readings of "¬" and "∨," DS fails it, and fails it in a way that *ex falso* also fails it. For, as with the case of DS, so, too, with *ex falso*; the relevantist cannot allow the nonexistence of I to constitute the validity of ⟨⌜P ∧ ¬P⌝, Q⟩. Here, too, there is the further requirement of satisfying K.

The whole business is knotted with question-begging. The breezy indifference of the parties is amazing – some would say appalling. If we were to reconstruct the to-ing and fro-ing of the past several pages as an dialectical exchange, we would get something like the following: (P is the strictist, A is the relevantist).

P: *Ex falso* is true.

A: Don't be silly.

P: Well, how about the Lewis-Langford proof?

A: It's invalid. Self-evidently preposterous, in fact.

P: Oh? What's wrong with it?

A: Disjunctive Syllogism.

P: What's wrong with *it*?

A: It's invalid.

P: You don't say. How so?

A: Well, to be precise, it's invalid for that meaning of "∨" for which the prior application of the rule of Addition was valid. On the other hand, there is another use of "∨" for which DS is valid but Addition is not. So either way, the proof is wrecked.

P: Are you saying that when "∨" is taken truth functionally (and "¬" too) DS is invalid?

A: I am.

P: But this can't be right. As long as we honor the truth functional interpretations of "or" and "not," it is trivial that there is no admissible interpretation on which ⌜P ∨ Q⌝ and ⌜¬P⌝ are true and Q false.

A: So?

P: So! So DS is valid.

A: It is not.

P: Do you doubt that there is no such interpretation, a counterinterpretation as we might say?

A: Certainly not.

P: So you *do* accept that DS is valid. You grant that no counterinterpretation exists; and that's just what validity is.

A: No, it isn't.

P: It isn't?

A: No.

P: Then what, pray, is it?

A: An argument is valid if and only if there exists no counterinterpretation and condition K is satisfied.

P: I can hardly bring myself to ask you what this mysterious K actually comes down to. Let me ask another question first. What's wrong with the old definition of validity; you know, the one according to which the nonexistence of a counterinterpretation suffices for it?

A: Use your head.

P: I'm doing my best.

A: Sorry. I didn't mean to be so sharp. Look at it this way. Suppose we pledge ourselves to the old definition.

P: Okay.

A: Then $\langle \ulcorner P \wedge \neg P \urcorner, Q \rangle$ would be a valid argument. In other words, *ex falso* would be true.

P: Sure.

A: But, you see, *ex falso isn't* true.

P: But what about the proof?

A: I told you, it's invalid.

P: Tell me again.

A: DS is invalid (assuming the proof's use of Addition is all right).

P: But it's not invalid. There is no admissible interpretation on which $\ulcorner P \vee Q \urcorner$ and $\ulcorner \neg P \urcorner$ are true and Q false.

A: So what?

P: So what?! DS is valid, that's what.

A: No it isn't. It satisfies the condition that there exists no counterinterpretation, but it does not satisfy the further validity-condition K.

P: Oh yes, it's coming back to me know. But, details aside, does not K wreck *ex falso directly*?

A: Well, I don't want to get too cute here. Certainly DS fails the validity test because, although it satisfies the no counterinterpretation condition, there is a further condition, which we've been calling K, which it does not satisfy.

P: Yes, but...

A: Hold on a minute. About *ex falso*, or anyhow its argumental counterpart, $\langle \{P \wedge \neg P\}, Q \rangle$, it too fails to be valid. Yes, it satisfies the no counterinterpretation condition, but there is a further condition on validity – call it K* for now – which it does not satisfy.

P: Are not K and K* the same condition?

A: It depends on whom you talk to. For some people, K, the condition that the *ex falso* argument fails, just is K*. But then K*, the condition that DS fails, cannot be K since DS does not fail K*!

P: I'm going home. I seem to have developed this whopper of a headache.

DOGMATISM

As our assessments show, both the Cologne objection and relevantist's objections reveal the *ex falso* dispute to have extremely wide consequences. If the Cologne objection stands, there are no valid *reductio* proofs – an intuitionist's dream come true, were the Cologne argument not so bad. Arguments against DS start rather narrowly, and they succeed only if a good deal more of classical logic is left in tatters. There are a further two things to be said about this development. For one thing, it is dramatic endorsement of the point that if we lose one classical connective we lose them all. On the other hand, it shows something further about the dialectical texture of conflict resolution measures in the abstract sciences. We seem to have it that as the consequences of a proposed resolution widen, the resolver finds himself stuck with commitments he must make the best of. He must now say things which at the beginning he would not have dreamed of saying,[5] for, at the beginning anyhow, they would have been thought easily as counterintuitive as the thing originally in dispute.[6] We see in this the makings of *dogmatism*. I will say that a party holds a thesis dogmatically to the degree that he holds it at the cost of accepting other things which, apart from the dispute at hand, he would have rejected for their counterintuitiveness, and whose acceptance now is driven solely by fidelity to his original position – that is, is *ad hoc*. *Ad hoc* or opportunistic espousal of what one does not believe for the sake of what one does believe is dogmatism if anything is. Dogmatism is the outcome of privileging an intuition at the expense of others apparently as strong. Lacking empirical check points, the abstract theorist must do what he can. It may seem obvious that he will get nowhere unless he consults and trusts his intuitions. But a conflict in an abstract science is little more than a clash of intuitions. The argumentative history of the *ex falso* dispute reveals a striking asymmetry between the strictist and the paraconsistentist. The strictist dishonors his original intuition that *ex falso* is a pretty strange customer for the benefit of having to dishonor no others. To do this, he invokes the explanatory device of *surprise*. His opponent is oppositely inclined. He honors his intuition about *ex falso* at the cost of the abandonment of his other intuitions on a scale that eviscerates what he used to take for logic. He, too, may plead the device of surprise, but if so he must tell us sooner or later how he could have been so massively and innocently mistaken about something like *logic*. He also must tell us a good deal more about the methodological role that he now reserves for intuitions. The asymmetry between the strictist and the relevantist discloses the finer workings of the intuitions methodology in its respective applications. For the strictist, suppression of one intuition permits him fidelity to the many others. For the relevantist, fidelity to the one intuition costs him fidelity to many others. If we see how the paraconsistentist is a dogmatist about *ex falso*, it is easy also to see something of the cost-benefit structure of the strictist's conservatism about it. I do not intend these remarks with any moral intent; their significance here is methodological. By the very structure of dogmatism, the dogmatic disputant is one who sacrifices general employment of the intuitions methodology for the sake of a privileged intuition whose status is usually unexplained.

As we have it here, dogmatism is largely a procedural affair. It is loyalty to a given belief at the expense of many others. On the face of it, there is nothing especially exotic about such situations. A wholly unanticipated experimental result can bring down an entire scientific theory. Recall the discovery in 1998 that neutrinos possess mass, and the bearing the daily press claimed it to have on the Standard Model.[7] True, there is in such situations a strain of what I have been calling "dogmatism." The conservative methodologist, when confronted by observational recalcitrance that threatens to cost his theory dearly, may seek to reinterpret the offending data, or to rebuttress – with epicycles, so to speak – the theoretical scaffolding that they place at risk. Even so, surprising discoveries do at times manage to dislocate large chunks, even the whole chunk, of theories previously thought secure and obvious. It is different with the abstract sciences; with disciplines that lack empirical checks. Recalcitrant data arise from the derivational mechanisms of such theories. In the absence of empirical checks, the measure of such recalcitrance must be taken in other ways. In the actual history of the dispute between relevantists and strictists over *ex falso*, two strategic models are broadly in evidence, one favored (though not exclusively) by relevantists and the other favored (though not exclusively) by strictists. One is what I have been calling the method of intuitions, and the other, the method of costs and benefits.

We may imagine the modern history of the disagreement between relevantists and strictists in the following way. Prior to their dispute, both parties shared a number of confident convictions – let us call them intuitions. In the sense in which I intend here, intuitions include beliefs held without argument and preinferentially, as well as other beliefs thought to be wholly and obviously secure. It bears on our question that throughout the history of this dispute both parties had access to, and even actively promoted, an epistemological view according to which *some* of a person's intuitions were epistemically privileged. They were "analytic," or objects of knowledge a priori, or some such thing.

Intuitions need not be active or occurrent. They may abide tacitly, awaiting occasion to surface. Let $\{B_1, B_2, \ldots, B_n\}$ be the set of predispositional intuitions shared by both parties. When it eventuated that *ex falso* was classically provable, both parties were given pause. There was a shared and almost certainly tacit intuition – B_3 say — which *ex falso* contradicted, and both parties were faced with the same pair of options. They could swallow *ex falso* and jettison B_3, or they could persist with B_3 and reject *ex falso*. Lewis and Langford chose the former course. Noting that the exclusion of *ex falso* required the rejection of other intuitions not otherwise in question, they asked their readers which of these they were prepared to do without. They did not ask, though they could have, which of these other beliefs they now believed false. Their reasoning therefore was dominantly cost-benefit, and markedly conservative. We could adjust to *ex falso* without dislocation of any intuition in $\{B_1, B_2, \ldots, B_n\}$ save B_3 itself. The relevantist was oppositely disposed. If the strictist's methodological bent was to minimize dislocation, and to avoid its costs, the relevantist's decision was to privilege B_3 no matter the cost. For this to be anything but a mad strategy, two things would need to be credible. The privilege extended to B_3 would have an identifiable rationale; and the now abandoned DS could be independently discredited. The derivability of *ex falso* placed both parties in a tricky situation. Both were faced with a strong challenge to an intuition they shared. Each would have to abandon that intuition or, if not that one, certain others. This abandonment should give us pause. How does one abandon something one confidently believes? In a rough and ready way, either one is seized by a contrary state of mind, which cancels the old belief, or one dissents from the old belief never mind that dissent does not, just so, cancel belief. The difficulty for the strictist was that he had to dissent from a single belief he may still have had, viz., that contradictions do not imply everything whatever, in order that his other beliefs B_i remained undisturbed, whereas the relevantist was required to dissent from many more things he believed in order to preserve his belief in B_3. The relevantist took the course of wide dissent in a field of enduring beliefs; the strictist opted for narrow dissent within the same field.

It may be thought that wide dissent in a field of persisting beliefs or dispositions to believe is pathological. Whether it is or not, it is common enough for there to have developed compensatory adjustments. One involves the effort of psychological displacement, in which serious attempts are made at discrediting the beliefs that one is already strategically pledged to have dissented from. Another is the dead opposite of this. It involves an attempt to abandon the realistic stance. It is an attempt to see that there is no prior fact of the matter about these things, and that a logical system creates a framework within which such facts as there may be secure such purchase as they may have. Of course, it is heuristically desirable that the logic we contrive is something we are actually drawn to, but it is misbegotten to give these feelings any role in the theory itself or in the methodology that constructed it. In a rough and ready way, the first means of adjustment flows from the methodology of analytic intuitions, whereas the second has more of a cost-benefit character.

Dogmatism carries dialectical consequences. Let P and A be our strict and relevant logicians, and let $\{B_1, B_2, \ldots, B_n\}$ be, as before, their shared intuitions prior to the dispute about *ex falso*.

P: My initial response was to think B_3 was right and there was something wrong with *ex falso*. It didn't take me long to come to my senses. *Ex falso* is fine; it's just that it came as such a surprise!

A: Whatever can you be thinking! B_3 is right. It is dead right. *Ex falso's* the problem.

P: But look, if B_3 *is* right, then B_4 is wrong.

A: Well, it *is* wrong.

P: You're kidding!

A: Not a bit of it. It's wrong.

P: What about B_7? Don't you lose B_7?

A: Absolutely. And a good thing, too.

P: !!

We have eavesdropped enough on this schematic conversation to see something of its dialectical structure. We see repeated attempts by P to marshal *ad hominem* arguments against A; and we see that P fails each time. It is not to be supposed that all propositions possess consequences that alone, or in conjunction with other concessions, discredit them, if only the opposing dialectician were shrewd enough to tease them out. Even so, parties to an *ad hominem* are wasting each other's time unless both accept the strategic possibility of there being discrediting consequences. If P is the *ad hominem*er, a search for discrediting consequences would be self-defeating if he were confident that there were none; and his target likewise would be a defective participant if his readiness to yield to a discrediting consequence were attended by the prior and continuing determination not to admit any such.

When is a consequence discrediting? At the level of intuitions there is an expectation (we might loosely call it a "rule") to the effect that if a consequence of a party's position contradicts another of its beliefs, the less counterintuitive of the pair will be surrendered. At the level of costs and benefits, the "rule" is that in such cases, parties will follow the course of least cost. It is well to emphasize that we are reconstructing the actual history of the modern conflict over *ex falso*. In the beginning, both parties shared the set $\{B_1, B_2, \ldots, B_n\}$. As the debate continued, and positions hardened, this same commonality of prior conviction was nowhere in sight. This matters in ways that bear on the *ad hominem*. In the situation we are describing not only do both parties initially subscribe to $\{B_1, B_2, \ldots, B_n\}$ but each presumes, perhaps tacitly, that his opposite number does. It is dialectically otiose for me to press you with consequences of your view that happen to offend my sensibilities. It is essential to the success of my project that they also offend your own. To this extent, *ad hominem* exchanges occur in communities of like-minded parties, communities in which this fact is known to its members. In our schematic account, we see that in his repeated determination not to "rollover," A is violating one, or perhaps both, of the expectation-rules. Certainly he is violating the cost-benefit rule, since in effect he is giving up on prospects for an extensional logic.

The person who finds himself the untutored guest of an alien culture will find himself encumbered or disabled in a number of ways. Assuming an intermediary who takes care of translation but who is unavailable for descriptions or explanations of local lore, he will find that there are classes of issues on which he has no purchase as a would-be *ad hominem*-maker. He may know that a native holds a view on a matter concerning that, say, there exists a good deal of cultural relativity with which the visitor disagrees, or thinks he does. But if the visitor does know what else his host believes, and believes that the visitor believes, he is dialectically disadvantaged. This is not to say that he cannot play a deduce-and-see-what-happens strategy, in which he seeks for consequences of which he himself would find discouraging, hoping for a like reaction from the other party. In the absence of such reactions, including lack of agreement about what the consequences actually are, our dialectician is growingly *alienated* in his encounter with his host.

As our schematic conversation unfolds, the dispute between P and A, which began as a tear in the fabric of shared belief, gapes alarmingly. It is increasingly apparent to P that he and A do not share enough in common to function as a fulcrum for P's probe. A's dogmatism compromises once-shared intuitions in $\{B_1, B_2, \ldots, B_n\}$ that P thought he could rely on, but now cannot. This places P in a condition of alienation that mimics that of the foreign visitor.

If begging the question is the use of a claim damaging to one's opponent's case but that the opponent has not conceded and cannot reasonably presume already to hold, alienation reflects what we have called a *converse of begging the question*. A denies a claim helpful to P's case that A either previously conceded or could reasonably have been presumed to accept. So we may say that in dogmatism there lies the disposition to commit *the converse of begging the question* (*CBQ*). It is the fallacy of illicitly *refusing* a virtual premise. And the degree of dialectical alienation varies proportionately to the number of times *CBQ* is committed. *In extremis*, where there is nothing bearing on the issue in question that P and A both agree on, alienation adumbrates a kind of incommensurability in which they have no common conception of the sort of thing they were wrangling about (no common conception of what logic is, for example), or a kind of methodological nihilism in which anything goes, since nothing in what initially divided them answers to a prior fact of the matter.

Our schematic conversation is more an expository device than a strict reconstruction of the fight between relevantists and strictists. We may safely suppose that no such conversation ever actually occurred between them. Let us remember that the Lewis-Langford proof was published in 1932, and the Anderson-Belnap papers did not start appearing until the early 1960s (and Read's criticisms were a later thing altogether). These temporal gaps are important. They afford the leisure of reflection and reconsideration and, in so doing, they integrate the assumption of shared beliefs precisely at the point at which the relevantist makes his attack. Historically, the relevantists themselves were more adept at teasing out the consequences of their own heterodoxies than were their classical *vis-à-vis*. All the same, something of the dialectical structure we have seen in our schematization of this dispute remains discernible in the real thing. This was Quine's point against parties to the debate between strong paraconsistentists and classicists about the issue of true contradictions. "They don't know what they are talking about," was Quine's observation. This has been taken as intemperate abuse of the paraconsistentist. Quine is no stranger to intemperate abuse, but this is not his purpose here. Rather, it appears that he was noticing the alienation that threatened to paralyze their interaction (I shall return to this point).

As we saw, the dogmatist is thus *procedurally* inconsistent. Except for his *idée fixe*, intuitions do not matter. The importance of this for us does not lie in the theorist's obsessiveness, nor does it overlook an occasional feat of "Copernican" insightfulness, in which an ambush actually comes off (e.g., Cantor's Diagonal Theorem, which established the indenumerability of the reals). Its significance for us is that the relevant logician is building his theories on methodological principles other than those he clung to at the beginning, except the one of them he clings to now. So we must ask, "Does he know what he is doing?"

It is known that the implication-negation fragment $R\overset{\sim}{\to}$ of the Anderson and Belnap system R of relevant logic gives a coherent intensional account of relevant proof, deducibility, and entailment. What is more, $\overset{\sim}{\to}$ is decidable and it goes over to a cut-free Gentzen formulation quite easily. It is notable, and intended, that in $R\overset{\sim}{\to}$, *ex falso* fails, as does the classical theorem

$$\Phi \to (\psi \to \Phi).$$

It is striking, and instructive, that if $\overset{\sim}{\to}$ is equipped with the full language of R itself; if, in particular the *extensional* connectives $\wedge$ and $\vee$ are added to the language of $R\overset{\sim}{\to}$, we lose all these features (Avron, 1992, pp. 261–2). Thus, $R\overset{\sim}{\to}$ cannot be a fragment of its own supertheory R without its ceasing to be a relevant logic.

The fragment $R\overset{\sim}{\to}$ was introduced by Church in 1951. Church notes that if $\ulcorner\Phi \to (\psi \to \Phi \wedge \psi)\urcorner$ is admitted to $R\overset{\sim}{\to}$, then $\ulcorner\Phi \to (\psi \to \Phi)\urcorner$ becomes provable. Concerning this discovery, Michael Dunn remarks that

For some mysterious reason, this observation seemed to prevent Church from adding extensional conjunctions/disjunction to what we now call $R\to$. (Dunn, 1984, p. 128)

To this, Arnon Avron makes just the right reply: "We believe that Church's reasons were not mysterious at all: He just has realized that it is impossible to add an extensional conjunction in a way which is consistent with the way he had constructed $R\to$ *and with its goals*"[8] (Avron, 1992, p. 280, n. 26, emphasis added).

This is a remarkable turn of events, and it gives rise to some interesting possibilities. One is that extensional connectives are somehow radically defective. Another is that relevant logic is radically defective. A third, rather more soberly, is that relevant logic and classical logic are incommensurable, so to speak. Whatever we may think of these alternatives in fine, it is clear that extensionality is a black hole for relevance of the Anderson-Belnap sort. The issue of extensionality connects with the question of *complexity*. A good theory of deductive reasoning, or of proof, should not be denied a fairly natural extension to mathematics. A theory of proof that precluded there being proofs for mathematical propositions would be a bad – even laughable – theory. By an obvious corollary, the same applies to theories of relevant reasoning or relevant proof. It is worth noticing that the systems of relevant logic that closely orbit the basic Anderson and Belnap system R fail this condition. The Anderson and Belnap approach is characterized by three main features: (1) R-proof is wholly transitive; (2) Disjunctive Syllogism is invalid; and (3) the set of propositional connectives contains a strong element of intensionality. If it is asked how a working mathematician might set out to relevantize his work, his efforts would be influenced by the fact that there is no proof that if a proof holds in classical logic then by certain procedures a relevant proof of the same thing also can be derived. In light of this fact, the relevant mathematician will have to proceed piecemeal, reproving his theorems one by one from required axioms. Then, as Kit Fine has shown (Fine, 1989), the semantics of the propositional fragment R will not go over to the full quantificational R in any naturally general way. Beyond that, the Gamma Rule fails for R#, which is the system of relevant Peano arithmetic derived from R (Friedman and Meyer, 1988).

The Gamma Rule is essentially the same as *modus ponens* taken as an *admissible* rule. Gamma provides that if Φ is a theorem of a system K and $\ulcorner \Phi \supset \psi \urcorner$ is also a theorem of K, then ψ is a theorem of K. The failure of the Gamma Rule in R# greatly encumbers R#'s proof-finding capabilities. It is a bad result for R#, and thus for R. Worse, and related, are some computational problems affecting R quite generally. As is well known, R is undecidable (Urquhart, 1984). Decidability can be recovered by deleting from R the distributivity axiom, which produces the system known as LR. But, though decidable, LR is not decidable by beings like us. LR has "an awesomely complex decision-problem: at best ESPACE hard, at worst space-hard, in a function that is primitive recursive in the generalized Ackerman exponential. From NP to ESPACE or worse, courtesy of relevance! – so much the worse then, for this brand of relevance"[9] (Tennant, 1993, p. 7).

In truth, the complexity objections may be less telling than they appear. It may be that the decision problem – the problem of *determining* whether an inference is valid – for relevant logic is ESPACE-hard or some such thing, but it does not follow that it is equally complex to *construct* an inference that conforms to the canons of relevance. Global searches are clearly out, but who – other than Quine – really imagines that they are part of the cognitive repertoires of beings like us? Similarly, Harman points out that reasoning in accordance with the rules of the probability calculus leads to a computational explosion. But what *is* reasoning in conformity with such rules? It is inferring the output of the rule from the input. The rule itself does not tell us anything descriptive about what happens "in between" – in the black box – so to speak, and hence nothing about its complexity. Saying so does not, of course, close the issue; it leaves it wide open and in need of attention. One possible answer to the complexity question is that reasoners do not reason in accordance with the principles of classical formal logic (**CFL**) or any decisionally more complex system, or in accordance with the equations of the mathematical theory of chance. Rather, they are guided to approximate fidelity to the outputs of such principles and equations by a battery of heuristics that combine accuracy with timelines. Among the so-called proof-heuristics, Newell's **LT** (Logic Theory) was an early contender. It was followed several years later by **GPS** (General Problem Solver), also conceived by Newell, in collaboration with Simon.[10] In the past twenty-five years, cognitive scientists have given much attention to the problems of deductive and probabilistic heuristics. We shall briefly return to this matter in Chapter 3.

Talk of complexity bears on a central question not yet fully posed in this book. The question is why one would *want* to impose relevance constraints on deductive arguments and proofs. The short answer is this: If we wish to *use* a deduction, whether as a proof or demonstration, or as an instruction-argument, or as a refutation of the adversary's thesis, the deduction must be processible by beings like us in a timely way (which is just another way of saying that it must be *usable*). On the score of efficiency alone, it would be better not to have to track and display boundless numbers of putative premises. Imposition of a relevance condition may seem to facilitate this task. It is arguably a first principle in any logic of discovery to bid the proof-seeker not to waste his time with irrelevant truths. What recent work on Anderson and Belnap systems makes plain is that systems such as R and their satellite systems cannot satisfy their own procedural motivation, namely, to facilitate the finding of proofs.

I do not wish to leave the impression that relevant logic is guaranteed to be a lost cause. It is nothing of the sort, as witness Neil Tennant's base system **IR**, which is a system of intuitionistic logic supplemented by rather different relevance conditions (Tennant, 1987). One might think that there is, in **IR,** hope for a logician like Aristotle, if not for Anderson and Belnap. In **IR**, deduction is not unrestrictedly transitive. The same is true of the relation "syllogistically implies"; it, too, is not unrestrictedly transitive in Aristotle's logical theory (Woods, 2001, Ch. 6). Pursuing the **IR**-similarities lies beyond the reach of the present work. I leave it as an open question.[11]

Relevant logic, like **QML**, was a good place to introduce the idea of the ambush-argument. Ambushes are arguments against what thitherto seemed secure against objection. We would expect them to have generally bad track records on that account alone. "Generally" here means "by and large," and it is a qualification of hard importance, as a Dedekind cut, or a Cantorian diagonalization or – the biggest ambush of twentieth century mathematical logic – a Gödelization makes us acutely aware. Even so, the quality of arguments by relevantists against *ex falso* is generally rather awful. I put this down to the propensity of ambushers to be dogmatists, together with their attractedness to procedural inconsistency. I have asked, without offensive intent, whether it is possible that relevant logicians do not know what they are doing. It was a genuine question. If the answer were Yes, it would explain the shabbiness of the relevantist campaign against *ex falso*. In our discussion of Quine's project against **QML**, we came close to saying not that Quine does not know what he is doing but, rather, that what Quine *is* doing as against **QML** is not to be found in those six silly arguments. If this is right, it would explain a further fact, namely, the near wholesale indifference of modal logicians to Quine's campaign against the modalities. If Quine's real case is not carried by those six arguments, they can safely be ignored. If the real case is that **QML** implies essentialism, which secures no purchase in Quine's physicalism, then it can still be safely ignored. No modal logician need swallow the idea that bodies are sets of quadruples of real numbers in arbitrary coordinate systems; and not even physicalism is protected by M^3, the maxim of minimum mutilation.

DOGMATISM AND PHILOSOPHY'S MOST DIFFICULT PROBLEM

As our assessments show, relevantist's objections reveal the *ex falso* dispute to have extremely wide consequences. It appears to be the case that as the consequences of a proposed resolution widen, the relevant resolver finds himself stuck with commitments he must make the best of. He must now say things that initially it would not have occurred to him to say, because initially he would have taken these to be easily as counterintuitive as the thing originally in dispute. We see in this the makings of dogmatism, as we have said.

We saw that a party to a dispute holds a thesis dogmatically to the degree that he holds it at the cost of accepting other things that, apart from the dispute at hand, he would have rejected for their counterintuitiveness, and whose acceptance now is driven solely by fidelity to the intuition in which his original thesis is lodged. The actual history of the *ex falso* dispute reveals a consequential asymmetry between the classicist and the relevantist. The classicist abandons his original intuition that *ex falso* is something of an oddity for the benefit of having to dishonor no others. So, he summons the explanatory device of *surprise*. His opponent is contrarily inclined. He holds to his intuition about *ex falso* even though it costs him his other intuitions. By the very structure of dogmatism, the dogmatic disputant is one who sacrifices general employment of the intuitions methodology for the sake of a privileged intuition whose status is usually unexplained. And this, I think, should give us pause about intuitions.

Dogmatism has an interesting structure. It engages *Philosophy's Most Difficult Problem* in an especially interesting way. It is important to bear in mind the *sort* of thing that triggers *Philosophy's Most Difficult Problem*. It is a derived consequence that strikes all concerned as *counterintuitive*, as giving the appearance of something awry. This produces *Philosophy's Most Difficult Problem* only if it occasions a disagreement about the import of this counterintuitiveness, with one party saying that it is *merely* counterintuitive and things have not really gone awry, and the other party saying that the counterintuitiveness *proves* that something has gone awry, that it is a counterexample or a refutation of the premises that imply it. The very structure of the difficulty we have been calling *Philosophy's Most Difficult Problem* provides that it takes a counterintuitive consequence to trigger it. There is no such problem for consequences that do not initially strike the parties as odd in the requisite ways. In the actual case of *ex falso*, this condition was met. *Ex falso* was provable in classical logic, and *ex falso* was counterintuitive to all concerned. It triggered the requisite pair of questions. Was this consequence a counterexample to classical logic, or was it a merely surprising truth? The Lewis-Langford proof was an attempt to answer this question in a non-question-begging way. Critics of the proof held not only that it did not answer that particular question, itself a case of *Philosophy's Most Difficult Problem*, but that it triggered such a Nasty Problem all on its own. For the critics now said that the proof discredited one of its own proof rules, DS.

Take a standard classical proof *Pr* of *DS* – it might be an elementary proof from truth tables. Details aside, dogmatism is the treating of that proof as if it occasioned a nasty problem, as if it called for a decision as to whether *Pr* is a sound but surprising proof of the validity of *DS*, or whether the fact that its validity derives *DS* shows *Pr* itself to be operating with a defective proof apparatus – say, for example, the truth table for ∨.

This is in effect what Anderson and Belnap actually claimed. The truth table for ∨ correctly endorses Adjunction, they said, but not *DS*, since *DS* is valid only for an intensional ∨. But if the truth table for ∨ is defective, then functional completeness is compromised. But there is a proof of functional completeness (and so on).

We can schematize the relationship between dogmatism and *Philosophy's Most Difficult Problem*, as follows.

1. *Stage one. The Trigger.* There is a valid proof of Y from X. All parties agree that Y is counterintuitive.
2. *Stage two. The Problem.* Party P takes this proof to be a sound demonstration of a merely surprising result. Party A take it to be counterexample to Y.
3. *Stage three. An Attempted Adjudication.* Party P offers a different proof Z of Y, emphasizing the intuitiveness of Z's proof-rules.

4. *Stage four. The Problem's Recurrence.* Party A take the conclusion of Z to be a counterexample of one or other of those rules.
5. *Stage five. Another Adjudication.* Party P offers a proof Z* of proof-rule W, emphasizing the intuitiveness of the proof-rules of Z*. Party A finds the proof to be a counterexample of one or more of Z*'s proof-rules.

.

.

.

I shall say that dogmatism enters the dispute at the stage (here marked with *) at which it fails to be the case that *both* parties find counterintuitive the conclusion of the proof constructed at the previous stage. So conceived of, dogmatism is *infectious*. Once it infects a stage, it affects every successor stage as well.

INTUITIONS

The dogmatist has lost control of (or perhaps has just given up on) two distinctions. He is unable or unwilling to distinguish in a principled way between intuitions that he is bound to honor, and intuitions that make no enduring claim on his fidelity; those that he can give up on when the going gets tough. He is also unable or unwilling to apply in a principled way the distinction between a proof of something counterintuitive and a valid *reductio* argument.

I am saying that the very fact of dogmatism and the very existence of *Philosophy's Most Difficult Problem* are methodologically important. Dogmatism, I am suggesting, is not moral pig-headedness but, rather, a kind of procedural inconsistency. It is not a mistake-free kind of inconsistency, since it involves the dogmatist in mispleading a diagnosis of counterexample with respect to a proposition the other party does *not* see as counterintuitive. This *is* a mistake; it is a dialectical mistake and a kind of question-begging, since the dogmatist operates on an assumption that the other party does not concede. Even so, the more important feature of the dogmatist's inconsistency is that it suggests, however tacitly, his abandonment of what I will call the *method of analytical intuitions*.

A dominant force in the abstract or nonempirical sciences,[12] the method of analytical intuitions is best known to philosophers as *conceptual analysis*. Intuitions are taken in Cohen's way, as untutored beliefs held without inference or evidence (1986, pp. 73 ff). They are analytic because they fix the very idea of what they are intuitions *of* – whether sets, truth, virtue, or whatever else. The method of analytical intuitions operates both positively and negatively. In its positive form, intuitions fix basic conceptual truths, and further truths are derived by logical and conceptual consequence. Not all theses in an abstract discipline are fixed in this way. As a matter of historical record, much of the content of such theories is fixed by starting points that are considered *plausible*, and further theses are derived by consequence, including what is sometimes called plausible consequence. Here is where the negative role of analytic intuitions secures a purchase. Let C be a consequence of a set of premises initially thought to be plausible. If C offends an analytic intuition, the consequence-structure is taken as a *reductio* of those once-plausible premises.

It is important to emphasize that on the method of analytic intuitions, intuitions are not beliefs, whether widely shared or deeply embedded. Intuitions are epistemically privileged. Intuitions are impeccable revealers of conceptual truths, of truths that flow with complete assurance from the very idea of the thing in question.

The method of analytic intuitions was robustly at work in the early efforts of Russell and Frege to get set theory up and running. Their efforts are known to this day as "intuitive set theory." With the notable exception of dialethic logicians, intuitive set theory was dealt a fatal blow by the Russell paradox, and with it the method of analytic intuitions. We return to this point in Chapters 4 and 8.

3

Managing Inconsistency

[T]he task is to find a system...which: 1) when applied to the contradictory systems would not always entail their overcompleteness, 2) would be rich enough for practical inference, 3) would have an intuitive justification..., the satisfactory condition (3) being rather difficult to appraise objectively.

S. Jáskowski, "Un Calcul des Propositions Pour Les Systèmes Déductifs Contradictoires," 1948

RECONCILIATION

Let us begin with a frank question: How is conflict resolution possible in the abstract sciences? So far I have been emphasizing the cost-benefit factors that attach to *ad hominem* maneuvers. Actual argumentative practice suggest five prominent forms of the *ad hominem*.

1. arguments *per impossibile*
2. *reductio ad absurdum* arguments
3. *reductio ad falsum* arguments
4. refutations of the Aristotelian sort
5. other arguments from cost.

Since each of these is a strategy for winning an opponent to the antagonist's view of things, or to get the opponent to abandon the defense and advocacy of his own version of those matters, it is appropriate that we give these five dialectical techniques the general name of procedures for *surrender*.

We have seen that *reductio* arguments, as they are standardly described by logicians, are caught in the maw of *Philosophy's Most Difficult Problem* (one disputant's *reductio* is another disputant's proof of a surprising truth). We shall have later occasion to see that not even the argument *per impossibile* escapes this snare.

Of the five, the last two – refutations and arguments from cost – stand out for the resistance they make possible to the Problems' indeterminacy. Their resistance is made possible by their respective *dialectical* characters. As we see, Locke is clear in saying that the function of the *ad hominem* is conflict resolution. Its aim is to win the agreement of one's opponent, or to embarrass him as to get him to quit the field and stop supporting his own thesis. By "concessions" Locke means what present-day dialogue logicians mean. A party's concession is a statement to which he has assented. By a "party's principles" Locke means what the party has pledged to as a principle. By "consequence," Locke means any relation strong enough to preserve assent. It is any deducibility relation under which concessions can reasonably be supposed to be closed. Operationally, a consequence of a party's concessions is any statement he is prepared to acknowledge as following from those concessions. Consequences, therefore, are *conceded* consequences.

It is not required that parties believe that to which they give their assent, although nothing precludes it either. Alternatively, Locke's account allows for the common distinction between belief as assent, in which belief is a social or sociolinguistic act, and belief as a feeling of certitude (or something approaching it), in which belief is a psychological state. It is clear, however, that belief-as-assent is the methodologically prior notion in this account.

Intuitions again reenter the picture. They are not the epistemologically privileged convictions of the discredited method of analytical intuitions but, rather, ordinary psychological states – either the belief that a statement P is true or the belief that it is worthy of assent, never mind that it itself might not be the object of direct belief. It is hugely important, even so, that while beliefs can enter the dynamic of *ad hominem* cut and thrust, Locke intends that it is perfectly legitimate to assent to the opposite of what you may chance to believe in the psychological sense. It is not that he thinks this is typical of routines of conflict resolution; it is rather that he recognized this as a possibility. In the more general case, assent brings psychological belief in its wake; so a standard way of changing a party's mind psychologically is getting him to perform the right speech act, and to get him to perform it thinking that it is, in the circumstances of the case in question, the thing that he should do.

At the core of this Lockean dynamic is a fundamentally economic standard of what is and is not appropriate for a party to concede or accede to. Most disputants do not want to saddle themselves with falsehoods, just as they would prefer to avoid escapable inconsistences. These are matters of a party's *preference* and are, in that sense, economic. As a first approximation, we can cite the

Cost-Benefit Rule: The *cost* of a transaction to a party is the net of his negative preferences in that situation. The benefit to him is the net of his positive preferences in that situation. The *price* of the transaction is benefits minus costs.

It is also important for the kind of account that Locke is sketching that preferences *pro* and *contra* need not be *standing* preferences. A standing preference with respect to a transaction is a preference the existence of which precedes the transaction. It is a preference a party brings ready-made to a transaction, and thus resembles what some theorists have called *pretheoretic intuitions*. For the Lockean account to work, it is necessary that latitude be given to two phenomena. One is instantaneous preference-formation, a situation in which a party now has a preference for something without ever having considered it previously. The other is a special case of the first. It is the emergence of a preference that countermands and puts into retirement a prior contrary preference, even a preference of standing. In the Lockean dynamic, preferences need not be the product of intense deliberation (although some surely are); rather, they can be things that befall us, like a sneeze, or the measles.

The actual struggle between classical and relevant logicians is a particularly apt test case of *ad hominem* strategies of conflict resolution. When Lewis and Langford offered their proof of *ex falso* they thought that they were pressing the opposition with consequences carrying too high a price to pay. One of the proof rules would have to go, concerning none of which was there any evidence of a standing *negative* preference. It lies in the nature of what we are describing that anything goes, provided one is prepared to meet the price. And it emerges from the real history of this struggle that classical and relevant logicians have different *values*, not just different beliefs. By the time the dust had settled, the price of relevant logic was the loss of extensionality, an unusually troublesome semantics, and horrific complexity problems. These are prices too dear for a classicist's pocket, and the price of their avoidance he sees as paltry by contrast. It is the acceptance of what once would have been a surprising consequence – *ex falso* – together with the rejection of any temptation to think that relevance between premise sets and conclusions must be delivered by something other than classical consequence itself.

It would be difficult to overstate the allure of intuitions in making and appraising abstract theories. Whatever else they are, they are psychological states, beliefs held with variable tenacity to the point of intractability, which the theorist brings to the shop before the conceptual scaffolding goes up. They are pretheoretical and untutored – held without inference or evidence, as Cohen says (1986, p. 75). Sometimes intuitions also surface posttheoretically when, for example, the theorist must consider the rebuttal of what he has wrought. There is nothing in this that contradicts the pretheoreticity of intuitions. On the temporal question of when to deploy, they are only contingently before or after.

Inherently, it is a matter of epistemic priority – without inference or evidence. My intuitions tell me that the defense of the Lewis-Langford proof is a success. Why, then, do I have no confidence that relevant logic will close itself down in a year or so? Whatever is to be said in greater detail about intuitions, it is plain that the design of cost-benefit conflict resolvers provides no inbuilt defense against stalemate. As we have described it so far, the avoidance of stalemate hangs entirely, and contingently, on what the parties cobelieve. Saying so is not to fault the design of such arguments; it simply registers a fact about conflict resolution. If parties do not share the appropriate beliefs, including beliefs about costs and benefits, resolution will not be possible. The same is true beyond an interest in what is the case. In a competition among a Deduction Theorem (for entailment), Simplification, and Contraposition, I may not have any intuition that makes it obvious to me that Deduction Theorem is false. But I may claim it as obvious that *of the three* it is the one that is best to let go of. If that appraisal is not shared, it is likely that I will find myself in a stalemate.

We have already noted the high rates of ecumenism across the field of logic. Perhaps "ecumenism" is not the word for it, but whatever it is, there is a good chance that it is born of a kind of dialectical fatigue. Given the absence of empirical checks, there seems little to go on but our intuitions, as I keep saying. If this is right, then by the very nature of disagreements in the abstract sciences, it is often the case that disputants have *nothing* to go on; for what is their disagreement but a clash of intuitions? Even when disputants did not simply throw in the towel and agree to disagree, resumption of the present pattern of cost-benefit dialectic is resumption of it with the confident expectation of concurrence much reduced. It is, as we saw, a form of *alienation* that is a natural response to dogmatism. It is a process that takes a narrow disagreement into an ever-widening one. The longer it goes on the wider it gets, short of what Aristotle calls "babbling" (recall that when you babble you just repeat yourself). It turns the dialectic in the direction of shots in the dark, provided the parties have something further to say.

It is apparent that at a certain point in the dialectic dogmatism, the point, namely at which one of the parties is prepared to reject an intuition that antecedently had been accepted – or anyhow not placed in question – by either party, the argument in process takes on the gravamen of the ambush argument, which we met with in Chapter 2. We see, then, that ambushes can *originate* a disagreement, as with Quine's out-of-the-blue attack on the very coherence of modal operators. They can also be triggered in *medias res* by an argument in which one (or both) of the parties is driven to increasing levels of dogmatism. It is worth repeating here not only that dogmatism is not a reliable indication of the inadequacy of a case that it invades, but that the same is true of ambush arguments. They, too, stand or fall on the merits of the particular cases that take shape under the auspices of ambush. If dogmatic and ambush arguments indicate trouble anywhere, it is at a more general remove usually from the particularities of the argument in dispute. The trouble, rather, is with a general methodological approach to abstract enquiry and to the resolution of conflict therein. It is the method of intuitions that these factors call into question. For an ambush argument is an assault on generally agreed intuitions, and a dogmatic argument is an argument that systematically retires received intuitions for the sake of some privileged insight. We develop this point in Chapters 4 and 8.

Parties who find themselves enmeshed in the alienation-dogmatism dynamic cannot long evade a sense of dialectical paralysis. Since dogmatism is not a virtue in the culture of *logos*, a certain discomfiture is unavoidable, if not *anomie*. Or so one would think. It is striking, in the natural history of conflict resolution in the abstract sciences, how often this is not what happens. As if to relieve paralysis and keep the dialectic moving along, disputants focus on what they can agree on, namely, that the contested claim Φ is true in system S and not in system S*. Now this point of agreement could not apply pertinently to the dispute at hand unless it were accompanied, tacitly, if not expressly, by a further agreement to surrender the presumption that there is a fact of the matter that settles the original question about Φ. It is a slope for the theorist to slide down to the point of a quite general system-relativity. Those for whom this maneuver constitutes a resolution, or at least a termination, of the disagreement, are precisely those not minded to press the question of which is the better system in relation to Φ. These are logical relativists, pure and simple. In their hands, logic has gone postmodern. My purpose is not to condemn logical relativism but to show how it arises and how, in somewhat surprising ways, it matters for the question of conflict resolution in the abstract sciences. Disputes, such as that over *ex falso*, often arise when and because the disputants are analytic theorists about entailment. When locked in a stalemate, two courses are open. One is to change the subject. The other is to embrace the relativist option, however tacitly. The two courses are identical. Analytically minded S5 or R theorists are not going to quit their respective views about *ex falso* just because, having found themselves in a stalemate, they decided to change the subject. All that changes is the reciprocal attempt at mind-changing over the *ex falso* question. Henceforth, each will pursue the logic of entailment in his own way, and as analytically as he pleases. The S5 theorist and the R theorist have nothing to disagree about since they have nothing to say to one another. There is a cost: explaining how such disagreements are possible for the analytically minded. For the analytically minded, there is a fact of the matter about *ex falso*, but no chance of joint access to it by these disputants.

In the present chapter, I shall examine mechanisms for overcoming stalemate. Parties are in a stalemate with respect to a disputed claim in an abstract science when the disagreement persists and neither party has confident expectation with respect to any proposition P, or with respect to the appraisal of any cost-benefit option E, such that if they shared a belief in P or made the same appraisal of E, their dispute would be ended. Stalemates correlate significantly with accusations of fallaciousness – question-begging, faulty analogy, *ignoratio elenchi* – and, of course, babbling. How do we break out of stalemates?

Perhaps the dominant break-out strategy is what we could call **Reconciliation**. *Reconciliation* takes antagonists in the opposite direction. If surrender is a win-lose game, reconciliation is a win-win game. It is what at first blush seems an attempt to find an interpretation on which apparently conflicting beliefs are both true. It involves proposing a crucial distinction in hopes that both parties might accept it. It is a variation on a common type of move made in philosophical argument, namely, a plea of ambiguity. Consider a stylized example. Someone believes

1. All A are B

and

2. No B are A.

So he is stuck with the consequence

3. All A are not A.

The ambiguation strategy bids the party who is not prepared, or not content, to give up (1) or (2) in the light of (3) to search for an ambiguity in the set $\{(1),(2)\}$. If he succeeds, for example, by determining say that (1) is true for such-and-so sense of "B" and (2) is true for some other sense of "B," then the argument in question is invalid, and the party's beliefs are not shown to be inconsistent. *Reconciliation* generalizes on this basic idea: when locked in a stalemate try to find something or other for each member of each pair of conflicted intuitions to be true *of*.

What are they talking about? It is possible to understand the dispute between the strictist and the relevantist as a good deal more than a parochial disagreement about *ex falso quodlibet* or Disjunctive Syllogism, and as something beyond the question of how widely consequenced the dropping of, for example, *DS* actually is. It is more generally, and more interestingly, a difference of opinion about the nature of inconsistency, both in theory and common belief alike, and about how its rational management is best achieved.

Much in the classical tradition suggests that inconsistency is simply catastrophic, that it disables, for useful work, any theory or belief stock afflicted with it. It is in this spirit that Frege wrote despairingly to Russell, following the demonstration of the inconsistency of set theory, that arithmetic was toppling. Count Koryé (1946, p. 344), sounded an even more general alarm in saying that, owing to the Liar, the most secure foundations of reason itself seem to be undermined; and, he added, this was *terrifying*. So we have **Koryé's Sorrow** about statements playing the role of **Frege's Sorrow** about sets.

It is obvious that this, the catastrophic view of inconsistency, turns on the conviction that simple negation-inconsistency coincides with absolute inconsistency, for without this connection there appears to be ample room for regarding simple inconsistency as a local and isolable disorder, an unattractive and unwelcome nuisance about which, all the same, something might be done. This was an opinion that occasionally tempted Wittgenstein, though it cannot be said that it was his settled view of the matter. It was sometimes Wittgenstein's

position that, contrary to Tarski, "a contradiction is not a germ which shows a
general illness" (1976, p. 211). For contradiction can be "sealed off" (1964,
p. 104), and so allowed to stand (1964, p. 168). Wittgenstein also was attracted
to the idea that, strictly speaking, the appearance of contradiction in a theory does not give rise to inconsistency, for a contradiction "is of no use; it is
just a useless language-game." But he had also said: "There is *one* mistake to
avoid: one thinks that a contradiction *must* be senseless: that is to say, if e.g.
we use the signs "p," "∼," "." consistently, then "$p \sim p$" cannot say anything"
(Wittgenstein, 1964, p. 171; cf. Woods, 1965). So it is clear that Wittgenstein
did not develop a consistent final position on this question.

The idea that inconsistencies, once safely sealed off, can be allowed to stand
gives offense to those of classical bent. Here is Quine on this point:

To turn to a popular extravaganza, what if someone were to ... accept an occasional
sentence and its negation both as true. ... [One] hears that ... a full-width trivializa-
tion could perhaps be staved off by making compensatory adjustments to block this
indiscriminate deducibility of all sentences from an inconsistency. Perhaps, it is suggested, we can so rig our new logic that it will isolate its contradictions and contain
them. My view is that neither party knows what he is talking about. (1970/1986,
p. 81)

It is evident that among the leading exponents of relevant logic and among its
technically most creative contributors, there has lingered the suspicion that,
quite apart from whether the proof of *ex falso* is "self-evidently preposterous"
and Disjunctive Syllogism obviously and transparently invalid, the classical
approach nevertheless seriously overstates and miscalculates the nature and
significance of inconsistency. This was the opinion of Robert Meyer in an
important paper of the early 1970s (1971).

Meyer's paper is a kind of landmark in the development of contemporary
dialethic logic from relevantist antecedents. I am not quite ready to characterize that evolution at the moment – this we will come to in due course –
but it is useful here to emphasize that relevant logic and dialethic logic are
different animals. For the present, it suffices to repeat that a relevant logic
is one in which *ex falso quodlibet* fails because of its failure to fulfill a relevance condition. A dialethic logic is one in which a contradiction might be
true (and so not merely derivable) and in which *ex falso* fails, but not necessarily because it fails to fulfill a relevance condition. Thus, one can have a
relevant logic that is not dialethic, a dialethic logic that is not relevant, and
also a logic that is both. The early relevant systems of Ackermann were not
dialethic (1956), nor were the systems of Anderson and Belnap collected in
Entailment. In an early paper by Anderson (1958), Wittgenstein is belabored
for what are, pretty clearly, dialethic proclivities. Anderson takes a dim view
of Wittgenstein's " 'so-what' attitude towards contradictions in mathematics"
(1958, p. 455). On the other hand, Jáskowski's paraconsistent systems, for example, his "Propositional Calculus for Contradictory Systems" [1969], are not

relevant, and several of the systems of Routley and his associates are both relevant and paraconsistent (Priest, Routley, and Norman, 1989, p. 152).

The importance of Meyer's paper for our present purposes is not that it forwards a relevantist approach that flirts with strong paraconsistency (though this it does do) but, rather, that it advocates the view that the strictist approach mishandles inconsistency. It misunderstands what inconsistency is, and it misunderstands how inconsistency is rationally to be managed. This, too, is a common complaint of dialethic logic.

It is Meyer's contention that the basic mistake of construing inconsistency in the manner of *ex falso quodlibet* is that this destroys the "*intuitive* parity" [1971], between a theory's negation-inconsistency, in which Φ and $\ulcorner\neg\Phi\urcorner$ are both theorems, and its negation-incompleteness, in which neither Φ nor $\ulcorner\neg\Phi\urcorner$ is a theorem. In the one case, he says, the strictist's view overdetermines Φ, whereas in the other case it underdetermines it. In each case the theory in question fails to honor the attractive ideal of proclaiming for each pair of its well-formed formulae $\{\Phi, \ulcorner\neg\Phi\urcorner\}$ which is the theorem and which the countertheorem. Meyer believes that the theory in question fails this ideal equally dismally, so to speak, in the one case telling us too much and, in the other, too little. So it is not reasonable that we should be more troubled by the one failure than by the other. Both are disappointments, but they are disappointments of approximately the same order of magnitude. The one is no more catastrophic than the other, and to suppose otherwise involves us in a kind of methodological hysteria.

Meyer is onto something important in claiming that the classical approach fails in some basic way to understand inconsistency. But it seems wrong to say that there is no coherent or convincing motivation for regretting inconsistency one jot more than we rightly regret incompleteness – metamathematical *tristesse* over theoretical shortfalls in each case. If we ask of a theory's sentences, Φ and $\ulcorner\neg\Phi\urcorner$, which is the theorem, the answer, "neither," will be disappointing if we prefer our theories to be complete. But if the answer to the question is not "neither" but "both," some will say that this would seem to be occasion to be not just disappointed but *dismissive*. Not only does there appear to be a natural and intuitive disparity between our reactions to incompleteness and inconsistency, it is recognized by what classical entailment theory says about negation-inconsistency. The shock of the answer "both" is represented strictly by the ensuing absolute-inconsistency and thus by the circumstance that contradictions "have a horrible way of infecting theories" (Priest, 1979, p. 226).

It is a mistake to hold that there is no intuitive disparity between inconsistency and incompleteness, and it is wrong, too, that there is no obviously natural motivation that the strict account possesses and its relevant *vis-à-vis* lacks. Admittedly, even if true, it does not show that the strict response to inconsistency is not an overreaction. It is a welcome reaction inasmuch as it is a reflection of the disparity between inconsistency and incompleteness. It is not a welcome reaction if it overstates the disparity or reflects it hyperbolically.

Considerations of "intuitive parity" to one side, Meyer is well aware that inconsistent and incomplete theories can be shown to be duals of each other. It may be, as Chris Mortensen has written, "that incompleteness has seemed easier to swallow than inconsistency," but, as he goes on to suggest, this is "something not so easy to justify given the duality results" (1995, p. 10; see also Routley and Routley, 1972; da Costa, 1974). Such dualities give us reason to think that "inconsistent and incomplete theories are deserving of equal respect" (Mortensen, 1995, p. 7), that they "are somehow equally reasonable" (1995, p. 125), since "inconsistent models can be simply transformed [by dualizing operations] to become incomplete and vice versa" (1995, p. 144). So it is still theoretically open to the relevantist to yield on the point of intuitive disparity and yet persist with the complaint that absolute inconsistency distorts it. He can still hold that there is room to argue that the strictist does not, after all, properly understand the nature and significance of simple inconsistency. I shall return to this point. Before doing so, there are two additional arguments lodged by Meyer against the strictist position which would repay examination. The one is worth looking at for its instructive mistake, and the other for its dialethic leanings.

The first of Meyer's additional arguments has to do with the question of how, rationally, to manage an inconsistency in a theory we accept. In places he interprets the strictist's account as requiring that everything whatever be believed in the face of a contradiction. He is quite right to think this absurd, but it is clear that this is not a requirement of the strict position soberly and realistically weighed. If, for example, Meyer is supposing that *ex falso quodlibet* is a (derived) rule of inference that requires that a contradictory belief give rise to the inference of everything whatever, it can quickly be dismissed as a faulty interpretation quite apart from its *quodlibet* clause. For it is not, as we will soon see, a rule of *inference* in the first place.

Perhaps we should understand Meyer as saying that *ex falso* embarrasses any theory of rational belief in which rational belief is closed under consequence. True, any such theory deserves all the embarrassment it gets, but there is nothing singular about the embarrassment given it by *ex falso*. Belief is not closed under consequence, never mind the special difficulties alleged on account of *ex falso*. The futility of this closure rule is transparent even without the complications of inconsistency. For where Φ is *any* proposition of my belief stock at t, given that Φ entails infinitely many propositions – relevantly or strictly, it does not matter – the present theory of belief would oblige me to add to my belief-stock at $t + 1$ that infinity of propositions. This is too much to bear. Not only are there infinitely many too many propositions for human believability, most of them are too "long" or too complex for representation in the belief-stock of a human being. Such are the burdens of our *cognitive finitude*, in Cherniak's apt words (1986). Since the closure rule is no good just as it stands, it is unnecessary to worry about the peculiarities deemed to importune its enforcement thanks to inconsistent beliefs.

Meyer sometimes writes of the consequence-closure rule so as to make it appear that it is a rule for commitments rather than beliefs. If commitments are just the beliefs that one is obliged to have as a consequence of the beliefs that one already holds, there is nothing good to say of closure under consequence. On the other hand, Meyer may understand by commitments the more general matter of keeping one's belief-stock consistent; that is, of avoiding the holding of beliefs which entail propositions which, even if we recognized them and could represent them in a timely way, could not consistently and without further ado be added to our belief-stocks. The damage caused by *ex falso* is then presumably that *some* among its infinite upshot will be accessible to inferential processing and that, once processed these will destabilize one's belief-stock. I doubt it. If some of the further inconsistencies implied by a contradiction were added to one's belief-stock, that would hardly improve the situation as regards its inconsistency. The belief-stock is already inconsistent, by virtue of its hospitality to a contradiction. Its inconsistency is hardly worsened by yielding to inconsistent consequences on a selective basis. No, the overarching requirement or *desideratum* on consistency is more plausibly honored by expunging or otherwise disarming the original contradiction and not worrying about what extra follies it commits us to.

Meyer's second additional argument is not an argument in all strictness. It is more of a rhetorical challenge. In making it, Meyer crosses the frontier between a purely relevantist approach and a strongly paraconsistent one. So the rhetoric is important. Why, he asks, are people so obsessive about consistency? It is true and important that there is quite a bit of consistency in our old world – my backyard fence is (probably) consistent, since for no property P, does it have both P and not P.[1] So, too, for ever so many of the things that clutter up my environment, including me. Although we may (or rather, certainly do) have inconsistent beliefs, it is unlikely that we ourselves are inconsistent objects. Even this is perhaps conceding too much, what with mind-body dualisms and the essential quantum physical component of our (at least bodily) makeup. Although there is indeed a good deal of consistency running through the realms of being with which we are acquainted, what reason, Meyer asks, is there to suppose that everything whatever must have it? Consider sets. Sets are rather strange entities. They are the kind of thing that we could quite reasonably expect to be rather different in their makeup from other things. Certainly *arbitrary* objects are a good deal different from the objects that they take as values. The arbitrary man is made interesting as much by his differences as his similarities to nonarbitrary, individual, men. Why cannot it be this way with sets? Might it not be the case that our expectation – no, our insistence – that sets always be consistent is really a prejudice arising from our affection for the commonplace? Why cannot sets be inconsistent? If, as Professor Higgins complains, a woman cannot be like a man, why cannot a set be unlike a fence?

This is dialethic rhetoric *par excellence*. It takes us from doxastic inconsistency to ontic inconsistency, that is, from the fact that beliefs are sometimes inconsistent to the possibility that inconsistent beliefs are sometimes true, and so to the possibility that sometimes objects or states of affairs are inconsistent. This is an issue to which we shall need to return. For now, we will take it as a transitional point in the contemporary repudiation of strict logic. For if inconsistencies of the latter, ontic, sort do exist and *ex falso quodlibet* is true, then beyond question the world would be absolutely inconsistent and, being so, would be catastrophic.

INCONSISTENCY

In Chapter 1, we briefly reviewed the inconsistency of intuitive set theory, and what likewise could be called intuitive semantics for natural languages such as English. Set theory falls under the weight of the Russell set, which is a member of itself if and only if it is not. Semantics lies exposed to the Tarski statement, which is true if and only if it is not.

By the lights of *Philosophy's Most Difficult Problem*, one way of understanding these paradoxes is to interpret them also as unwelcome and exotic surprises. They are held to be exotic in the sense that they exploit certain technical resources of the theories that generate them at, or close to, the limits of their intuitive applicability. So, for example, the comprehension axiom of naive set theory is strongly intuitive in its general use, but that intuitiveness is sharply diminished in limit applications, as in the case of sets that are their own members, something that Cantor would not stand still for. It is important for this, the "standard treatment" of the set theoretic paradox, that the counterintuitiveness of these limit applications be discernible entirely apart from whether paradox can be derived with their aid. If a paradox *can* be derived, the oddness or counterintuitiveness may be thought to give way to counterexample, and the axiom in question must be dealt with. But since such axioms and principles are highly intuitive in their general use, the style of their abandonment which is required by the standard treatment needs to be consistent with that fact. Thus, the principle in question is *qualified*, not dispensed with altogether.

Similarly for the semantic paradox. Principles permitting self-reference ("I can utter at least one English sentence") and the mingling of object language and metalanguage ("The sentence, 'It is raining' may be true in Groningen and false in Bonn") are, in their general use, intuitive and quite harmless. But they also admit of counterintuitive applications, as in "This sentence is true" (*which* sentence?), the counterintuitiveness of which obtains independently of paradoxical results. Since at their limits they also beget paradox, as in "This sentence is untrue," then counterintuitiveness again appears to give way to counterexample, and the principles must be redrawn.

The modern history of set theory and semantics in the aftermath of the paradoxes is the history of the redrawing or recalculating of basic principles in such ways as prevent paradoxical recurrence. The reworking of set theory and semantics proceeds against the background of three important assumptions, all of which the paraconsistent logician rejects.

(1) The inconsistency of mature and well received theories is a rarity, an occasional aberration that is always a setback. (2) Paradoxes falsify even the strongly intuitive general principles that beget them. (3) Consequently those principles must be reexpressed by adjusting those features of them that lead to the (nonparadoxical) counterintuitiveness of their limit applications. Hence, set theory is probably best adjusted by constraining set-membership options, and semantics is probably best adjusted by constraining self-reference and semantic closure, the comingling of object language and metalanguage.

Apart from the pivotal fact that the dialethist need not take an inconsistency-generating paradox to be anything other than a sound demonstration of the inconsistency it implies, neither will proposition (1) brook muster. Inconsistency is not an isolated, uncommon quirk of the occasional mature theory already predisposed to trouble due to its embrace of principles that are intuitive in their general use but counterintuitive in their limit applications. On the contrary, inconsistent theories abound (Priest et al., 1989, p. 152). Newton's theory of gravitation is inconsistent with Galileo's law of free fall and Kepler's laws. The Newton-Leibniz calculus is inconsistent. By the requirement of division by infinitesmals, where α is an infinitesmal, $\alpha \neq 0$. But by the requirement that infinitesmals and their products are discounted in the final value of the derivative, $\alpha = 0$. Statistical thermodynamics is inconsistent with the second law. Wave optics is inconsistent with geometrical optics. Bohr's model of the atom is inconsistent with Maxwell's equations. In Bohr's model, an electron can orbit an atom's nucleus without radiating energy. But according to Maxwell's equations, which are essential to Bohr's theory, an orbiting electron does radiate energy. The special theory of relativity was persisted with notwithstanding the contrary experimental results of Kaufmann in 1906.

It is easy to see that not all these consistencies are *self*-inconsistencies. The calculus is self-inconsistent, whereas Newton's theory is inconsistent with Galileo's theory, and relativity theory is inconsistent with some experimental data. However, each of these adventitious inconsistencies is transformable into self-inconsistency under the rule of Adjunction. If we hold with relativity theory and hold with the relevant experimental data then, by Adjunction, we hold the conjunction of relativity theory and the relevant experimental data; and the augmented theory is indeed a self-inconsistent theory. Similarly, if we want to hold Newton's gravitational theory and Kepler's laws, then the conjoined theory got by Adjunction is self-inconsistent.

Unless we can think of independent reasons for abandoning Adjunction, then very many of our most mature and authoritative scientific theories are inconsistent in *something* like the way that naive set theory and naive semantics are inconsistent. This is getting to be quite a lot of inconsistency.

It is striking that so much of what we are encouraged to count as the most secure and authoritative branches of knowledge are, in their most natural or most widely received formulations, simply inconsistent. But this is far from the end of the story. We might take our lead from Harman (1986), and liken inference to a function that takes belief-stocks (or data-bases) into belief-stocks under the guidance of such regulative principles as, "Take note of the entailments" and "Strive for the acquisition, preservation, and restoration of consistency." Then, although this account recognizes that inconsistency may often dog our beliefs, it pictures this phenomenon as, for the most part, a transitory and tractable affair. Consider an example. We might suppose that at time t my belief-stock $\Gamma = \{\dots, A, (A \to B), \dots\}$. Part of my task as a rational agent is to decide whether, and if so, how to adjust Γ at $t + 1$, that is, to decide on what the successor belief-stock Δ will be. Since I honor the regulative principle of taking note of applicable entailments, one of the options that I might consider is that of detaching B. But now that I am actually considering B, I might come to see that B is false. We can represent this state of affairs, though in a idealized and overreflective way, as my falling into momentary *cognitive dissonance*, for Δ is at $t + 1$ $\{\dots, A, A \to B, B, \neg B, \dots\}$. If I am on my toes, I will adjust Δ at once. In an attempt to relieve the cognitive dissonance I will honor the other regulative principle which instructs me to restore consistency, and so our function f will take me to Σ at $t + 2$, where Σ might be $\{\dots, A, (A \to B), B, \dots\}$, whereupon consistency is recovered. In this picture, inconsistency is something that occurs at the fringes of the transition from one belief-stock to another. Inconsistency, now manifested as momentary cognitive dissonance, can be understood to be that which motivates my maneuvers from Δ to Σ, that is, from inconsistency to consistency. It can be seen then as a basic driving force of the deduction engine, and so as a cognitive commonplace.[2]

It also bears on our question that our belief-stock also might be routinely afflicted by deep, central, and more or less abiding, inconsistencies that have little or nothing to do with momentary dissonances that drive the operation of hum-drum belief revision. If belief-stocks were inconsistent in these rather more permanent and endemic ways, then the sense in which what we believe is inconsistent would resemble the sense in which set theory, say, is inconsistent, what with inconsistency a permanent or next-to-permanent feature of each. If belief-stores were then inconsistent in this way, we would be simply awash in inconsistency, and there would be little left to say in defense of (1).

Is there reason to think that belief-stores are routinely, deeply, and more or less permanently, inconsistent? Recent work on human memory offers an interesting counterfoil to the picture of rational belief-store management suggested by Harman's account of inference. It is not by any means incompatible with Harman's account, but it fills it out, so to speak, in dramatic ways. If we imagine a database to be an inventory of what is perceived, believed and remembered at a given time, then it must be said at once that there is far too much in long-term memory for anything like exhaustive retrieval. Complete systematic inventories are not possible objects of search and rehearsal. Even if we were to allow that our retrieval capacities are far more potent than they actually are, the very extensive plumbing of our memory stores would take paralytically long, and so would derange the business of rational belief-management, and of life itself. We would never find the time to figure out how to make our way to Central Station. Human memory is sectoral, much in the way that governments are compartmentalized by ministerial portfolios (Klatzley, 1975). Beliefs tend to cluster in more or less independent sectoral subsets, the contents of which are routinely inconsistent. Sectoral scrutiny and comparison is difficult to bring off and it is, therefore, a wholly natural consequence of the organization of human memory that intersectoral doxastic inconsistency is rather easy not to notice.

The psychological literature on memory convincingly distinguishes between *occurrent* memory and *dormant* memory, answering also to a distinction between short term and long term memory (see Howe, 1970; Collins and Quillam, 1969; Lindsay and Norman, 1977). These two reminiscential operations work quite differently from one another. Occurrent memory presents beliefs that are, then and there, ready for action, for driving inferences and influencing behavior. Beliefs stored in dormant memory are not available for premissory action in inferences; they do not contend or interact with one another; and some say that they do not influence conduct.

It is natural therefore that occurrent memory should be governed by tougher requirements than dormant memory. Occurrent, or short-term memory, is and should be, troubled by inconsistency, but inconsistencies lurking in dormant memory are virtually harmless. It is unreasonable to impose on dormant memory rationality conditions that are appropriate only for occurrent memory, for once again there are too many beliefs in dormant memory for comprehensive searches and global scrutiny.

Beliefs and memories are not the only things held to consistency expectations in rational life. Desires are expected to be consistent, too: "beliefs and desires can hardly be reasons for action unless they are consistent" (Elster, 1985, p. 4). In much of contemporary social science, consistency is expected in preferences. The minimal condition on preference is transitivity. If agent X prefers A to B and B to C then he can be expected also to prefer A to C.

The idea of preference extends in a natural way to sequences of options. A time preference is definable as the importance an agent at a given time gives to later times. Time preferences usually discount the future, that is, they give less importance to future rewards than to the rewards of the moment. Consider what economists call a "plan of consumption," which is made at time t_1 for the allocation of consumption between times t_2 and t_3. Then, for the preferences that the plan embeds to be consistent, the plan for the interval $t_2 - t_3$ should remain in effect when t_2 arrives, given that the agent's set of feasible options has not changed in the meantime. The person whose time preferences are inconsistent fails to stick to his plan. If an agent's plan were to spend only half his vacation money between t_2 and t_3, and yet when t_2 arrived he spent nearly it all on a shopping spree, his preferences would have lapsed from consistency. What is required to maintain consistency for time preferences is that the "future must be discounted at a constant ratio" (Elster, 1985, p. 7), that is, time preferences must be *exponential*.

It is also widely held that preferences, if they are to serve as reasons for action, need to satisfy a condition of continuity. Thus, if X prefers A to B and A changes slightly, this would not suffice to reverse the preference. So, if I prefer *The Times* to *The Sun* and the price of *The Times* rises by five pence, with no change of price in the other, I should be expected to continue preferring *The Times*. A further condition to which preferences are held is the condition of completeness. Preference is complete for X if for any pair of options for X, A or B, either X prefers A to B or B to A, or X is indifferent to A and B alike.

It is fair to ask why preference is held to these conditions and what it is about them that makes noncompliance give rise to inconsistency. Why should transitivity be a consistency-condition for preferences? If X prefers A to B, he must be prepared to pay something (money will do for the example at hand – never mind for now the problem of its diminishing marginal utility), perhaps only a very small amount of it in order to exchange B for A. Similarly, for the exchange of A for C and of C for B. In the end this leaves X with B, which is what he started with, *and* X is now out of pocket. If X continues to act on this intransitive preference the above exchange profile recurs and recurs again, and so on, until X has put itself in the poorhouse (Raiffa, 1968, p. 78). One might suppose that human beings, at least those whom we are prepared to count as rational agents, will more or less routinely fulfill these conditions on consistency of desire. True, we might say, there will be occasional defections – no one is perfect – occasioned by fatigue, inattention, or carelessness, but inconsistency in rational life will be the exception, not the rule.

There is reason to doubt this assumption.[3] While it is quite true that people usually try to adjust their preferences if intransitivities are pointed out to them, the fact is that adopting intransitive preferences is far from uncommon. Moreover, even after intransitivities have been identified, some people persist in their preferences (Raiffa, 1968, p. 75). However, Davidson's experiments indicate something quite baffling about human preference. Preference is frequently intransitive in individual episodes of choice, but in sequences of choices the intransitivities tend to be leached out. Preference is somehow self-correcting, but we know not how (Davidson, 1976, p. 107).

Time preference inconsistency is also a common occurrence in human life. People are strikingly prone to nonexponential time preferences, that is, to time preferences in which the future is discounted at nonconstant ratios. However, here, too, it seems that remedies exist that enable the nonexponential time preferer to deal with this problem. George Ainslie has suggested that by collecting "several future choices the chances are increased that in each of them one will take the option with a later and greater reward" (Elster, 1985, p. 8; cf. Ainslie, 1982).

The principle that preference should be continuous is breached by what are called non-Archimedean preferences, such as lexicographic preferences attached to normative hierarchies. Here is an informal example, which I quote from (Elster, 1985, p. 9):

> If I am starving and am offered the choice between an option involving one loaf and listening to a Bach record and another involving one loaf and listening to Beethoven, then my love for Bach may make me prefer the first option. If, however, from the first option is subtracted even a very small crumb of bread, as small as you please, then I switch to the second because at starvation level calories are incomparably more important than music.

We see then that the continuity principle is not just breached, it is refuted. We might wonder why it was proposed in the first place. The answer is that such an assumption helps streamline theory. For if preference is held to the conditions of transitivity, completeness, and continuity, preferences are representable as real-valued utility functions and the mathematics of preference manipulation takes on a certain power and elegance.[4]

To return briefly to belief, there is evidence to indicate that people are bad probabilists, as in the work of Nisbett and Ross (1980, p. 146). Experimental results suggest that subjects are prone to numerous errors and misinferences arising from defects in "the cognitive mechanism" (Elster, 1985, p. 26). "Recent empirical studies of behaviour under uncertainty has [*sic*] brought out what has appeared to be systematic inconsistencies in the evaluation of risk and in the comparative assessment of alternative decision. Many of these results have been interpreted ... as simply "mistakes" in perception or reasoning. Even if that view is fully accepted, the *prevalence* of such behaviour indicates the case for making room for departures from the usual requirements of "rationality" in understanding actual behaviour" (Sen, 1987, p. 69). Here is one of the classic examples. The proposition that X is a Republican was judged by subjects to be highly likely and that X is a lawyer was found to be unlikely. When asked to assess the proposition that X is both a Republican and a lawyer, this was found to be middling likely. In so saying, the subjects were taking the probabilities additively rather than multiplicatively, in violation of the Conjunction Axiom of the probability calculus.

The significance of the past several pages is this: Inconsistency is no *rara avis*, not the defection from logical rectitude of an occasional abstract theory. Inconsistency routinely dogs us in the management of our beliefs, the manipulation of our memories, and the organization of our desires. There is no reason to think that such inconsistencies are anything other than something to be avoided if possible, or corrected once noticed. If we like, we can say that human beings have internalized something like a strictist's bias toward inconsistency. But the strictist's canons are not, or not always, the ones that people resort to as strategies for inconsistency-*management*. The last thing that a competent reasoner would do is to submit himself to *ex falso* as such a strategy. Point to the nonstrictist.

It is also quite true that not all of these cases of inconsistency are representable as logical contradictions. However, again with the aid of supplementary assumptions, for example, the rule of Adjunction, some certainly are. Nonexponential time preferences yield nothing in the form $\ulcorner \Phi \wedge \neg \Phi \urcorner$ without further premissory supplementation, and, as we have seen, non-Archimedean preference ranking is not inconsistent in any sense, save in ideal models designed to simplify the mathematics.

It might be said that the case of the Republican lawyer provides us with a logical contradiction, inasmuch as the judgment of moderate likelihood breaches the Conjunction Axiom of the probability calculus. Conservatives might jib: The probability calculus is not logic. Say what we like, if the axioms of the probability calculus yield up a satisfactory analytical reconstruction of probability, perhaps we could say that the judgment of middling probability regarding the Republican lawyer is analytically false. It should be mentioned, however, that among some probability theorists there exists considerable confidence that there are concepts of probability that the probability calculus does not faithfully describe. As Jonathan Cohen points out, "... though aleatory probability [i.e., the probability involved in games of chance] always requires a complementational principle for negation and a multiplicative principle for conjunction, there are contexts in which credibility conforms to non-Pascalian [i.e., nonaleatory] principles" (1989, p. 13). Thus, it may be that the subjects in the Nisbett and Ross test were, for one reason or another, *not* making a mistake. They would not have been if in the experimental context in which they found themselves they were invoking a nonmultiplicative concept of probability and were right in invoking it.[5]

At any rate, there is a good deal in these psychological disclosures about memory, preference, and the like to give comfort to the dialethist. It is not only that we have many scientific accounts of things, mature and well-considered accounts too, that are inconsistent; and it is not only that the engines of inference are driven by repeated momentary and transitional cognitive dissonance. We also have it that memory is dotted by intersectoral inconsistency, that deep or dormant memory is pervasively inconsistent, and that preference ordering is routinely inconsistent as well. What is more, there appears to be, in this empirical evidence from psychology, reason to think that inconsistency is not disabling, and that even when, once noticed, an inconsistency is set up for repair, it does not seem that *ex falso quodlibet* plays any role. Are we then at the point at which it is appropriate to admit something like a paraconsistent logic as (part of) a theoretical reconstruction of what the rational management of inconsistency might come to? Let us see.

DIALETHIC BELIEF-MANAGEMENT?

Here is the principal motivation for the ideas of this section. The notion that there is something catastrophic about inconsistency is hardly borne out by our reactions to it. That this is so suggests, although it is far from proving it, that in large part it is the pervasiveness of inconsistency in human life that makes it possible to deal with it sensibly, and certainly not in general by getting rid of it no matter the cost. The empirical record alone suggests that inconsistency is just too intractable for efficient surgical removal as the only permissible recourse. We are met, obviously enough, with two competing psychologies of belief-inconsistency. We might characterize their difference roughly as that between a pain in the neck, the end of the world.[6] If we take seriously the metaphor of a pain in the neck, it is natural to ask what might its remedy be? The remedy will be one that attempts to remove or minimize or block the pain, but not at *any* cost, certainly not at the cost of removing the neck!

In order to examine a paraconsistent remedy for belief-inconsistency or, as we might also say, a paraconsistent recipe for the rational management of inconsistency, let us briefly visit a zero-degree system of dialethic logic, which I shall call **PR**; it derives from an account presented by Priest and Routley, which, in turn, rests substantially on work by their colleague Ross Brady (Priest et al., 1989, pp. 168–70). The name "**PR**" reflects the fact that the system it denotes is a strongly paraconsistent logic that also is relevant. So **PR** represents an interesting technical linkage in our passage from relevant systems to paraconsistent ones.

PR's point of departure is relevant entailment, that is, an entailment relation (here, denoted by "→") that satisfies condition **Ent** as interpreted by condition ****Rel**. Thus ⊢(A → β) only if A is relevant to β, and A is not relevant to β unless A and B share propositional letters.

It is clear already that **PR** will evade the "paraconsistently execrable" (Priest et al., 1989, pp. 168–70), *ex falso quodlibet* and so is a promising candidate for a paraconsistent logic. What else is required that will make **PR** paraconsistent, that is, at least, minimally tolerant of the *possibility* of deriving an inconsistency therein? It is something which, following Brady (1984, pp. 63–74), I shall call a "depth relevant logic," which is actually a weaker system than Anderson and Belnap's system R.

We begin with a language L. If Φ is a sentence of L then [Φ] is the *sense* of Φ, and ≤, a partial ordering, is the relation of *sense-containment*. Thus, [Φ] ≤ [ψ] if and only if the sense of ψ contains the sense of Φ.

We also introduce the three functions, ∪, ∩ and *: [Φ] ∪ [ψ] = [Φ] ∨ [ψ]; [Φ] ∩ [ψ] = [Φ] ∧ [ψ]; and [Φ]* = [¬Φ]. By these operations we express the fact that in **PR** the sense of a compound sentence is a function of the senses of its constituents.

With these operations at hand, it would appear that the partial ordering of senses goes over into a *distributive-lattice* for which ∪ is the *join*, ∩ the *meet* and * an *involution*, that is, a function obeying the conditions.[7]

  i. $\Phi ** = \Phi$; and

  ii. if $\Phi \le \psi$ then $\psi* \le \Phi*$

So, in particular, the sense of $\ulcorner\Phi \to \psi\urcorner$ will contain the sense of Φ and the sense of ψ, and anything containing both these senses will also contain the sense of $\ulcorner\Phi \wedge \psi\urcorner$. Thus, we are given a *De Morgan lattice of senses*.

Our DeMorgan lattice of senses enjoys a far from arbitrary motivation. It allows us to define entailment in such a way that an entailment is true precisely when the sense of the antecedent contains the sense of the consequent. We can notice, in passing, that such a definition also honors (for what it may be worth) the preanalytic intuition that entailment is a matter of a *connection of meaning* between antecedent and consequent: Thus, $\ulcorner\Phi \to \psi\urcorner$ is true if and only if $[\Phi \to \psi]$. And so we may imagine that T is a *prime filter* on the lattice, that is, that we have the following conditions:

  i. $\Phi \cap \psi \in T$ iff $\Phi \in T$ and $\psi \in T$

  ii. $\Phi \cup \psi \in T$ iff $\Phi \in T$ or $\psi \in T$

  iii. If $\Phi \in T$ and $\Phi \to \psi \in T$ then $\psi \in T$

PR provides as logical truths

$$\Phi \to \Phi$$

$$\neg\neg\Phi \to \Phi$$

$$\Phi \wedge \Phi \to \Phi$$

$$\Phi \to \Phi \vee \psi$$

and as valid rules of transformation

$$\Phi \to \psi, \psi \to \chi \vdash \Phi \to \chi$$

$$\Phi \to \psi, \neg\psi \vdash \neg\Phi$$

$$\Phi \to \psi, \Phi \to \chi \vdash \Phi \to \psi \wedge \chi$$

$$\Phi \to \chi, \psi \to \chi \vdash \Phi \wedge \psi \to \chi$$

$$\Phi, \Phi \to \psi \vdash \psi.$$

PR serves the basic purposes of a dialethic logic based on relevant entailment. It is relevant; the Curry paradox fails;[8] and negation is recognizable and well-behaved (it obeys Double Negation, Contraposition, and de Morgan). Moreover, as Brady shows (1989), it is possible to absorb naive set theory and naive semantics into a logic such as **PR**, in which inconsistencies are allowable and yet the resulting system is provably nontrivial, that is, noncatastrophic. Absolute inconsistency is avertible.

Disjunctive Syllogism fails in **PR**. "Indeed the relevant approach holds the losses from classical logic to a minimum at the zero degree level. This is the only loss on the relevant paraconsistent position. Moreover, the loss of the disjunctive syllogism is not as great a blow as might be thought." In fact, it is "generally valid" (Priest and Routley, 1989, pp. 169–70). It will prove important when, in Chapter 5, we examine the claim that, short of catastrophe, **PR** is capable of absorbing a theory of sets and a theory of semantics each of which is intuitive and manageably inconsistent. For now, our more immediate preoccupations are with the question whether a logic such as **PR** offers helpful clues about how to describe efficient and rational strategies for inconsistency management.

Let us set the scene once more. The paraconsistentist is attracted by the following argument.

(1) Inconsistency is endemic. Not only is it a persistent feature of many of our best theories, it is also an habitual visitor of our ordinary beliefs, memories, and desires. (2) One thing in particular is quite clear. This unabatable inconsistency is not paralyzing and is not catastrophic for belief and action. (3) This means, therefore, either [a] that although our theories, preferences, beliefs, and memories commit us, as the strictist insists, to the triviality result, somehow in actual ratiocinative practice we are able to outmaneuver these inconsistencies and to leave the triviality-consequence uncashed; or [b] that, as the paraconsistentist insists, some at least of these inconsistencies are true, or can be taken as true in such a way that the triviality-consequence collapses. (4) It is therefore appropriate that we look to a paraconsistent logic such as **PR** for models of the situations reflected in (3) [a] and (3) [b].

(3)[a] suggests that we are, to some extent, dogged by unexpungeable inconsistencies that somehow can be maneuvered around, short of having to cash the triviality-consequence. Whether **PR** models inconsistency management in an interesting way will turn on the extent to which it honors what is known empirically. There is little in the psychological literature about how cognitive agents manage inconsistent belief-sets. Even so, broad patterns of their handling are evident in common practice. Such evidence as there is indicates something like the following routines of inconsistency-management. Faced with a (noticed) contradiction in his belief-stock, a person may:

(I) Quickly forget that it is there if it seems harmless and inessential to any of his cognitive tasks at hand. This attitude is reflected in Emerson's couplet that "a foolish consistency is the hobgoblin of minds, adored by little statesmen and philosophers and divines" (1967, p. 37).

(II) If he recognizes that it may be important for his cognitive agenda to resolve the contradiction sooner or later, and if he lacks the means of doing it now, he may quarantine the contradiction and keep it out of premissory action for the time being. (This is reminiscent of Wittgenstein's supposition that a contradiction that we do not know how to dissolve can be "sealed off" and allowed to stand.)[9] A case in point is Frege's achievement in the *Foundations of Arithmetic*. Appreciating the achievement requires that we attend to the distinction between principles *conceded* by *Foundations* and principles *used* in the generation of arithmetic. The set of principles conceded is inconsistent, whereas the set of principles used is consistent; and those principles deliver the goods for arithmetic. As Boolos sees it,

A piece of mathematics carried out in an inconsistent theory need not be vitiated by the inconsistency of the theory: it may be possible to develop the mathematics in a suitable proper subtheory. (Boolos, 1998a, p. 182).

(III) If an agent's cognitive agenda presses for resolution then and there, he may attempt to discern an ambiguity in the inconsistent proposition and so discover that the sense in which it is true is not the same as the sense in which it is untrue. Something like this occurs in the kind of naive, semantically closed system developed by and large independently by Kripke, Gupta and Herzberger (see Kripke, 1975; Herzberger, 1982; Gupta, 1982).[10] Such systems – KGH systems, we might call them – possess valuation procedures that provide that a paradoxical sentence, such as **T** ("This sentence is untrue"), is not both true and untrue "in the same respects," that is, at the same valuational levels. We shall return to KGH systems in Chapter 5.

(IV) Depending again on the urgency of his cognitive agenda, he might decide to "split" the inconsistent sentence and put each of its conjuncts to noncontending premissory work, this, too, on sufferance, until more stable remedies suggest themselves (Rescher and Brandom, 1980, pp. 44–81). And so he will be well advised to refrain from the Adjunction rule, the failure of which, although a principal feature of some systems of paraconsistent logic, is not by any means always particularly paraconsistent in its motivation. For example, it is quite possible to be a skeptic about what used to be called the Unity of Science, and so a skeptic about unqualified resort to the Adjunction rule, without having any particular inclination to abandon strict logic (Cartwright, 1983, pp. 101–04).

How well does **PR** model this range of responses? The question is set by the more carefully worded (3)[a] above: Is it the case that "although our theories, beliefs, and memories commit us, as the strictist insists, to the triviality result, nevertheless, somehow in actual practice we are able to out-maneuver these inconsistencies and to leave the triviality-consequence uncashed, thus, in a sense, exemplifying Wittgenstein's idea that contradictions can be allowed to stand?" Notice that in taking the position of (3)[a],

(4) we agree to leave uncontested whether *ex falso* is (really) true, and acknowledge that, whatever its status, it holds in classical logic and fails in **PR**;
(5) we agree that **PR**, never mind its standing as a theory of entailment, might model or mimic actual rational practice concerning the management of inconsistency.

and

    (6) this said, we now ask, "Does it do this in acceptable and illuminating ways, if at all?"

Concerning (IV): Does **PR** reflect the fact that people confronted with an inconsistency sometimes withhold the Adjunction rule? No, the Adjunction rule is valid in **PR**.

    Concerning (III): Does **PR** capture the strategy of contradiction-removal by way of disambiguation? It is moot on the question. Although it does not, and does not have appropriate occasion to, disapprove of such a maneuver, its main interest lies in the claim that inconsistencies do not have to be parsed out of existence. They can be "allowed to stand." This is not to say that **PR** actually encourages persistence with "inconsistencies" that are not really inconsistent, but there is nothing in **PR** to support as a general strategy the attempt to make inconsistencies go away by trying to find interpretations under which they are not inconsistent.

    Concerning (II): Does **PR** approve the strategy of quarantining contradictions, of keeping them out of harm's way? No, since there is no question of their being "in harm's way." In particular, the assumption of (II) is that the harm's way that inconsistency is in is that it gives the triviality result. So it is natural, on that assumption, to want to disarm such things. The practitioner of maneuver (II) supposes that a contradiction can be disarmed by not triggering some of the consequences that it actually has. The dialethist holds that since it does not have those consequences there is no need for selective constraint.

    Concerning (I): Does **PR** model the fact that human reasoners sometimes just do not bother with inconsistencies that play no central role in the cognitive agenda at hand, that is, that they do not give such things any serious place in their cogitations? To the extent that such a strategy suggests the "uselessness" of contradictions, **PR** does not honor it. But, in so saying, a more important question is forced on us. If, as in strategy (I), we sometimes do not bother to make inferences based on unimportant or irrelevant contradictions or on any of their conjuncts, it is necessary to ask, "What inferences do the rules of **PR** license us to make, and is the strategy of (I) compatible with those inferential procedures?"

    In asking what are the inferences that **PR** licenses us to draw, we hit on the central weakness of **PR**. Like R and like classical logic, *PR is not a credible theory of inference*, never mind its special provisions for such inferences as it thinks are allowably got from inconsistencies. In the first place, **PR** offends against that regulative rule for inference that says that noticed inconsistencies should always be removed from one's belief-stock. Strategy (I) all but complies with this in its decision to disable unimportant contradictions for premissory action and for the guidance of conduct. **PR** dishonors it flatly.

Of course, the *dialethic* paraconsistentist has a reply to this. He can say, and does, that **PR** admits precisely those contradictions that we know to be true or have good reason to think true, and so **PR** honors the more basic regulative ideal, of which ours is a special case, of retaining in our belief-stocks only those things that are *true*. This, the most distinctive feature of **PR**, does not, however, figure in a situation of the type that we are now examining, that is, in a situation of the type described by (3)[a]. So we shall leave the point for now, returning to it in Chapter 5 when we take up the discussion of situations of the type described by (3) [b].

The other key respect in which **PR** fails as a theory of inference is that, like classical logic and R, it possesses perfectly uncontroversial transformation rules that, although they are correct for entailment, will not do as f-rules; as rules of rational belief-stock modification. An example will make the point. In **PR** we have the rule Hypothetical Syllogism,

$$\Phi \rightarrow \psi, \psi \rightarrow \chi \vdash \Phi \rightarrow \chi.$$

If this is given its weakest interpretation as a rule of inference, then it is always permissible to place into one's belief-stock

$$\Phi \rightarrow \chi$$

if

$$\Phi \rightarrow \psi, \psi \rightarrow \chi$$

are already there. But this is untrue. It is not always permissible to do this. As we have said, one might notice that $\ulcorner \Phi \rightarrow \chi \urcorner$ is false, at which point it becomes necessary both to erase something that is already there, as well as refuse entry to $\ulcorner \Phi \rightarrow \chi \urcorner$.

We have been asking ourselves whether **PR** is an acceptable and illuminating model of empirically indicated strategies for the efficient management of inconsistency in situations of the sort described by (3)[a]. We have been asking, whether or not **PR** is a good theory of entailment, could it nevertheless be the case that it is a good theory of rational inconsistency-management in situations in which, whatever the entailments really are, they are in some way or another not cashed, not acted on strictly, so that major difficulties might be averted?

The answer is No. The four strategies I to IV for the management of cognitive dissonance are not preserved well, or at all, in **PR**. Moreover, quite apart from what inferences it is competent to draw under circumstances of inconsistency, **PR**, like classical logic and R, is not a good theory of inference in general. Rules for entailment, whether paraconsistent, relevant, or strict, are not good rules of inference; they are not f-rules for the rational adjustment of belief-stocks in the light of (noticed) entailments.

INFERENCE AND IMPLICATION

It will help us to recognize a distinction between belief and acceptance. Belief is a *biopsychological state*, and acceptance, also, can have a psychological mooring. Often enough, when one accepts what one does not believe, one roots the acceptance in the belief that, all things considered, accepting what one does not believe is a good thing, or that there is something to be said for it. In its pure form, acceptance is better thought of as *assent*, that is, as a *social act*, something that I pledge myself to publicly under dialectical constraints involving some other agent. If my belief that Φ is a matter of how my psychological states converge on Φ, my assent to Φ is a matter – in pure form – of what I *say*.

As we have been conceiving of it here, inference is a function f from belief-stock or data-bases to belief-stocks or data-bases, hence a function that takes biopsychological states into other biopsychological states. There are conditions on f that we could think of as *state-conditions*. When I infer ψ from Φ, a biopsychological event displaces a biopsychological event, under state conditions that permit this to happen, or which, as the case may be, causally require it. Perhaps there is a state condition resembling *modus ponens*. *In abstracto*, it provides that a biopsychological state constituting a belief that $\Phi \rightarrow \psi$ when adjoined by a biopsychological state constituting a belief that Φ will, all else being equal, *produce* a biopsychological state constituting a belief that ψ. Much is borne by the qualification "all else being equal." Such is the purport of Harman's insistence that it is not invariable that a belief that ψ will come, or should, from the prior belief that $\Phi \rightarrow \psi$ and the concurrent or subsequent belief that Φ. As Harman points out, the belief that Φ might get us to attend to ψ in ways that make us think again about whether $\Phi \rightarrow \psi$.[11]

Arguments are interpersonal exchanges of speech-acts under rules that constitute them as such and other rules that help regulate to and fro in appropriate ways. Inference is governed by state conditions on biopsychological events, but arguments are in the ambit *rules*, which a party can comply with or disobey. Perhaps here, too, there is a rule suggestive of *modus ponens*. If you have assented to $\ulcorner \Phi \rightarrow \psi \urcorner$ and also to Φ, then you must now assent to ψ as well. This is something you will be *obliged* to do, unless you do something else to compensate. In strictness, there are no *rules* for inference. State conditions are causal conditions, which can be neither obeyed nor flouted. Argument rules are social conventions. They are prescriptive rather than causal. They do not *make* us assent or dissent in certain ways. They *oblige* us to.

, *Modus ponens* is connected to entailment. It appears in logical theory in two ways; but they are the ways of confusion. We may see *modus ponens* as condition on entailment – as a *truth condition*, as when it is said that entailment is such that the conjunction of Φ and $\ulcorner \Phi \rightarrow \psi \urcorner$ entails ψ. We may also see it as a transformation "rule," which tells us that when marking out what follows from a set of statements we may enter ψ if the set contains Φ and $\ulcorner \Phi \rightarrow \psi \urcorner$.

There is a twofold confusion. Truth conditions on the entailment relation are taken for rules, and even in the technical sense in which logic gives us trans- formation "rules," these are rules on derivations that are mistaken for ar- guments. Of course, a derivation is a proof if it satisfies the appropriate ax- iomatic conditions; and a derivation is a conditional proof if it satisfies certain natural deduction constraints. Proofs and derivations are not arguments in any straightforward sense. They are not arguments on the hoof. They lack the requisite dialectical features. Mill was right to regard them as a kind of bookkeeping device, a register or re-presentation of the antecedent labor of thinking them up (Mill, 1839–1845/1974).

With the distinction between belief and assent now in hand, it will be useful to introduce the notion of *concession*. I will say that

$$\text{concession} = \langle \text{assent} + \text{commitment} \rangle.$$

Behind this primitive idea is another even more basic one. If we assent to a statement under certain circumstances then (i) we are stuck with it in those circumstances, and (ii) there are additional things we are also stuck with.

Answering to condition (i) is the idea of *initial commitments*.[12] Suppose that in a discourse of type D someone assents to Φ. Then he is initially committed to Φ in D just in case dissenting from Φ or assenting to Φ's negation, without first having cancelled one's assent to Φ, would be D-pathological. It would be a *defective* move in D, a move that *frustrates realization of D's goals*.

Answering to condition (ii) is the idea of *derivative commitments*. Sup- pose that someone is initially committed to Φ, and suppose that Σ is the set $\{\psi_1, \ldots, \psi_n, \ldots\}$ of Φ's deductive consequences. Then is he committed to all the ψ_i in Σ, or only to some? If the latter, what is the principle of selection?

We may favor the initially plausible suggestion that we are not committed to every consequence of anything to which we chance to be committed (and, correspondingly, that in conceding something, we do not concede every con- sequence of it). Consider the case in which concessions Σ are inconsistent. Then Σ implies everything, if classical logic is correct. Yet for most discourse- types D, if not all, the concession of every proposition is D-pathological. It is a defective move in D as it frustrates the purposes for which we engage in D.

Here are two options to consider. (A) We want not to say that commitments (hence concessions) have nothing to do with the entailment relation; so the first option is to constrain the entailment relation. We try to do this in such a way that *real* implication is that relation that delivers from a set of conces- sions all and only its derivative concessions. So we retool entailment in such a way that the relation of "being a derived concession of" coincides with the re- lation of "being a consequence of" on sets of concessions. Against this it might be objected that the present option allows the tail of derived concessions to wag the dog of entailment. (B) On our second option entailment is left alone. The basic idea is that we should try to keep it classical for as long as possible. Then, concerning derived concessions, we introduce defining conditions that are appropriate to what they are conditions *of*. It is well to emphasize that the idea of a derived concession is this: Let Φ be a logical consequence of a set of concessions in a discourse D. Then Φ is a derived concession in D if and only if not conceding it without compensatory adjustment would be D-pathological. Thus, what is needed is the specification of *new* constraints on derived concessions, rather than the standard ones that define the D-*independent* relation of propositional entailment. The same holds for inference when considered as *belief*-revision.

If we decide for option (B), we are acknowledging that commitment is different from entailment. There is a standard definition of entailment. What is needed now is a definition of commitment:

For agent S and statements Φ, ψ, S is *committed to* any consequence ψ of what he concedes Φ, just in case $\ulcorner\neg\psi\urcorner$ entails $\ulcorner\neg\Phi\urcorner$, and the entailment of $\ulcorner\neg\Phi\urcorner$ by $\ulcorner\neg\psi\urcorner$ does not depend on the modal status of either Φ or ψ (in other words, the entailment in question is modally independent).

Now,

The entailment of Φ from ψ is *modally independent* just in case the modal status of neither Φ nor ψ is sufficient for the fact that Φ entails ψ; or the modal status of Φ or ψ does suffice for this fact, but there is a true entailment schema χ entails χ' for whose truth the modal status of neither χ nor χ' is sufficient, and the entailment of ψ by Φ is a case of this entailment-schema.[13]

Concerning the first clause: "x is red" entails "x is colored," but the entailment depends on neither the necessary falsehood of "x is red" nor the necessary truth of "x is colored." Concerning clause two, consider the statement

$$A \wedge \neg A \Vdash A.$$

A follows from $\ulcorner A \wedge \neg A\urcorner$, but it also is a case of the Simplification-schema

$$\Phi \wedge \psi \Vdash \Phi$$

which itself is a true entailment by classical lights. The relation of "case of" is truth-preserving; so

$$A \wedge \neg A \Vdash A$$

holds by Simplification.

A certain stratagem re-presents itself. It is to *download* complaints against implication to the commitment relation. We may or may not like it that a contradiction implies everything whatever, and we may or not like it that a necessary truth follows from everything (and nothing, too); but there is nothing to be said for the proposition that a person is committed to concede any statement implied by what he inconsistently concedes. The strategic principle, therefore, is that it is best to press our complaints where they stand the best chance of being incontrovertibly correct.

Do commitments serve as a general closure condition on concessions? If we put it that an agent S concedes any commitment of any of his concessions, we would have it that S actually dissents from a commitment of his concessions. We might seek relief by postulating tacit concessions, and the possible concurrence of an agent's tacit concession of ψ and his express dissent from it. On the whole, we are better served by imposing a recognition condition on concession: If S concedes that Φ, and S is committed to ψ on account of Φ, and if S *recognizes* this commitment, then S concedes that ψ. An improvement, it is still not good enough. It neglects the fact that sometimes an agent will *not* concede, and will expressly dissent from, a statement he recognizes he is committed to. Better yet, we should abandon the idea of general closure conditions for concessions, and that we "normicize" the concept of closure. Whereupon, if S concedes that Φ and recognizes that in doing so he is committed to ψ, but does not concede that ψ, it can straightforwardly be said that conceding ψ is something he is *obliged* to do, that doing so is directly within his power, and that not doing so is a violation of the cooperative norms that undergird the very concession-commitment exchange in which his defection is mired. So instead of general closure conditions, we have a quite general *performance* condition, or, as we might say, a pragma-dialectical *norm*. Concession is both pragmatic and dialectical. It is dialectical because it is always directed *to* someone. It is pragmatic because concession is always concession *by* someone.[14]

If inference is belief-revision or belief-updating, it is not pragma-dialectically specifiable. It does not obey – it is *unable* to obey – the pragma-dialectical norm. If we thought of inference as a kind of concession-revision, it would fall under the norm. Consider a case in which an agent violates the norm by failing to concede ψ that he recognizes that he is committed to concede. Taking inference as concession-commitment, his not conceding ψ in these circumstances just *is* his not inferring ψ. On the other hand, there are lots of cases in which what an agent fails to concede is what he has come to believe, hence something he has inferred. A person may not want to lose an argument he is involved with, or he may be shy about admitting what he is committed to concede and what, as in the present example, he actually believes.

In other cases, what the party does not concede (and therefore does *not* infer) is identical to what he does believe (hence *has* inferred). We could try to remove the contradiction by pleading ambiguity, itself a kind of the reconciliation strategy. We could say that the party *did* infer ψ in the biological sense and yet did *not* infer ψ in the social sense. I am not much drawn toward the ambiguation strategy in the present case. There is a different and better strategy to which we could appeal.

Armed with the distinction between commitment and entailment, we can say more of the distinction between commitment and concession. Consider:

If *S* concedes that Φ and recognizes that Φ commits him to ψ then *S* is *required* to concede that ψ.

This, as we say, is the fundamental pragma-dialectical rule. It regulates *argument*, a pragma-dialectical exchange, a kind of social interaction. We also have seen that belief-revision is not governed by rules. We cannot intelligibly command someone just to *be* in a biopsychological state that he is not now in, since belief cannot be commanded by the will. Inference always has been supposed to have something to do with implication. This is the source of considerable confusion, because argument is also held to have something to do with implication. This has produced two mistakes: (I): Inference, argument, and implication are "effectively" identical, the topics of a common theory (called logic). (II): Although inference and argument are not the same as implication, they sufficiently resemble each other so as to be the objects of a common theory (also called logic). When we reflect on the fundamental pragma-dialectical rule, it is easy to see how concessions (and commitments, too) differ from implications. What is now needed is a characterization of inference that sets it apart from implication, and also from argument.

If S believes that Φ and that Φ commits him to ψ, then S *will* (as opposed to "is required to") believe ψ, provided that S's doxastic devices are functioning properly.

This is the *fundamental condition on inference.*

I lack a theory of proper functioning for doxastic devices. Perhaps something like Ruth Millikan's account will do (Millikan, 1984). If so, an agent S's cognitive devices are functioning properly, or as they should, if they are operating in such ways as to facilitate the replication of those devices in the descendent-chains of which S is a member. Not everyone is satisfied with this line of approach (Plantinga, 1993), but it does not matter for our purposes here. It suffices that somewhere there is a true account of proper functioning (perhaps in Plato's Heaven). I assume that there is, and I appropriate it into the service of the fundamental condition on inference.

Implication, argument, and inference are *three* things:

(1) Implication is fixed by *truth conditions*. Φ implies ψ if and only if it is not possible for Φ to be true and ψ false.
(2) Argument is regulated by *pragma-dialectical* rules, which are rules of social interaction.
(3) Inference is characterized by *state-conditions* that "predict" output biopsychological states from input states.

It is greatly to be doubted that a common theory will do for all three in any commonly systematical way. But it is precisely this that modern logic has taken more or less for granted. It was a huge mistake.

The strictist has an important advantage in the controversy with the dialethist. He can say that the superiority of **PR** over classical logic cannot be established by its more credible representation of the rational dynamics of cognitive dissonance-management or of inference in general, for they both fail in such matters. He can also say that the pragma-dialectical dynamics of argument fails to give enough of a nod to **PR** for it to matter in the rivalry between it and classical logic as a theory of argument. This leaves as the only serious question on which the adjudication of their differences can sensibly turn: Which is the better theory of *entailment*? Once again, we are back where we started. Is it possible to avoid the stalemate in determining the superior claim on entailment?

It may be that strongly paraconsistent systems are, when considered as models of rational belief management, too strong for their own good. But we would do well at this juncture to flag a different reservation and resolve to give it our attention before chapter's end. The other possibility is that what counts against **PR** as a theory of rational belief-management, or even that part of it we could call a theory of inference, is that **PR** represents itself as a logic and indeed is enough of a logic to merit the following objection. No *logic* is a theory of *inference*. I shall come back to this point below. For the present I want to return our discussion to two of the strategies for the management of cognitive dissonance, strategy III (the ambiguation strategy) and strategy IV (the split-the-contradiction strategy). Although a strongly paraconsistent analysis of the **PR** sort gives inadequate guidance for these maneuvers, it may be that *weakly* paraconsistent systems will do better.

WEAK PARACONSISTENCY

Whatever else it takes for something to be a logic, serious contenders turn out to share a property. The attribute in question is *preservation*. We look to our system's consequence relation to preserve target properties of premise sets. In classical logic (and the several alternatives and rivals that orbit it), we want the consequence-relation to be *truth*-preserving (or, syntactically, *consistency*-preserving). Let $\Vdash$ and $\vdash$ be our semantic and syntactic consequence relations on a set of sentences Σ. The decision to close Σ under $\Vdash$ and $\vdash$ gives highly conservative extensions of it. Let Δ be a subset of Σ. Then any consistent or satisfiable extension of Σ is also a consistent or satisfiable extension of Δ; so $\Vdash$ and $\vdash$ are also monotonic.

Now let Σ be inconsistent or unsatisfiable. Since $\vdash$ and $\Vdash$ operate in ways that deny Σ consistent or satisfiable extensions, then two facts coobtain. One is that under classical construal $\vdash$ and $\Vdash$ (trivially) cannot preserve the consistency or satisfiability of Σ; and the other – equivalently – is that anything and everything must follow from Σ. *Ex falso quodlibet.*

The basic insight of weakly inconsistent systems is that, although $\vdash$ and $\Vdash$ cannot deliver consistent or satisfiable extensions of an inconsistent or unsatisfiable Σ, it is desirable that there nevertheless be a nonempty class of "acceptable" extensions of inconsistent Σs. Contriving such extensions is the business of what Bryson Brown calls "*faute de mieux* reasoning" (Brown, 2000). One way of achieving acceptable extensions is to pick some classically inconsistent sentence-sets and rig them with properties making them "desirable" and then contrive consequence relations *on them* that preserve these desirable properties. A case in point is the *discussive systems* of Jáskowski (1969, pp. 143–57). Imagine a disagreement between two parties (it might be the diagreement between relevantists and strictists over *ex falso!*). In a Jáskowski-game each party's store of commitments is deemed consistent at each stage of play. When parties disagree, their positions are inconsistent with one another, but the total net of commitments is free of contradiction. In da Costa's C-systems, a negation operator is introduced by which both Φ and $\ulcorner\neg\Phi\urcorner$ can be assigned T (1974). As we have seen Priest and Routley adopt a three-valued logic of which the third value is the paradoxical value $\{T,F\}$. Dunn adapts this to a four-valued logic by addition of the value $\{\neg T, \neg F\}$ (1976). Relevant logics of the more or less standard sort use a multiworld semantics in which negations are evaluated at a world bearing a particular link to the different world in which other sentences are evaluated. Rescher and Brandom introduce the operations of superposition and schematization. These operations serve to sever the link between $\ulcorner\Phi \vee \psi\urcorner$ and $\ulcorner\Phi$ or $\psi\urcorner$, and $\ulcorner\Phi \wedge \psi\urcorner$ and $\ulcorner\Phi$ and $\psi\urcorner$.

In all these systems $\Vdash$ and $\vdash$ are truth-preserving and consistency-preserving. This exposes them to the objection that in their unacceptable extensions of unsatisfiable and inconsistent sets, that is, those extensions that are satisfiable *in a sense* and *consistent in a sense*, negation (i.e., *real* negation) has been trifled with (Slater, 1995). One is tempted to think, "Here we go again! A clash of semantic intuitions about 'not.'" I note the temptation, only to reobserve the tenacity of the method of analytic intuitions in disputes such as these. My greater interest is to show how one can be weakly paraconsistent in ways that avert contentions about negation.

Systems of this second sort have come to be known as *preservationist logics*. There are two main versions of such logics, each of interest for the business of thinking about the rational management of cognitive dissonance. In the first variation, which I will call *aggregative*, new properties are introduced and a new consequence relation that preserves them. In the treatment of Schotch and Jennings (1989), the property preserved by their consequence relation is that of the *level* or *degree of incoherence* of a set of sentences. A set Σ's degree of incoherence is a measure of how aggressively we must subdivide Σ to make all its subdivisions classically consistent. The same approach can be taken with semantic incoherence. The degree of Σ's semantic incoherence is the number of subdivisions needed to make each a classically satisfiable set. The Schotch and Jennings consequence relation preserves the degree of a set's incoherence, and adapts so as to preserve the degree of a set's semantic incoherence (Johnson and Jennings, 1983).

Suppose now that Σ is a set of sentences in which I have an interest. It might be my own current belief-stock, or it might be the axioms of intuitive set theory. For various reasons, I may wish to reason from this Σ without being able or knowing how to expunge its inconsistency. Even so, in the business of reasoning from this inconsistent set, it is fair to assume that I would wish to satisfy the condition that Schotch has dubbed *Hippocratic: primum non nocere* [1993]. In Schotch's usage, the injunction, "First, do no harm" carries the meaning, "In reasoning from an inconsistent Σ, make things no worse than they already are, and in one respect at least make them better. So do *not* preserve triviality" (Schotch, 1993). So, if I began my reasoning with a set of premisses whose degree of incoherence is k, *Hippocratic* requires that the set of my inferences from that set have no higher degree of incoherence than k.

Systems of the present sort are aggregative systems, or systems of *aggregative logic*, because their consequence relations aggregate premise sets differently from how this is done under classical consequence. If $\Gamma = \{P_1, \ldots, P_n\}$ and $\Gamma^* = \{Q_1, \ldots, Q_n\}$ then classical consequence aggregates Γ and Γ^* fully, that is, as $\{P_1, \ldots, P_n, Q_1, \ldots, Q_n\}$. In aggregate logics premise-aggregation is more fine-grained.

Let $C_1, \ldots, C_n$ be the properties we want our bespoke consequence relation to preserve. Let Γ be a set of sentences. Provided that inconsistent and/or unsatisfiable sets exist and have C_i for some i, then there will be a nonempty set of supersets of Γ that likewise have C_i. This makes possible the definition of a non-trivializing consequence relation.

$$\Gamma \blacktriangleright \alpha \text{ if and only if, if } C_i\,(\Gamma \cup \Delta), \text{ then } C_i\,(\Gamma, \alpha \cup \Delta).$$

Schotch and Jennings define:

Con (Γ, ξ) if and only if there is a family of sets $A = \phi, a_1, \ldots, a_i$ for $i \leq \xi$

such that:

    i. $a \in A \Rightarrow a$ is consistent.
    ii. $\forall\,(\gamma\,) \in \Gamma, \exists\,(a) \in A \colon a \vdash \gamma$

This notion generalizes on consistency. Let T be a set of tautologies. Then Con (T, ξ) holds for $\xi \geq 0$. For consistent but non-tautologous sets, X, Con $(X\,\xi)$ holds for $\xi \geq 1$. Then the *degree of incoherence* of a set Γ, $i(\Gamma)$, is:

$$i(\Gamma) \overset{\text{def}}{=} \min \xi \mid \text{Con}\,(\Gamma, \xi), \text{ if this limit exists}$$
$$\overset{\text{def}}{=} \infty, \text{ otherwise}$$

Again, Γ's degree of incoherence is a measure of how much we must subdivide Γs in order to make all the subdivisions consistent.

In the Rescher and Brandom theory, Adjuction or $\wedge$-intro does not hold universally. It is not a consequence of $\{p, \neg p\}$ that $(p \wedge \neg p)$ since $i(\{p, \neg p\}) = 2$, whereas i ({p, ¬p, (p ∧ ¬ p)} = ∞. However, unlike the system of Rescher and Brandom, aggregative logic has a valid aggregation law.

A second approach to preservationist logic is that of Bryson Brown (2001) and Apostoli and Brown (1995). As Brown says, the motto of the preservationist turn in logic is: "Find something you like about your premises, and then fashion your consequence relation so as to preserve it." Let Σ be an inconsistent set of sentences. Let l be the level of its ambiguation, that is, the number of formulas in Σ whose postulated ambiguity would render Σ consistent. Then we want our consequence relation to be l-preserving, that is, consistency-under-l-ambiguation-preserving.

Brown defines

Amcon: Amcon (ξ, Γ) if and only if Γ can be made consistent by treating ξ atomic sentences (sentence letters) as ambiguous.

Then *lamb* (the level of ambiguity) of a set Γ is denoted $l(\Gamma)$. $l(\Gamma) = \text{Min}_\xi \mid$ Ambcon (ξ, Γ). And now the desired consequence relation $\triangleright$ drops out.

$\Gamma \triangleright \alpha$ if and only if every l-preserving extension of Γ is an l-preserving extension of Γ, α.

It follows that whenever $l(\Gamma) = 0$, $\Gamma \triangleright \alpha$ if and only if $\Gamma \vdash \alpha$. We also have it that when $l(\Gamma) > 0$, closing Γ under $\vdash$ produces triviality, whereas closing under $\triangleright$ may avoid this outcome. Furthermore, as Brown remarks, $\triangleright$ does not trivialize all inconsistent sets, and it preserves the classical consequence relation for consistent sets.

For all their similarities, the aggregative approach of Schotch and Jennings differs from the preservationist approach of Brown. The technicalities of this difference are not needed for the purposes at hand.[15] It suffices to note that the Schotch and Jennings approach has displayed an affinity for our Strategy III in which if one's belief-stock contains Φ and $\ulcorner\neg\Phi\urcorner$ (or even $\ulcorner\Phi$ and $\neg\Phi\urcorner$, Φ and $\ulcorner\neg\Phi\urcorner$), they may be used separately in subsequent inferences, but not either jointly or conjunctively. Similarly, Brown's preservationist approach resembles Strategy III according to which inconsistencies in one's belief set should be disabled by ambiguation of the offending elements before further inferences can be drawn. Even so, there are differences. In Strategy III, an effort is made to find ambiguities that actually exist, hence which show that the set in question is not really inconsistent. In Brown's approach (at least in its most distinctive application), there is no doubt that the set is inconsistent. The question is *how* inconsistent? The answer is that its degree of inconsistency is the number of times an atomic sentence would need to be *assumed* ambiguous.

In Brown's logic, the consequence relation is allowed to operate as normally as possible under minimal ambiguation of classically inconsistent sets. Likewise, with Schotch and Jennings; the consequence relation is allowed to function as normally as possible under the most minimal aggregation that preserves degrees of incoherence.

It is easily seen that the weakly paraconsistent logics here under review are designed to be user-friendly. They are developed with actual real-life reasoning in mind. They do this in such a way that requires us to see that what the competent reasoner often must not do is seek for consequences under a relation that guarantees truth-preservation. For sometimes pursuing such a goal lands the reasoner in triviality. It is also of some importance that although weakly paraconsistent logics will accommodate a dialethic truth value, they do not require it. If dialethism is true, these are candidate logics for it; but dialethisms need not be true for such logics to have a competent rationale. Therefore, for reasons we have already canvased, I think that it must be said that weakly paraconsistent logics are in general better models of strategies for the management of cognitive dissonance than are their dialethic cousins. Even so, there are reasons for dissatisfaction with preservationist systems, as I shall now attempt to demonstrate.

Both aggregative and preservationist logics in the manner of Brown make a larger claim on the sympathies of those interested in more inference-friendly logics than the strict or classical alternatives. For one thing, they honor the downward principle. They acknowledge the existence and provenance of classical entailment. They acknowledge that classical entailment will not deliver the goods they want delivered, and they define a second consequence relation designed to deliver the goods in question. In this, there is an interesting similarity to Aristotle. Aristotle wanted a wholly general theory of argument. He was convinced that his ambitions would be dashed – as they were for all the Sophists – unless he could contrive a *logic* that would serve as the indispensable core of the theory of argument. This subtheory is the theory of syllogisms, occasioned Aristotle's boast that its likes had not been seen before in all of human history. It is important that in Aristotle's logic there are two relations of entailment. One of these gives ordinary validity: An argument is valid if and only if its premises entail or necessitate its conclusion. Syllogisms are not just valid arguments. They are *restrictions* of valid arguments; that is, they are valid arguments that satisfy additional constraints. The most important of these are that conclusions of syllogisms not repeat a premise, and that there be no idle or superfluous premises. When a set of premises interacts with a conclusion in such a way that they constitute a syllogism our second conception of entailment is now in play. The premises in question *syllogistically entail* or *syllogistically necessitate* the conclusion in question. Aristotle fashioned the relation of syllogistic entailment precisely because it would help deliver the goods for a wholly general theory of argument, and in precisely those respects, ordinary entailment could not deliver. Aristotle has nothing against ordinary entailment (although he never got beyond regarding it as a theoretical primitive in the logic of syllogisms). He never said that *real* entailment is syllogistic entailment, and that nonsyllogistic, ordinary entailment is a fraud, or a disaster, or whatever else. Ordinary entailment is indispensable to the theory of syllogisms, but it is not fine-grained enough to serve the peculiarities of a theory of argument.

The similarities with preservationist logics are striking, and we need not go on and on about them. However, it is not surprising that preservationist logics inherit a problem that also caused trouble for Aristotle. Aristotle wanted a logic for its contribution to the general theory of argument. He also wanted the same logic for its contribution to a general theory of deductive reasoning, or inference (it is *not* a loosely contingent matter that translators render *syllogisms* as both *deduction*, in the sense of argument, and *reasoning*). The syllogism was made to be argument-friendly. It was also made to be inference-friendly. In *neither* case, however, do the conditions or syllogisms serve in pragma-dialectical conditions on argument or as state-conditions on inference.[16] The best that can be said of them in this connection is that the conditions on syllogisms, unlike the condition on ordinarily valid arguments, adapt more easily than they, and adapt *differentially* to the different requirements of a theory of argument and a theory of inference.[17]

Aggregative logic and preservationist logic also differ from Aristotle's approach. Aristotle wanted a logic that left the ordinary entailments alone and provided a bespoke consequence relation, syllogistic consequence, which, under suitable supplementation, would serve the ends of a theory of argument in general and for reasoning in general. Aggregative and preservationist theories also leave the classical entailments alone and they engineer a consequence relation, ► or ►, which is designed to deliver the goods not for reasoning in general, but for reasoning from inconsistent premises or inconsistent hypotheses. This leaves the question of what consequence relation delivers the goods for reasoning from consistent sets. Their answer is that such reasoning is entirely classical. This is a setback. We have already seen reason to doubt that such reasoning is classical; that it is regulated by the principles of classical logic serving as "inference rules."

We suggested pages ago that theories of argument and theories of inference are not logics. Perhaps this is an overstatement. No one doubts that in axiomatic setups the principles of logic validate steps of axiomatic proofs, and that in natural deduction setups, the principles of logic validate steps of conditional proofs. Moreover, if one reasoned about a matter in the form of a good proof of it, one would surely have reasoned correctly (which is why these logical rules are called "rules of inference"). Usage being what it is, we must yield. There is a technical sense of "argument" and a technical sense of "inference-rule" according to which it is quite right to think of a logic as a theory of argument and a theory of deductive reasoning. What it is not, of course, what it is nowhere close to being, is a theory of argument in any sense in which the aim is to unearth and systematize sets of pragma-dialectical rules, or a theory of inference in any sense in which the aim is to specify and systematize the state-conditions on belief-revision. Once they are released from these false presumptions, classical logics are free to run, and to do what they do best; that is, to serve as theories of implication, consistency, and satisfiability. As Aristotle saw, the implication relation is too latitudinarian for (real) inference and (real) argument. There is more to inference and argument than truth-preservation. Aristotle tried for a more inference-friendly and argument-friendly implication relation. It was a move in the right direction. But the theory of even this relation could not be the theory of the pragma-dialectic constraints on arguments on the hoof, and could not handle the state-conditions on real-life, real-time belief-revision. Modern logicians such as Schotch, Jennings, and Brown pursue like strategies for even more limited purposes. Even so, they may claim two of the virtues of the Aristotelian paradigms, just as they must share one of its difficulties. The virtues are these: One is that ordinary (or classical) implication is of no use in the theory of reasoning from inconsistencies. The other is that, unlike the relevantist, preservationist logicians see no need to say that classical implication does not exist, or that, if it does, it is not implication. The inherited difficulty is that no implication relation that is a restriction on classical implication is such that its truth conditions will serve as the pragma-dialectical *regulae* of argument and/or as the state-conditions on belief-revision. *This* is the meaning of the claim that theories of argument and theories of inference are not logics.

For all that separates them, and not withstanding their individual failings and technical difficulties, the generically paraconsistent theories we have examined in this chapter can all be reconciled to the strategy we have been calling *Reconciliation*, a strategy that bears an interesting similarity to our ambiguation strategy III. Given a disagreement ⟨Φ, ¬Φ⟩ of the sort we have been examining in these chapters, *Reconciliation* bids both parties to try to find different things of which Φ and ⌜¬Φ⌝ are true. The strategy produces a *resolution* if and only if (a) both parties are agreed on the different things of which Φ and ⌜¬Φ⌝are true, and (b) they also now agree that there is no single thing in the ambit of their shared interests of which Φ and ⌜¬Φ⌝ are both true. We could say that when one half of the first condition, (a), is satisfied, and the second, (b), not, *Reconciliation* has produced a *potential resolution*. I take it that the best that *Reconciliation* has been able to do in the case of the dispute between the strictist and the relevantist is a *potential resolution*. Both parties can agree that *ex falso* is no principle of argument or of inference on the hoof, but beyond that they have not been able to go. Still, this is not to denigrate the importance of *potential resolutions*. They offer recalcitrant parties resolutions that save their face. On the other hand, the weakly paraconsistent logics of aggregation and preservation fit *Reconciliation* like a glove when the issue is *ex falso*.

The view of inference sketched in these pages is in the spirit of Harman. Like him, I am concerned to discourage the misconception that classical logic is a theory of inference, except where "inference" is a technical term designed to facilitate the peculiarities of the logician's interests. To do this, it is unnecessary to have a better alternative theory up and running. This has not stopped people regretting the sketchiness of Harman's positive account, a fate that my own approach also tempts. Perhaps, then, it would be useful to have a slightly fuller appreciation of the kind of approach to inference that Harman's own, and mine too, are rejecting. It is an approach to inference exemplified by what have been called "AGM theories" (Alchourrón, Gärdenfors, and Makinson, 1985; Gärdenfors, 1987).

AGM theories are an influential version of Classical Belief Revision Theory, a theory that accounts for changes – revisions and updates – in sets of beliefs. Such changes are postulated as being made by, or as happening to, an ideal doxastic agent. They are changes in his belief-set. Belief-sets are closed under deduction; they are negation-consistent, hence absolutely consistent; and somewhat in the manner of Aristotelian demonstrations, they are epistemically ordered. The ideal inferer's belief-set is infinite; he believes all deductive consequences of what he believes; and hence any deductive consequence of what he believes is a member of his belief-set. A belief-set is the set of its own deductive consequences. In the AGM approach, an agent's belief-set is a logical theory. As is widely recognized (Hintikka, 1962), the identity of an agent's belief-set and the set of its deduction consequences constitute the ideal inferer as deductively omniscient. Some theorists, not content to mitigate the counterintuitiveness of such a consequence merely by pleading the ideal agent's ideality, displace talk of beliefs with talk of commitments (see Fuhrmann, 1988, cited by Girle, 1997, p. 10 of preprint). It will not work. Certainly it will not work for the notion of commitment developed in this chapter. Nor will it do for any conception of commitment on which a commitment is in any way presumptive with regard to inference.

Classical Belief Revision Theories have difficulties with belief-justification. Gärdenfors writes that:

the idea of including justification for the beliefs seems fruitful. This aspect of belief systems seems neglected in other models of epistemic states. (1987, p. 35)

The problem with justification in classical theories is the closure of beliefs. It is obvious that in such systems justification could not be the material conditional. If we put it that, for beliefs Φ and ψ, $\ulcorner \Phi \rightarrow \psi \urcorner$ expresses the justification of ψ by Φ, then, since it is a logical truth that $(\Phi \rightarrow \psi) \vee (\psi \rightarrow \Phi)$, every belief would justify or be justified by every belief (Girle, 1997, p. 11). It is hardly surprising that the truth functional conditional fails the burdens of belief-justification so spectacularly; it was not a serious classical candidate in the first place. A more serious problem is this: If it is belief rather than commitment that is closed under consequence, then it falls to the ideal reasoner to believe vastly more than he is justified in believing; that is, it is a requirement of his ideal rationality that he do so. Yet, if deductive closure applies to commitments, if beliefs are a proper subset of these, and if justification is a finite operation on beliefs, then, even so, the ideally rational agent will be committed to all sorts of things that he could not be justified in believing. The point has a particular pinch when applied to the commitments that flow from an agent's justified beliefs. Let J be a consistent set of such beliefs and Φ an arbitrary selected statement form the sprawl of J's deductive closure. Let Φ not be something that agent S believes but that, since belief (in J) is also commitment, is something to which he is committed. Suppose we take commitment not in the manner of our suggestion of pages ago, but rather in some stripped-down way, such is this:

S is committed to Φ with respect to J if and only if (a) J ∪ {¬Φ} is inconsistent, and (b) it is not rationally permitted that S concede ⌐¬Φ⌐.

Anything that would be rationally impermissible for S to concede would be unjustified for S to believe. Accordingly we have it that, in the absence of additional procedural guidance, the ideally rational agent is required to be an agnostic about each of the abundance of the commitments of his justified beliefs that exceeds the reach of his belief-justification mechanism, whatever it is. For every such Φ, S's situation is that he is not justified in believing Φ and he is not justified in believing its negation. The whole force of his commitment to Φ is that he not deny it, hence not disbelieve it. Let K be the set of S's total commitments in J. K will include as a subset of K^a, S's agnostic commitments in J. The task of the theorist is to characterize mechanisms that produce S's justified beliefs in K, that is, which specify $K-K^a$.

It is not to our purpose to dwell in the coils of this problem, which, in any event, is part of the ongoing research program of Classical Belief-Revision theories. It is enough to remark that whatever else is required to accommodate the problem of the justification of belief in such accounts, justified belief cannot be closed under consequence. That commitment is allowed the deductive latitude that must be denied justified belief counts for little apart from the commonplace that the inferer, ideal or everyday, is not permitted a concession to, hence a belief in, anything that contradicts what he is already consistently committed to.

In Chapter 1, we reviewed a system of dialogue logic, in which commitment displaces belief in a principled way. Systems such as Girle's **DL3** are logics (or as we must now say, "logics") of conversation, a central purpose of which is the regulation of series of exchanges of speech-acts. In these setups, conditions on the commitments of parties regulate what speech acts may be performed and when. Belief is recoverable in such systems under an assumption of sincerity for speech acts and for responses to the mechanisms of challenge. It is open to question as to how faithfully **DL3**, a theory of conversation, mimics inference, a process of psychobiological adjustment. Even so, there are features of **DL3** that one would want rather centrally placed in a theory of inference. One is the failure of deductive closure. Another is the failure of the presumption of consistency for commitment-stores. Still another is that the mechanisms for conflict resolution are realistic; in particular, they allow for conflict resolution devices to produce inconsequential results. The operational rules of **DL3** were reviewed in Chapter 1. They are summarized in Table 3.1.

Of particular note for our present interests is that **DL3** is paraconsistent. It is paraconsistent in a sense that lies between our own notions of strong and weak.

TABLE 3.1 *Dialogue Logic DL3*

S LOCUTION at Step n	S STORE	H STORE	H RESPONSE
categorical statement: P	$+P$	$+P$	
reactive statement: $\therefore R$	$+R$ $+If\ R\ then\ S$ $+\langle \dots Why\ S\ ?\rangle$ $-Why\ S\ ?$		
term declaration: t	$+Ft$	$+Ft$	
withdrawal $-P$	$-P$		
withdrawal $-(\exists x)Fx$	$-(Qx)Fx$ $-(\exists x)Fx$	$-(Qx)Fx$ $-(\exists x)Fx$	
tf-question $P\ ?$	$-P$		one of $P, \sim P, \iota P$ or $-P$
wh-question $(Qx)Fx$	$+(\exists x)Fx$	$+(Qx)Fx$ $+(\exists x)Fx$	one of $t, \sim(\exists x)Fx$ $\iota(\exists x)Fx$ or $-(\exists x)Fx$
challenge $Why\ S\ ?$	$-S$ $+Why\ S\ ?$	$+S$ $+Why\ S\ ?$	one of *acceptable* $\therefore R$ or $\sim S$ or $-S$ or a resolution demand as in (v)
resolution *Resolve P*	$+P$ $+Resolve\ P$	$+P$ $+Resolve\ P$	withdraw part P or state part P as in (*vii*)

DL3 requires not that the conversationalist be consistent in his commitments but, rather, that he keep his commitments free from *challenged* inconsistencies, hence free from the localized disturbance they create. So there is nothing dialethic about **DL3**. On the other hand, with respect to a party's consistency-obligations, it is not enough that he decline to yield on every proposition whatever when his commitments fall into inconsistency. In other words, **DL3** is more than weakly or generically paraconsistent.

We see in **DL3** an attractive disposition to honor the inconsistency-management strategies reviewed in this chapter. Strategy I is to overlook, and to leave standing, an inconsistency that seems harmless; that has not gotten in the way of the cognitive (or as we can now also say, the dialectical) task at hand. **DL3** sanctions the strategy. A noticed inconsistency is all right until *challenged*. Strategy II permits inconsistencies not covered by Strategy I to be sealed-off and denied a premissory role until such time as the means of its resolution becomes clear. **DL3** does not deal with this case directly, but it allows for the option by the consent of the parties. The third strategy, to dissolve the inconsistency with a plea of ambiguity, is allowed by the **DL3**. Strategy IV is to split the inconsistency and to allow its consistent parts premissory work in separate argumental environments. In effect, the strategy denies us the unconditional use of Adjunction, but Adjunction fails independently both as a state condition on inference and as a rule of dialogue.[18] In this respect, **DL3** does conspicuously better than systems of paraconsistent *logic*, such as **PR**. There is a reason for this event apart from **PR**'s dialethic character. **PR** is at bottom a logic of a paraconsistent *implication* relation. **DL3** is purpose-built for *argument*. Argument is not inference, of course; but they share properties that it is expensive to ascribe to the implication relation. Inference and argument are nonmonotonic; they require special constraints on premise-inconsistency; they cannot abide multiple conclusions except under tight constraints; and they are bound by relevance conditions of some kind. A theorist is free, if he wishes, to impose these properties on the implication relation. It will cost him to do so, as we saw in the case of relevant implication. The logic "goes" intensional and the decision problem becomes hyperintractable, in a manner of speaking. Besides, since the truth conditions on the implication relation are neither state conditions for belief-revision, or inference, nor rules for human conversation, it is unnecessary to retrofit the implication relation so that it might better bear the traffic of inference and argument. Other devices bear that traffic better, and even when rejigged, the truth conditions on implication will not bear it well enough to justify the effort.

DL3 helps us see that we are misconceived in our efforts to construct theories of inference, and argument, too, on the model of what logic, for technical reasons, recognizes these things to be. Any realistic attempt at getting inference and argument right will reserve a central place for paraconsistency, but not everything that is paraconsistent will be a contender for such work. This was the moral of **PR**.

We have been for some time now pondering over the classical wherewithal of logic. In the case of *ex falso*, we have seen it the object of the observation that, since it will not do for inference and for argument, then (in effect) it will not do for implication either. It has been the burden of this chapter to show how misgotten this line of reasoning is. If that burden has adequately been met, it should have offered a certain promise of success to our reconciliation strategy, which was to find something for conflicting intuitions about, for example, *ex falso* to be true *of*, although not the same thing. All the same, it is classical *logic* that comes up trumps if what we are saying about the logic of inference and the logic of argument is also true. It is not just that classical intuitions are true of the implication relation; classical *logic* is also true of it. The intuitions of relevant logicians – certainly the central ones – are true of inference and of argument, but relevant *logic* is not a good theory of these things. This reveals *Reconciliation* to be a bit of a letdown (though not for the classicist), and perhaps something of a fraud. It is not just an attempt to find for the paraconsistentist something that he and everyone else sees his intuitions as true of; but in the things that *Reconciliation* find them to *be* true of, we find occasion to propose the abandonment of paraconsistent *logic*. In short, in our use of *Reconciliation*, we have found reason to accept the paraconsistentist's intuitions and yet deny the theory of which they were a central motivation.[19]

4

Semantic Intuitions

[A] *space* is nothing but the verbal substantialisation of mutually compatible spatial relations.

Georges Lechaleas, *Étude sur l'Espace et le Temps*, 1896

We should again call to mind the frequency with which it is claimed by people who do not like it that *ex falso* is counterintuitive. With some theorists, that is the whole case against *ex falso*. It is too slender a case, but the reverse mistake is equally bad. It is the mistake of ignoring the near indispensability of intuitions and counterintuitions in the abstract sciences. Let T be any abstract theory – of sets or of truth, for example – for which it is presumed provable that there are no intuitive concepts of its target properties. Then how, with respect to the adjudication of rival accounts of such properties, can appeals to intuitions and counterintuitions be in *any* sense "indispensable"?

We will go only a little way toward appreciating what counts as conflict resolution in the abstract science until we have an understanding of the *dialectical epistemics* of conflict resolution. It is "dialectical" inasmuch as conflict resolution, minimally a two-party affair, is inherently a consensual arrangement, whatever else it is. It is "epistemic" inasmuch as the conflicts whose resolution-strategies the theory specifies and pronounces on are disagreements about what people say they know, what they claim or concede to be true, and whose resolution therefore may be expected to inherit something of this epistemic cachet. We may see an abstract theory T as a set of claims K and a projective mechanism M, which derives new sentences. Normally M's projective propensities will make it appropriate to speak of T as the closure of K under M. Where M is not a purely deductive mechanism, it is understood that some sort of deductive device is a proper submechanism of M. So M will force out of K the deductions appropriate to the kind of theory that T is. What

120

makes T abstract is that the selection of K, as of M, is independent of both T's particular empirical design (if any) and of T's empirical consequences (if any). Independence, here, may be taken discretely or holistically. Discretely, K and M alike are proof against recalcitrant observations. Holistically, they are the last, or close to last, to submit to revision in the web of belief, or in the laddered fabric of theory.

Abstract theories can be approached either *formally* or *analytically*. In the first case, the specifications of K and M are independent of the theorist's belief that the K-sentences are true and the M-mechanisms sound. Sometimes they are chosen because they have properties that make them attractive things for the theorist to "play with" (van Benthem, 1983, p. 1). For example:

Now that we all understand the virtues of a model-theoretic semantics satisfying general Montagovian standards of rigor and clarity, there is joy in *playing around* with virtually every specific detail of Montague's original paradigm. The following... illustrate various aspects of this new wave of free speculation. (van Benthem, 1979, p. 337, emphasis added)

Motivations for a theory can be numerous and complex, but one is surely logical *curiosity*. Suppose one takes claims about the underlying Hegelian dialectic (as opposed to classical logic) at face value; can a rigorous logical system be formed incorporating just these claims? (Compare Heyting's axiomatization of such a seemingly *esoteric* subject as intuitionistic logic.)

On the other hand, when a theory is approached analytically there is antecedent presumption of the theory-independent truth of K and the projective soundness of M, and for the view that jointly they constitute an analysis of some target concept, such as *set* or *consequence*. Such theories, when they are nonlogical axiom systems T, aim at what Hintikka calls *descriptive completeness*. T is descriptively complete when all models are intended models (Hintikka, 1996, p. 91). However, Hintikka at times is not wholly at ease with the role of intuitions in achieving descriptive completeness, which he describes as "fallible" and "prejudices" (Hintikka, 1996, p. 19). Analytical theories in the sense I am trying to characterize are less skeptical of intuitions. Even so, Hintikka is drawn to the main idea:

Even abstract mathematical theories can be thought of as explications of certain intuitive concepts, topology as an explication of the concept of continuity, group theory of the idea of symmetry, lattice theory of the notions related to the idea of ordering, and so forth. (Hintikka, 1996, p. 10)

See also John P. Burgess:

There may be principles so central that anyone who rejects them and nonetheless uses the usual name for the concept may be said to be using not the usual concept but another one of the same name. Such principles may perhaps be called "conceptual," and by those who accept them as truths, "conceptual truths." We may perhaps be hypothetically obliged to accept them as true *if* we accept the concept. But we are not categorically obliged to accept them as true, since we are not obliged to accept the concept. We may even on the contrary be obliged to *reject* the concept – and are *obliged to reject it if the principles lead to inconsistencies.* (Emphasis added in the last instance.)[1]

There is good reason for thinking that Frege,[2] and Russell, too, in the early days, took the axioms on sets to reflect a conceptual analysis in this sense, and that much of the contemporary disagreement about consequence or entailment is a disagreement about how the concept of entailment is to be analyzed. "Formal" here is a term of art, a neologism portending neither logical forms nor uninterpreted calculi. In my use, a formal theory's formality denotes a doxastic property of the *theorist*: *He need not believe its assertions.*

It should not be supposed that nonempirical ventures formally entered into are mere play. A good deal of pure research is highly and unapologetically formal. Riemann's geometry was shown consistent (if Euclid's is) long before relativity theory furnished an empirically commanding model of it, and that alone effectively ended the *à priori* imperial pretensions of the old geometry.

Various disciplines, as we have said, are distinguished by their brisk and pacific pluralism, in which a multiplicity of generally inequivalent, and often incompatible, theories and would-be theories coexist at levels of discord and cross-examination that are noticeably low. Pluralism of this pacific and tolerant kind is a valuable type-indicator of its constituent incompatible theories. But pluralism is not always so gentle. It is sometimes a matter of high vexation and spiky noise. This, too, is an indicator of theory-type analytic theories for noisy pluralism, formal theories for gentle. Paragraphs ago, we wondered how intuitions could in any way be indispensable indicators of properties of which demonstrably there is no intuitive conception. There is a related question about what might be called *analytical pluralism*: How is analytical pluralism possible? Ambiguation is one of the received answers. If T and T′ are incompatible analytical theories of a target concept C, then T and T′ marshall incompatible C-intuitions, each "indispensable" to the respective analytical account. If we ask how T and T′ could both be true, it is said that they capture different senses of C. If we take pluralism to be the view that T and T′ *can* both be true, our question now is whether an analytic theorist can coherently espouse pluralism about what he himself is analytically minded about. Suppose he can. Then he must be antecedently disposed to regard his own intuitions about C as showing C to be ambiguous. But in general this is not true of analytic rivals. This being so, the ambiguation strategy for conflicted analytical theories T and T′ is not as easy a strategy for the T-theorist and the T′-theorist as it is for some third party who, let us note, need not himself be analytically minded about C. So I conclude that pluralism, tailor-made for formal theories, can be problematic for analytic ones.

Our distinction between analytic and formal is rather slack, and certainly neither exclusive nor exhaustive. There are those who judge the naive theory of sets to have been an analytic account (and a failed one). Concerning the pathological biconditional on which the Russell Paradox rests, Quine notes, apocalyptically, "the thundering heptameter that shattered naive set theory: the class of all those classes not belonging to themselves" (Quine, 1987, p. 146). If the naive theory were not an analytic theory, Quine's description would be overblown; nor would there be occasion to say that "desperate accommodations were called for" (Quine, 1987, p. 148). It is widely believed, but also contested, that successor theories (e.g., ZF) are not, and do not purport to be, analytic theories, but hardly anyone seriously conceives of ZF on the historical model of Riemann's own highly formal entertainments. Where intuitive set theory can be seen as an attempt at an analysis of sets, ZF can be seen as a *strategic stipulation* on sets, judged at bottom by its mathematical fruitfulness and, as Russell himself said, by its propensity *to be believed*.[3] Seen this way, it becomes explicable how doing set theory was possible in the wake of Russell's paradox for anyone who thought of the naive theory as an analytic theory. At least it becomes partly explicable. If ZF is not an analytic theory, it does not matter that there is no intuitive concept of set for it to be a theory of.[4] Why does this not place ZF in the camp of purely formal theories in which there is no presumption that anything there is true?

When we stipulate that α has Φs we make $\lceil \Phi(\alpha) \rceil$ **true by stipulation**. Many people suppose that stipulated truth stands to truth as fool's gold stands to gold. I propose a contrary assumption, on sufferance: that one way for a sentence to *be* true is to be made true by stipulation. If this is so, a distinction falls out between the pure forms of formal and stipulative theories. Purely formal theories leave nothing true that was not true before. Stipulative theories leave new truths, true by stipulation. This leaves a difference between how a formal theory's entertainments and a stipulative theory's stipulations are grounded and justified. The pledge of truth by stipulation needs to be redeemed. This is undertaken later in this chapter and in Chapter 6 and the Epilogue. For the present it is proposed that stipulative theories be understood casually enough to make of "stipulative" itself a stipulation. Let us simply say that a theory is stipulative to the extent that it makes posits that there was no inclination to make prior to the theory, and that it holds that making them is making them true.

In real life and natural history of theories, the distinction between analysis and stipulation, often in any event a matter of degree, divides its terrain diachronically. Sometimes we see that what starts out as stipulation takes on an analytical gravamen in the fullness of time. For some mathematicians, this has been the fate of ZF. To some extent, it is a matter of learning conditions. Someone immersed in the culture of intuitive set theory is likely to have, at most, a stipulative appreciation of ZF. But if from the beginning one's exposure to sets has been ZFish, the greater the likelihood of thinking of ZF-structures analytically. We ought not be surprised by the *generational* shift from Black to Shoenfield and Martin. Here is Black:

...once we abandon our recourse to our intuitions and "resort to myth mak-ing"... the need to bolster our myths by some justification other than pragmatic considerations of convenience for the technical needs of mathematics and the sci-ences becomes imperative. (Black, 1943)

On the other hand, the Russell Paradox

does not really contradict the *intuitive* notion of a set [...since for any set A,] A is not one of the possible elements of A; so the Russell paradox disappears. (Shoenfield, 1967, p. 238, emphasis added)

Indeed, sets from the ZF perspective just are

the standard conception of set. (Martin, 1970, p. 113)

There is a moral to this. It is insufficient to disqualify a theory for its failure to be, or to attempt to be, an analysis of its target concepts. I would say that any modal theory in which the Barcan formula[5]

$$\Diamond \exists \alpha \, \Phi \to \exists \alpha \Diamond \Phi$$

is embedded is a defective analysis of the concept of possibility, never mind that the formula is embedded in a number of successful modal theories, notably in the quantificational extension of S4. This is telling. It suggests a slope slippery enough to slide down. Short of giving up on a substantial part of modern modal logic, we must say that the goodness of no good formal theory depends wholly or dominantly on its being an analytically good theory of its target concept. But what is this, if not the stipulation of the negation of a narrow fact for the wider good?

In Chapter 1, we encountered an asymmetry problem. How is it, we asked, that conflict resolution in the abstract sciences is an antirealist enterprise, whereas (as many people assume) ordinary theorem-proving is a realist ven-ture? Russell's answer – although it is one that no analytic philosopher should be happy with, and is one that Russell never characterized in the way that I do – is that this is *not* the way it is. Russell in 1902 is headed for strong logicism, writing thirty years before the Gödel incompleteness proof. By his logicism, all of mathematics is translatable without relevant loss in what would be the logic of *Principia Mathematica*. But after 1902, Russell's position is that set theory has to be stipulated. This is enough to "irrealize" mathematics if logicism is true. After Gödel, logicism is a crimped affair, but it remained the case, and still is, that set theory is a very large part of modern mathematics. Thus, a very large part of modern mathematics is antirealist, although this is not a fact that Russell emphasized. How, then, does this correct the antirealist-realist asymmetry of Chapter 1? It turns it into an antirealist-irrealist *symmetry*. Conflict resolution is antirealist in the ways already noted; and ordinary theorem-proving is made antirealist by stipulations necessitated by the death of the intuitive concept of set.

Readers will have noticed that analytical theories need not be seen as theories that give the *meanings* of terms for its target concepts.[6] For a theory to give an analysis of a concept, it suffices to specify truth conditions for sentences invoking or applying the concept in appropriate ways. What is required for analytic status is that the truths conditions that the theory specifies are conditions that obtain independently of their recognition by theory. Meanings are not required for analysis unless truth conditions are meanings. What is essential is not that truth conditions constitute meanings but, rather, even if they do, that they be theory-independent, and in that sense *objective*. Objectivity enters the picture in two different ways, as we said in the Prologue. If T is an analytic theory, T's objects exist independently of their recognition by T. If T is an analytic theory, T's truths Φ are true independently of the fact that $T \vdash \Phi$. And if T is an analytic theory whose domain of interpretation is D, then T's objects exist independently of their derivability in T. Theories objective in the second way are semantically objective. Theories objective in the first way are ontically objective. In standard model theoretic semantics, realistically construed, the two kinds of objectivity coincide. Thus, abstract theories fulfill realist presumptions when they are coincidentally objective.

It is a matter of controversy whether analytic theories could be theories of meaning. What is not in doubt is that if there are any analytic theories worthy of the name, they are theories rooted in realist presumptions. They are theories in which it is intended to tell the objective truth about matters that objectively instantiate target concepts. It appears to be distinctive of abstract theories that their Ks are heavily populated by intuitions. When a theory sanctions a consequence judged to be counterintuitive, it is striking how readily one inclines to the view that the theory is wrecked or, more soberly, that it has something to try urgently to explain away. If we take a dominantly analytic view of theories, it is easy enough to specify a concept of *counterexample*; for what would a counterexample to a theory be if not a counterintuitive consequence of it? But if our approach is dominantly formal, counterintuitiveness can be a more or less routine concomitant, and the idea of counterexample becomes all the harder to get a grip on. One is left to wonder whether there is any serious methodological role to be played by considerations of intuitiveness in the dynamics of any theory that pretends to be dominantly nonanalytic.

SEMANTIC INTUITIONS

We see in these considerations something of the methodological designs of philosophical skepticism about meanings. To admit meanings, the skeptic says, is to assign them load-bearing roles in theory. If we suppose that a sentence is counterintuitive if it violates the meaning of embedded or implicationally adjacent terms, then meaning-violation can be expected to play an explanatory role in the dialectics of the appraisal of counterintuitiveness. If a relevant logician attributes counterintuitiveness to *ex falso quodlibet*, a classical logician may accept the attribution or reject it. If he accepts it, he is unlikely to give to the counterintuitiveness any mind other than that it is surprising. But whatever the details of his view of what the counterintuitiveness positively consists in, his negative view will be dialectically recalcitrant: it is *not* a counterexample. On the other hand, if the classical logician refuses the attribution, his refusal is correct or incorrect. If correct, nothing more here needs saying. If it is incorrect, then by the assumption of the present case the classicist does not know what "entails" means. The relevantist's invocation of counterintuitiveness is *dialectically impotent*. For if the classicist does not know what "entails" means, how can pointing this out get him to see that *ex falso* is not true? This is not to deny that when confronted with *ex falso* a theorist may reject the theory of entailment that sanctions it. Nor is it to deny that, if asked, the repenting logician may explain himself by saying that his now disabled theory let him down by implying something counterintuitive. This is not, however, the case under review, for which the would-be defeater is not

(1) Your theory implies *ex falso*

but rather

(2) Your theory implies something *counterintuitive*.

Philosophy's Most Difficult Problem looms here, and it leaves the urger of (2) with no dialectical slack to cut. A complaint along the lines of (1) *may* get a theorist to lighten up, but if he does not there is no prospect of gains to be got by pressing (2) even *if it is true*. For, again, if it is true, anyone holding *ex falso* does not know what *ex falso* means, that is, does not know what

$$\ulcorner \Phi \wedge \neg\Phi \text{ entails } \psi \urcorner$$

means. And if that is the case,

(3) $\ulcorner \Phi \wedge \neg\Phi \text{ entails } \psi \urcorner$ is counterintuitive

cannot serve as dialectically telling against anyone for whom (1) is not already a counterexample. We can generalize on this: Let T be a theory and Φ a counterexample to it. The T-theorist either recognizes that Φ is a counterexample or not. If he does, that is the end of the story. If he does not, then the perfectly true assertion, that Φ is a counterexample to T, cannot be availing. I conclude, then, that a charge of counterintuitiveness as presently conceived is dialectically underdetermining.

It is possible that I have played the theme of meaning-violation rather too lightly (and some would say too tightly). Certainly, there is a substantial tradition in analytic philosophy that offers sharper purport for our metaphor. On this view, it remains true that if Φ is counterintuitive, Φ involves a violation of meaning; but in its turn, if Φ involves a violation of meaning, then Φ is *false-by-meanings-alone* (or *m*-false), hence inconsistent with a statement true-by-meaning-alone (or *m*-true).

What now are the dialectics of a charge of counterintuitiveness directed to someone who resists the attribution? For the charge to be dialectically serious (something other than the exasperation of, "It's obviously false; why cannot you see it this way, too?"), the maker of the charge must specify the truth, true in virtue of meaning, with which he takes the counterintuitive sentence to be inconsistent. If it is Φ that is counterintuitive, then a sentence with which it is inconsistent is ⌜¬Φ⌝. But it is dialectically otiose to cite it, for it makes the attributor's case circular: Φ is *m*-false because ⌜¬Φ⌝ is *m*-true. On the other hand, if ψ is some sentence distinct from ⌜¬Φ⌝ with which Φ is inconsistent, the attributor might cite ψ. In this he might succeed in winning his opponent's concession, or he might not. If he does succeed, he has a "starter" for his complaint, and it suffices for this that his opponent grant ψ's truth (and its inconsistency with Φ). For this to happen, it is unnecessary for the opponent to agree that ψ is true *by meanings*. If, on the other hand, the opponent resists ψ, cannot see that ψ is true, or cannot see that it contradicts Φ, this resistance will not be overcome by citing ψ's truth by virtue of meaning. If the opponent fails to see that ψ is true, he will fail to see that ψ is true by meanings even if it *is* true by meanings. And if it also happened that ψ could not be seen as true unless it were seen as true by meanings, it does not come to pass in the general case that attributing counterintuitiveness to Φ will get an opponent to see that ψ, with which Φ is inconsistent, is true by meanings.[7] The same is true for the inconsistency of Φ with ψ. If Φ is inconsistent with ψ by virtue of meanings, and the opponent fails to see the inconsistency, the attribution of counterintuitiveness to Φ will not get him to overcome this failure. Perhaps his dialectical *vis-à-vis* will offer him a proof of Φ's inconsistency with ψ. If so, he may have established to his opponent's satisfaction Φ's falsehood on ψ's truth. But it is not necessary to the success of the proof of inconsistency that the opponent be made to see that the inconsistency obtains in virtue of meanings. I conclude that where it is not circular, a case against a sentence based on its counterintuitiveness stands or falls *independently* of an opponent's recognition of the truth *m*-truth (or *m*-falsehood) of any sentence germane to the construction of the case. In this, the meaning-skeptic is vindicated. Meanings fail the methodological conditions imposed on them by theory. They do not carry enough weight to bear fruitfully on conflict resolution in the abstract sciences.

There can be a similar problem with intuitiveness. It is widely conceded, and quite true, that we try to get our theories to conform to our intuitions, and want their provisions to be intuitive. Sometimes we accept intuitiveness as if it is understood on the model of counterintuitiveness: Φ is intuitive when *it is true by meanings*. Consider a case. Theorist S registers his claim that Φ. S believes Φ and he may believe that Φ is true by meanings (and he might be right). Suppose that S is queried by S′ who doubts that Φ. Suppose that S has no reason to give for believing Φ, that is, that Φ is *probatively inert* for S, except for his belief that Φ is true by meanings. S says so to doubting S'. It is no use. If the doubter really does doubt that Φ, S's reply must be unsuccessful. Since S' does not believe Φ to be true, he can hardly believe it to be true by meanings, even if it is.

Invocation of truth by meaning is dialectically otiose. If Φ's truth by meaning has any real work to do, it is work done only within a community of Φ-*believers*. Truth by meanings can there be invoked to explain why Φ has been proclaimed without proof or evidence. Truth-by-meaning triggers the transformation of no evidence into self-evidence, and that is something, but not much. It makes self-evidence the conspicuous runt in the litter of evidence. Evidence is probative even across the grain of intersubjective disagreement. Self-evidence can survive no such journey. Consider a proposition Φ that I believe and you disbelieve. If you are reasonable and open-minded you might ask for evidence and I might furnish some. It is not that we are antecedently agreed that "my" evidence *is* evidence or that, if it is, it is adequate to sustain Φ. The point, rather, is that where the only evidence for Φ is self-evidence, no one who doubts that Φ would or should ever consider the question of Φ's self-evidence, except negatively. In doubting that Φ, the doubter is committed to the view that self-evidence is precisely what Φ does not have.

As we have been conceiving of them so far, intuitions and counterintuitions are so much methodological dead weight. They inhibit the dynamics of conflict resolution. They also bankrupt at least one historically prominent conception of analytic theories, according to which the business of analytic theories is the analysis of concepts, and the analysis of concepts consists in partitioning in the right way a set of sentences into those that are m-true, and those that are m-false. In particular, they make it impossible for analytical theories to recover from the devastation of paradox. Even so, no serious abstract theorist, save those devoted to pure play, thinks that he can proceed *without* intuitions, or that he can succeed in utter indifference to counterintuitiveness. So, an important question for such a theorist is, what does he think he is talking about in saying such things?

DIALECTICAL ECOLOGIES

If what has been said just now is so, it becomes apparent that the significance, and value, of invoking intuitions is deeply a function of dialectical circumstance. By the lights of Chapter 1, this is as it should be. We would do well to flesh out the dialectical character of conflict resolution beyond what we did in that chapter, although, partly for want of space and partly for want of insight, everything offered here will be an approximation of modest ordinality. The notion of a dialectical community will prove a useful starting point. Dialectical communities are what artificial intelligence theorists call *distributed environments*. We begin by retailoring the idea of counterintuitiveness. Counterintuitiveness is defined for dialectical communities.[8]

Counterintuitiveness: Let Φ be derivable in a theory T. Then Φ is counterintuitive in a dialectical community **dc** if and only if (i) except for its T-derivability, or some like support, Φ would not be accepted in **dc**; and (ii) in the absence of further considerations Φ's derivability together with the fact stated in clause (i) disposes **dc** to divide its opinion over Φ, which places **dc** in a state of cognitive dissonance with respect to the following triple of statements:
(1) That Φ is derivable in T is some reason to accept Φ.
(2) Φ is not (really) derivable in T after all.
(3) That Φ is derivable in T is some reason to reject, or cancel acceptance of, T.

Next we have

Cognitive dissonance: Let Σ be a set of statements $\sigma_1, \sigma_2, \ldots, \sigma_n$. A dialectical community **dc** is in a state of cognitive dissonance with respect to Σ if and only if for each σ_i, if σ_i is accepted (or rejected or suspended) in a subcommunity s(**dc**) of **dc**, and its acceptance (rejection, suspension) is *troubled* in s(**dc**).

And now

Trouble: An acceptance-rejection stance toward a statement σ is troubled if and only if it is accompanied by *psychological doubt, and the doubt* is not compensated for by assent toward σ even of the kind that is compatible with concurrent disbelief.

Ex falso helps in illustrating these notions. Let **dc** be the community of logicians, let T be a system of modal logic such as S5, and let Φ be *ex falso quodlibet*. Φ is derivable in T. It follows directly from T's definition of implication. Historically, Φ was counterintuitive in only a proper subset of **dc**, and then only for a time. We may assume a (perhaps small) subset of **dc** in which Φ's derivability was no problem. In other subcommunities, confidence equally untroubled ran in the other direction. For hard-core relevant logicians, *ex falso* was not counterintuitive in our present sense of the term; it was *false* or *absurd*. Lewis himself may be supposed to have fallen, for a time anyhow, into the subcommunity of logicians for whom *ex falso* is counterintuitive. In the absence of the Lewis-Langford proof, we can suppose divided and cognitively dissonant opinion with respect to the triple:

(1) That *ex falso* is derivable in, for example, S5 is some reason to accept it.
(2) *Ex falso* is not (really) derivable in S5.
(3) That *ex falso* is derivable in S5 is some reason to reject or cancel acceptance of S5.

There is reason to suppose that Lewis himself was initially in the camp of the cognitively dissonant to the extent that he looked on the subsequent Lewis-Langford proof with relief. The proof was consequential for the subcommunity for whom Φ was troubling; for many of whom, it transformed counterintuitiveness into surprise.

A proposition counterintuitive in **dc** is one judged to violate the community's "pretheoretical intuitions," as is sometimes said. "Intuition" is an unfortunate turn of phrase for our present purposes. It suggests knowledge by special license, or knowledge with exotic foundations, for example, from meanings. These suggestions should be disarmed. An analytic theory's pretheoretic intuitions in a community are the propositions in the theory's domain that members of the community are prepared to be realists about. They are propositions judged objectively true independently of their place in theory. They constrain, in turn, what will count as a satisfactory theory.

Unexpected disclosures are the stock in trade of interesting theories. When the theory is empirical, that it produces a counterintuitive result is, in general, nothing to complain of. It is widely recognized that Nature is full of surprises. It is with abstract theories of the formal sort that counterintuitiveness seems to matter distinctively. Dissonant counterintuitiveness leaves the surprise/counterexample question unresolved. For empirical theories, an unexpected result is either a counterexample or a surprise. In the end, the matter is determined by Nature, in principle at least. But for theories we have been calling abstract, counterintuitiveness is not unexpectedness. It is unexpectedness plus cognitive dissonance. It is quite true that in the complex dynamic between epistemic communities and their theories, counterintuitiveness is often a transitory affair, something that resolves itself and peters out. When this happens, the surprise/counterexample question has been resolved.

Counterexample: Let Φ be a consequence of a theory T. Then ψ is a counterexample to Φ in **dc** if and only if ψ is judged in **dc** to be inconsistent with Φ, and for those reasons, either Φ is believed false in **dc** or $\ulcorner \neg\Phi \urcorner$ is accepted in **dc**.

This is the general case, of course. When a counterintuitive consequence transforms into a counterexample, then the role played by ψ is played directly by Φ's negation. In the transition, a cognitively dissonant disposition to accept Φ (ditto for rejection and suspension) gives way to a judgment of Φ's falsity.

Surprise is the dual of this:

Surprise: Let Φ be a consequence of a theory T. Then Φ is judged a surprise in **dc** if (i) Φ is initially counterintuitive in **dc** and (ii) it is *subsequently* determined in **dc** that the truth of (i) constitutes no adequate reason for supposing that there exists a counterexample to Φ.

What is it for something to be recognized or believed or accepted in an epistemic community? Universal consent is not a prerequisite, certainly. Italians can love garlic without Luigi having to do so as well. There are two likely candidates, neither of them maturely developed to the point of settled consensus: (1) It is believed in **dc** that Φ if most members of **dc** believe that Φ. (2) It is believed in **dc** that Φ if **ec**-members believe that Φ. Case (1) offers us *plural quantification* in the manner of "Most tigers are four-legged" (See Sher,

1991; van Benthem and ter Meulen, 1985). Case (2) reflects the use of *generic* statements in the manner of "Tigers are four-legged."[9]

We have chanced on a distinguishing feature of abstract theories. Abstract theories can give rise to counterintuitive consequences; empirical theories cannot. This is so in the pure cases, of course. Empirical theories whose predictions await an empirical check function in obvious ways as abstract, and, as such, are open to the production of dissonantly counterintuitive consequences.

As we have described them here, counterintuitions and counterexamples have something in common in the absence of the empirical check. Both offend against a community's intuitions. It is tempting to see their difference as one of degree, with counterexamples rooted in an intense feeling of violation, and counterintuitions attended by resistance of correspondingly lesser intensity. It is preferable to see the difference in question in a community's dialectical behavior. If it judges a consequence Φ to be counterintuitive, its conversations and modes of enquiry will be different from those attending the judgment of counterexample. The difference is clear enough as a general idea. The question is whether it will bear any weight; whether it can be made to do onerous and useful work. Conflict in the abstract sciences is often conflict about whether a theory's consequence is a counterexample or a counterintuition. In the history of *ex falso*, relevant logicians have opted for a verdict of counterexample, whereas strictists would concede nothing more than counterintuitiveness, if that.

It is here that our medical metaphor again takes hold. Diagnostically there is something on which the counterexamplist and the counterintuitionist agree. *Ex falso* (or whatever else) offends their intuitions, or those of the epistemic communities of which they are members. Where they disagree, triage is the issue. How bad is it? Given the lexical refinements of the present chapter, there are three options, not just two. In ascending orders of probative seriousness they are: surprise, counterintuitiveness, and counterexample. Strictists opted for a verdict of surprise precisely because they thought there was a proof of *ex falso* deemed valid in the opposing epistemic community, as well as their own. In supposing so they were wrong, of course. If what we were saying in Chapter 2 is so, the supposition was wrong, and innocently so, precisely because it failed to take into account the possibility of dogmatism on the part of the generic paraconsistentist. Where surprise is not judged to be an option in an epistemic community, contention as between judgments of counterexample and counterintuition is an especially difficult thing to arbitrate. The primary fact of the disagreement is that *ex falso* (or whatever else) offends intuitions that both parties share. Call these intuitions *IN*. It is a difference of opinion concerning the significance of the violation. We see at once a limitation on the utility of the methodology of intuitions in the abstract sciences. Where do we send triagic disagreements for adjudication? It cannot be to their intuitions *IN*, since their triagic disagreement can be seen as the one that *IN* failed to settle. Whatever else we may say of them, the *IN*-intuitions are what the parties antecedently believe about the matter at hand. If *those* beliefs fail to establish a triagic accord, it is difficult to see what further intuitions might turn the trick. It can sometimes be doubted whether the parties have of *any* intuitions to guide them in their triagic judgments. The consequence in question was unexpected. Indeed, it was always possible to see it as a surprise. Not only is the consequence unexpected, but so, too, is the fact that parties are in triagic disagreement about. It is highly unlikely that these facts are preceded by untutored beliefs relevant to the adjudication of triagic rivalry. Even in the unlikely event that such beliefs were available for higher-order service, if they failed to produce a resolution, the utility of the methodology of intuitive consensus would surely have exhausted itself.

The juncture at which the methodology of intuitions reveals itself as having stalled is where costs and benefits need to take over. Of the options – surprise, counterintuition, counterexample – a terminal judgment of counterintuition is rarely an affordable cost. If the consequence in question is of moment enough to generate cognitive dissonance rather than indifference, a decision to take the matter no further than a judgment of counterintuitiveness is a decision to remain in cognitive dissonance about something we care enough about to have engendered the cognitive dissonance in the first place.

The standard ways of going beyond, or doing better than, a judgment of counterintuition is displacement by surprise or a verdict of counterexample. In the case of the dispute about *ex falso* itself, we have seen that the surprise option is the lesser cost; but it would be naive to imagine that in a contest between surprise and counterexample, the nod always or typically goes to the surprise option, or even that one rather than the other will always be seen in the relevant **dc** as having the decisive edge.

Epistemic conflict at its most interesting is conflict in communities in which a rivalry exists as between the verdicts of surprise and counterexample. It is well to emphasize that there is no prospect of recourse to empirical checks. Epistemic communities, like all social organizations, are ecologies. Essential to the structural integrity of such ecologies is affordable access to strategies of conflict management. One of the most economical of those strategies is an agreement to disagree. As we have said, it is not a psychologically realistic strategy in those cases in which uncertainty is attended by cognitive dissonance. In other words, it is a psychologically unnatural strategy when at least one of the parties finds the consequence in question to be counterintuitive in our present sense.

A conflict in **dc** about whether the derivation of Φ in T is the derivation of a surprise or of a counterexample might prompt division of **dc** into **dc′** and **dc″** in one of which the nod goes to surprise, and in the other to counterexample. If these are highly abstract issues, disagreements will be far removed from the practical necessities of life. These will be subcommunities that can afford to pay one another no mind. Alternatively, they will beg one another's questions, and no one will much care. Of course, this is not a course of action for theorists of every stripe. For the playful theorist, there is so little to the very

idea of theoretical disagreement as to make resolution strategies other than
live-and-let-live an exercise in overexcitement. Formal theorists – made so by
their readiness to accept what they do not believe – have a way of bridging the
chasm between agreement and disagreement, although it is certainly not guaranteed always to be available to them. For they can sometimes, if not always,
bring themselves to accept what the other party believes and they themselves
do not.

Genuine trouble awaits the analytical theorist. In a competition between
surprise and counterexample, he is bound to be brought up short. Consider *ex
falso* again. To judge its derivability in a theory you have up to now liked, a
surprise is tantamount to subscription to the proposition that the negation of
ex falso fails to inhere in the very idea of implication. However, whoever opts
for a finding of counterexample is committed to the opposite view, that it does
indeed inhere in the very idea of implication that *ex falso* is false. Here is the
making of stalemate.
So,

Quandary: Let Φ be a consequence of T. Then a dialectical community **dc** is in a
quandary with respect to Φ if and only if (1) it is unable to displace the counterintuitiveness felt toward Φ with a judgment of either surprise or counterexample; and
(2) T is a theory that matters in **dc**.

And,

Stalemate: Let Φ be a consequence of T. Then dialectical subcommunities **dc′** and **dc″**
are in a stalemate with respect to Φ if (i) for every argument offered by **dc′** as a proof
of Φ, and for every argument offered by **dc″** as a *reductio* of Φ, **dc′** sees it as a proof
of Φ; and (ii) each party is disposed to press its respective cases; that is, T matters in
both **dc′** and **dc″**."

These conditions will suffice for stalemate only if they leave the contending
parties with nothing else to do to achieve consensus. If T is a playful theory in
dc′ and **dc″**, there is, in pure cases, no occasion for a judgment either of surprise
or of counterexample, by the nature of playfulness. In pure cases, therefore,
stalemates are undefined for playful differences.

Formal theories – again in pure cases – are another matter. The purely
formal theorist is one prepared not to believe the things he accepts in T.[10]
Typically, the nonbelief-acceptance matrix is rooted in cost-benefit considerations. A decision to assent to what one does not believe is cynical (or naively
hopeful) unless secured by the second-order belief that, all things considered,
it is preferable in the case at hand to accept what one does not believe rather
than reject it. For such a theorist a judgment of surprise is wholly reconcilable
to his formalism; it is a judgment to accept what he does not, or did not, believe. A judgment of counterexample also can be squared with the theorist's
formalism. For a pure formalist to judge a consequence Φ of a theory T as a
counterexample, it suffices that he believes that it contradicts something
he accepts, and for that reason he accepts that it is preferable to reject Φ.

Inasmuch as formal theorists are capable of verdicts of surprise and of counterexample, they are capable of stalemates, because they are also capable of
reciprocal persistence with their disagreement, fueled by their second-order
beliefs about what, all things considered, it is preferable to accept.

As could only have been expected, stalemates are a special liability for
analytical theorists. An analytical theorist can be understood as one for whom
assent and dissent coincide precisely with belief and disbelief, and suspension
with the absence of belief and disbelief. The analytical theorist – in pure form,
anyhow – lacks a resolution device open to the formal theorist. He cannot
accept what he does not believe.

It is one of the more interesting features of the natural history of abstract
theories in this century that the development of the theory of sets, from
Russell's failed attempt to present day versions of ZF, was by and large a development unafflicted either by quandary or by stalemate, whereas in the history
of entailment theory from 1912 to the present, we see very little *but* stalemate.

Stalemates are nothing to despair over. They can be removed in the light
of a subsequent discovery or a future clever insight – something that had not
been thought of before. Often this happens when one of the parties sees a
way of *repositioning* the original disagreement. The process, under the name
of *Reconciliation*, was explained in the preceding chapter. Of our triple of
explanatory concepts, {surprise, counterintuitiveness, counterexample}, perhaps surprise is philosophically interesting in ways that the others are not.
Consider a case. In the community of mathematicians who accept the axiom
of choice, the Tarski-Banach result is highly counterintuitive, but is acquitted
of the charge of counterexample. It is a surprise. This is the surprise: Let us
say that objects Σ and Γ are congruent by finite decomposition if they par-
tition finitely into sets $\Sigma_1, \ldots, \Sigma_n, \Gamma_1, \ldots, \Gamma_n$, and $\Sigma = \cup\Sigma_i, = \cup\Gamma_i$, and Σ_i is
congruent with Γ_i (for i a non-zero whole number). The Tarski-Banach the-
orem establishes that all *spheres* are congruent by finite decomposition. It is
not surprising (no pun) that in some quarters the theorem is known as the
Tarski-Banach Paradox (Jech, 1977, p. 35 ff).

Surprise, as we have it here, shares structural features with Moore's
Paradox, the notorious "Φ, but I don't believe it." This alone tends to make us
think that a surprising consequence counts against itself probatively. When the
conditions of *Surprise* are met, the "surprisee" is committed to an utterance
in the form

Moore: Φ, but I don't believe it.

This is Moore's Paradox in benign form. This means that uttering *Moore* does
not constitute a blindspot for anyone to whom it is directed.

Blindspots: "a proposition p is a blindspot relative to a given propositional attitude
A and a given individual a (at time t) if and only if p is consistent but a cannot have
attitude A towards p ... Given the constraints imposed by certain desiderata of belief,

I cannot believe that 'It is raining but I do not believe it' even though it is a consistent proposition." (Sorenson, 1988, pp. 24, 52–3)

In the context in which *Moore* is uttered, it is not impossible for the addressee to know either what the speaker is asserting, or what he believes. Context makes it clear that what the surprisee does not believe is something to which he nevertheless assents.

The factor of surprise calls seriously into question the utility of purely analytic theories. We said that T is an analytic theory when, for its initial claims K and its projective mechanisms M, T is the closure of K under M, and K is a set of sentences believed objectively true. If there is to be *any* methodological role for counterintuitiveness with respect to formal theories, acquittal must be possible in principle. But with acquittal comes surprise, and with surprise comes sentences in the form *Moore*. If, as we said, a theory counts as an analysis of its target concepts, the analysis is given by the sentences of T believed objectively true. By these lights, it will be entirely commonplace that successful nonempirical theories will not qualify as analyses of their target concepts. This will be so to the extent that T carries consequences that are assented to but not believed.

Once the distinction between belief and assent is allowed to bear on consideration of such things as counterintuitiveness and surprise, it is easy and natural to grant to it a broader provenance. In particular, it is quite straightforward that the K of a theory need not be restricted to sentences believed objectively true, and that the M of a theory need not be confined to projection mechanisms that are believed objectively sound (think of M as an *abductive* device, for example). It becomes apparent that abstract theories seriously worthy of the name will be hybrids of what is believed and what is accepted. Such theories cannot be analytic theories, although it may be that they contain analytic fragments. In this connection, we might mention our former distinction between analysis and stipulation, which is what hybrid theories appear to be hybrids of.

Formal theories also become easier to characterize. In pure cases, they are theories whose K-members are neither believed objectively true, nor accepted as true, and are so in sufficient numbers as to make it powerfully odd to think of them as giving accounts of how target concepts really are. When Riemann constructed his brilliantly peculiar geometry, he constructed something within which, and about which, all sorts of things are objectively true, or rationally acceptable. What Riemann did not think he was doing was giving a rationally believable or rationally acceptable account of real space. In this he was wrong, of course, but that does not change the fact that the theory he thought he was constructing was a formal theory in the sense at hand.

Stipulative, hybrid, and formal theories are themselves natural antidotes to intractable disagreements. They afford disputants the luxury of not minding overmuch their deadlocked beliefs, and they make possible some fruitful poaching in the opponent's preserve. It is a good thing, too. Often a theorist's creative play with a theory he detests will cause him to lighten up. He may come to respect his opponent's insights. He may even make theoretical strides on his opponent's behalf, proving new theorems or streamlining old axioms, for example. The playfully motivated proofs of new theorems in the enemy camp may eventuate in the poacher's enlarged acceptance of the rival theory. They might even precipitate a full scale conversion.

If acceptance is what stipulation seriously contends for, it becomes important to specify conditions on a successful outcome. Such conditions are notoriously difficult to pin down with any exactitude or promise of exhaustiveness. I will mention only the standard ones: internal coherence, propensity to solve otherwise unsolvable problems, predictive facilitation, simplicity, reshaping of the research program, and so on. Overall, these are economic rather than objectively probative considerations. Their importance is such that some philosophers have been prepared to rewrite the definition of "probative," by pragmatizing truth and depsychologizing belief. I will not comment on these deviations except to say that the revisions they encompass attest to the importance of economic considerations in the construction and appraisal of theories.

Defense: For cognitive agents S and S′ and some set of nonempirical sentences Σ proclaimed by S′ and attacked by S, S′ makes a satisfactory *economic defense* of Σ against S's attack by making the success of S's attack too costly for S.

Cost: S's attack on Σ is *too costly* for S if and only if its success commits S to propositions he is not prepared to accept or to the rejection of propositions he is not prepared to reject, or to methodological adjustments he is not prepared to make.[11]

Thought of this way, defenses are pure cases of Lockean *ad hominem* maneuvers, as we have seen. I will say that an S′-defense is *secure* rather than merely economical if and only if what S′ considers costs would also be considered costs by S were he relevantly situated; if, in other words, their argument is *cooperative* in the sense of Chapter 1. A case in point is Quine's defense of classical logic in the face of intuitionistic defections. We lose the law of Double Negation, said Quine, and therewith "classical negation" (Quine, 1970/1986, p. 74). But "[i]t is hard to face up to the rejection of anything so basic" (Quine, 1970/1986, p. 85). We may agree that Quine is citing costs that he himself is not prepared to bear, and that this may be said to be a sort of *autodefense*. But inasmuch as the rejection of classical negation is precisely what the intuitionist earnestly wishes for, Quine's remark is no defense of classical logic *against the intuitionist*, and is still less a secure defense. Indeed, we may say that autodefense stands to defense in something like the dialectical relation in which self-evidence stands to evidence.

I am not able to be very specific about what dialectical communities are. They are, even so, handy things to make theoretical use of. Apart from the fact that they are, so to speak, the natural habitats of dialectical transactions, they also appear to be the natural medium for the changing of minds about conceptual matters. It is exceedingly difficult to see how it came to pass that from the collapse of set theory in 1901, ZF came to be, as Norman Martin says, the standard conception of set, except by reference to mechanisms of adjustment and accommodation that are essentially social.

IDEALISM

It is time to enlarge on an idea sketched in the Prologue. Frege was devastated by Russell's news of the Paradox. It led him to proclaim what we are calling

Frege's Sorrow: There is no intuitive concept of set.

We have been asking, in effect, how it is possible for a theoretical setback to qualify for *Frege's Sorrow*. The history of mathematics since the Russell paradox (and well before) displays a remarkable poise in the face of such difficulties, and a readiness to soldier on. This suggests that Frege's reaction was a psychological peculiarity and a methodological anomaly. In fact it was neither. Frege introduces what he calls "Hume's Principle" (HP).

HP: The number of Fs equals the number of Gs if and only if the concept of F is equinumerous with the concept G.[12]

When HP is annexed to second-order logic, the result is Frege arithmetic, FA^2. Crispin Wright has established that within second order logic, Peano arithmetic, PA, can be got from HP (Wright, 1983). So there is a consistent system of arithmetic in Frege's *Foundations*, and that this is so, that is, that PA derives from HP, was shown by Frege himself (Boolos, 1998a). It is true that Frege's concession to Basic Law V commits him to an inconsistency about sets (i.e., courses of values). And it is true that, in its *sole essential* use in *Foundations*, an instance of inconsistent Law V was used to derive *consistent* HP. Since HP is provable without Law V, there is a huge mathematical achievement in *Foundations* unaffected by the inconsistency about sets. What Frege seems unaware of here is the suitability of our strategy II for the management of inconsistency, the strategy that bids us to isolate the inconsistency and keep it from harmful consequence.

Even so, while Frege was wrong to say that the inconsistency discovered by Russell threatened to topple arithmetic, by his own methodological lights he was not wrong about *Frege's Sorrow*, since by those lights he had lost the concept of set.

To ask how Frege's reaction was possible is just to ask how it could be that a mathematician takes an analytic approach to his target concepts. It is to ask, "How is an analytic theory of sets explicable?" As it happens, there is a fateful concurrence between, in Jena, the death of psychologism in logical theory and, in Cambridge, the death of idealism in philosophy. As for the latter, it is convenient to mark the years between 1897 and 1903 as the period when Moore was spiriting Russell away from idealism and to something that was to be called "philosophical analysis." As we saw, 1897 saw the publication of *An Essay on the Foundations of Geometry*, Russell's first, and last, overtly idealist book. In 1903 there appeared *The Principles in Mathematics*, the first, and in some ways the most aggressive, work of his conversion (1956). Here are six years of such mathematical fruitfulness, especially in Germany, that it is easy to overlook philosophical developments at home, in Cambridge. Among the mathematical and dominantly German results of the period are: Frege's own masterwork, *Grundgesetze der Arithmetik* (Frege, 1893–1903/1964); Cantor's "Beiträge zur Begründung der transfiniten Mengenlehre" (1895–1897/1915/1952); Burali Forti's "Una questione sui numeri transfiniti" (Burali-Forti, 1897/1967); Hilbert's *Grundlagen der Geometrie* (1899/1902/1971); and just a year away was Zermelo's "Beweis, dass jede Menge wohlgeordnet werden kann" (1904/1967). Meanwhile, in Cambridge, Moore was revolutionizing the course of English-speaking philosophy.

If we are to answer the question of how it is that the collapse of set theory could have given rise to *Frege's Sorrow*, it will be necessary to solve the following puzzle, which we might call the

Riddle of Analysis:[13] Russell discovered the paradox that bears his name in June 1901.[14] Doing so should have *either* (1) preempted entirely his conversion to Moore's conception of philosophy *or* (2) been the occasion of his abandonment of the philosophy of mathematics.

Our puzzle is set by (i) the philosophical changes that Russell underwent between *The Foundations of Geometry* and the *The Principles of Mathematics*,[15] and (ii) the discovery of 1901.

In the *Foundations of Geometry*, Russell attempts to reconcile two fundamental insights. The first is that there is but one true geometry, which, in turn, is the one true theory of space, some of whose axioms are known *à priori*, and some by experience. The second is the paradigmatically idealist claim that the concept of space is inherently and incorrigibly contradictory, and hence, so too is geometry. Russell writes:

After hypostatisizing space, as Geometry is compelled to do, the mind imperatively demands elements. . . . But what sort of elements do we thus obtain? Analysis, being unable to find any earlier halting place, finds its elements in points, that is, in zero quanta of space. Such a conception is a palpable contradiction. . . . A point must be spatial, otherwise it would not fulfill the function of a spatial element; but again it must contain no space, for any finite extension is capable of further analysis. Points can never be given in intuition, which has no concern with the infinitesimal: they are a purely conceptual construction, arising out of the need of terms between which spatial relations can hold. If space be more than relativity, spatial relations must involve spatial relata; but no relata appear, until we have analyzed our spatial data down to nothing. (Russell, 1956, p. 128)

Noteworthy is that the irremediable inconsistency of geometry prompted in Russell nothing like *Frege's Sorrow*. Contradictions can be dealt with "dialectically" in Hegel's sense. If geometry is afflicted with contradiction, then, like any theory so fated, geometry demands supplementation (with what Aristotle called "further qualifications"),[16] under which the contradiction disappears. The supplementation of an old theory involves changing its subject matter to a different, but linked, subject matter. If the old theory is inconsistent, then a satisfactory supplementation of it must involve a change in subject matter that removes the inconsistency. In the case of geometry, the contradiction is subdued by abandoning the abstraction that underlies the old theory. The abstraction is that of conceiving of space as the mere possibility of diversity, that is, as the abstraction to empty space from the actual diversity of things in real space (Russell, 1956, p. 128). "[I]t is empty space ... which gives rise to the antinomy in question; for empty space is a bare possibility of relations, undifferentiated and homogeneous, and thus wholly destitute of parts or thinghood" (Russell, 1956, p. 191). In fact, "the relativity of space ... renders impossible the expression of ... [any theorem] of pure Geometry, in a manner which is free from contradictions" (Russell, 1956, p. 128).

Given that the inconsistency arises from our conceiving of space as an abstraction from real things in space, we must, says Russell, resolve "to give every geometrical proposition a certain reference to matter in general" (1945, p. 190), itself "a peculiar and abstract kind of matter, which is not regarded as possessing any causal qualities, as exerting or subject to the action of forces" (1956, p. 191). Here is the first glimmer of logical atomism, not fully born until 1918. In its adumbration here there is nothing to recommend the postulation of acausal atoms save for their contribution to a rehabilitated geometry. True, "the mathematical antimonies ... arise only in connection with empty space, not with spatial order as an aggregate [of acausal atoms]" (Russell, 1945, p. 196), but, even so, we are driven to our inconsistent concept of space by "*an unavoidable psychological illusion*" (Russell, 1956, p. 196, emphasis added). Inasmuch as we are driven to our inconsistent notion of space by an unavoidable psychological illusion, it may be said that it is our *intuitive* conception of space that suffers the inconsistency.

Because a dialectical resolution of the inconsistency requires that we change the concept of space, the new concept will not be an intuitive concept and will involve the invocation of factors that we have no antecedent reason to accept and that we accept now because doing so (a) removes the contradiction and (b) preserves a link with the old concept of space. It will be as intuitive as consistency allows, namely, not very. Readers familiar with the reconstitution of set theory in the aftermath of Russell's paradox will notice the similarity to what Russell is proposing for the reconstitution of geometry, but I daresay that it may safely be supposed that they would be surprised to learn that Russell conceived of such reconstitution as a working out of the Hegelian dialectic (in McTaggart's somewhat peculiar, but interesting, understanding of it).[17]

The author of *Foundations of Geometry* was an idealist. He engineered the rescue of geometry from its intuitive inconsistency in ways entirely faithful to his idealism. Russell thus subscribed to the four defining conditions on what we have been seeing as *minimal idealism*, that is, idealism stripped down and shorn of its Romantic puffery. First, human knowledge is at least partly *creative*; the constitution of the human knower is at least partially constitutive of what the human knower knows. Second, truth is *not absolute*. It is comparative (and so some claims are more true than others); it is partial (and so nothing a human knower can know is completely true); and it is mutable (what is true today may be false tomorrow). Third, with the exception of a special kind of thinking, which idealists called "metaphysical," all (human) thinking is *defective*: incoherent, contradictory, and incomplete. Fourth, something is real only to the extent that it is self-sufficient, that is, not dependent on other things, and, failing this test, the objects of human knowledge are *less than fully real*.

It is perhaps worth noting in passing that the ideas for which idealists seem to have received persistent (and alarmed) attention are corollaries or appendages of those listed here. "Psychologism" is a word for the first claim, that human knowledge is at least partially constituted by the human knower, but it should be emphasized that Bradley was at least as hostile as Frege to psychologism in logic.[18] Moreover, the infamous attack on internal relations is misnamed. Bradley's view was that all relations were unreal, external and internal, never mind that he made special arguments against internal relations. For consider the general case: if R is a relation such that for all objects x and y, which satisfy the condition that x bears R to y, the existence of R is bound up with the existence of its relata. Relations thus fail the self-sufficiency test. It is rather delightful that, if relations are taken extensionally, that is, as sets of n-tuples of *relata*, the dependency of relations on their *relata* is entirely obvious.

Idealists' exception to the law of Excluded Middle can be seen as a straightforward consequence of the relativity of truth. Here, too, much nonsense has been written about the role of dialectic in idealistic thought. But in its barest essentials it is little more than the rescue of an inconsistent theory by the contrivance of a consistent nearest-thing. Then there is the Absolute, also the subject of more florid description than it ever deserved; it is best seen as that which is described by the limit on which dialectic converges.[19] Suppose, then, that all descriptions of the world, after much successive refinement, were free of inconsistency, provably coherent, and fully compliant with the law of Excluded Middle; then the totality of such descriptions could be said to be an utterly faithful record of Reality and thus *absolute*. There is ample opportunity to agitate over the conditions whose fulfilment constitutes this "end of dialectic," but the main idea is not at all exotic.

Our minimal idealism is one of those philosophies in which most of the fun is in the details. The basic ideas are not quite as common as dirt, as the saying has it, but they are entirely without shock value: the human condition guarantees only limited and problematic access to how things really are and, in their congress, such as it is, with the world as it *is*, there is the constitutive press of the human cognitive apparatus and repertoire.

Moore's conception of analysis, although not itself expressly defined, involves fundamentally and essentially the unqualified repudiation of our four basic conditions on idealism and of their corollaries and supplements. This is the negative characterization of analysis: (1) How the world is owes nothing whatever to any presuppositions of our knowledge of it. Hence, the objects of knowledge are wholly nonmental.[20] (2) Truth is not relative, and Excluded Middle holds without exception. (3) The concepts of ordinary thinking are not as such defective; they are not as such inconsistent. (4) That of which humans have knowledge – when they do – is fully real.

Analysis also has a positive dimension. The world is a totality of propositions, of which the constituents are concepts. Concepts are either simple or complex. If complex they decompose into simple concepts. Complex concepts are thus *analyzable* into simple concepts. Simple concepts have no analyses and are directly and accurately intuited by the knower. There is no difference between a true proposition and a fact; so Moore espouses not a correspondence "theory" of truth, but an identity "theory" (1899, 1898, reprinted in 1903).

This is a view that persists to this present day. It is a view of concepts that divides them into two types: simple and complex.

Complex concepts are definable in terms of simple concepts, which are indefinable. We can use simple concepts to analyze other things, but they are too fundamental to be analyzed themselves. Rather, they constitute our starting point in analysis. We have no choice but to take them for granted. (Soames, 1999, p. 21)

Russellian Analysis. By 1903, Russell appears to have been disabused of his former idealism. It was Moore's doing:

On fundamental questions of philosophy, my position, in all its chief features, is derived from Mr. G. E. Moore. I have accepted from him the nonexistent nature of propositions (except such as happen to assert existence) and their independence of any knowing mind; also the pluralism which regards the world, both that of existents and that of entities, as composed of an infinite number of mutually independent entities, with relations which are ultimate, and not reducible to adjectives of their terms or of the whole which these compose. *Before learning these views from him, I found myself completely unable to construct any philosophy of arithmetic, whereas their acceptance brought about an immediate liberation from a large number of difficulties which I believe to be otherwise irreparable.* The doctrines just mentioned are, in my opinion, quite indispensable to any even tolerably satisfactory philosophy of mathematics. (Russell, 1903/1937, p. xviii, emphasis added)

Russell credits Moore with the idea that a *philosophical definition* of a mathematical concept "professes to be, not an arbitrary decision, to use a common word in an uncommon signification, but rather a precise analysis of the ideas which . . . are implied in the ordinary use of the term. Our method will therefore be one of *analysis*, and our problem may be called *philosophical* – in the sense, that is to say, that we seek to pass from the complex to the simple, from the demonstrable to its indemonstrable premisses" (Russell, 1903/1937, p. 2, emphasis added). Thus, the method of philosophy as regards mathematics is analysis, and analysis decomposes target concepts into simple concepts that the knower directly intuits, or of which he has a direct "nonsensuous" perception. On the face of it, the paradox of set theory is a genuine catastrophe. It means that we cannot treat classes philosophically. This appears to bring Russell alarmingly close to *Frege's Sorrow*. There is no concept of class, or anyhow; "I have failed to perceive any concept fulfilling the conditions requisite for the notion of *class*" (Russell, 1903/1937, pp. xv–xvi). Now we have an explanation of the **Riddle of Analysis**. The Riddle, we said, is that on discovery of the paradox Russell was in consistency bound either to revoke his commitment to philosophical analysis or to abandon all hope for a philosophy of mathematics. If reality is consistent, and if knowledge is knowledge of reality, and if knowledge of reality involves the direct apprehension of simple consistent intuitive concepts, then there can be no knowledge of classes.

Of course, Russell dissembled. He invoked a distinction between *philosophical analysis* and *mathematical analysis*. Having conceded the philosophical intractability of the concept of class, Russell makes room for a procedure that gives, for something that we might loosely and inaccurately call the concept of class, "merely a set of conditions insuring its presence."[21] This "mathematical sense of *definition* is widely different from that current among philosophers" (Russell, 1903/1937, p. 15). In fact,

it is necessary to realize that definition, in mathematics, does not mean, as in philosophy, an analysis of the idea of be defined into constituent ideas. This notion, in any case, is only applicable to concepts, whereas in mathematics it is possible to define terms which are not concepts. (1903/1937, p. 27)

Moreover

of the three kinds of definition admitted by Peano – the nominal definition, the definition by postulates, and the definition by abstraction – I recognize only the nominal.[22]

Russell opts for nominal definitions in mathematics in part because "definition by abstraction, and generally the process employed in such definitions, suffers from a wholly fatal formal defect: it does not show that *only one object* satisfies the definition" (1903/1937, p. 114, emphasis added).

A mathematical definition of something specifies a set of conditions that, if satisfied at all, is uniquely satisfied. The definition of Fregean sets cannot have been a philosophical analysis, because there is no concept of set. It cannot have been a mathematical definition either, because the paradox guarantees that the conditions on classes are not satisfied, hence not uniquely satisfied. For the Fregean set there is no "set of conditions" insuring its presence. Small wonder, then, that Russell's remedial treatment of sets via the theory of types should strike him as "harsh and highly artificial" (1903/1937, p. 500). Even so, the method of mathematical philosophy can be seen as bearing correctively on the perception of indefinables. It can get us to recognize that "we are trying to perceive the *wrong thing*, and so redirect our attention" (Hylton, 1990, p. 234, emphasis added).

As we now see, Russell has a sort of solution to the *Riddle of Analysis*. Sets cannot be analyzed philosophically; since there are no sets, there is nothing for philosophers to be realists about in that connection. Even so, the *term* "set" (or "class") is amenable to mathematical "analysis," even when it produces results that are harsh and arbitrary. The arbitrariness inexplicable by the fact that mathematical analyses are nominal definitions, and the harshness is explained by the unexpected difficulty of hitting on a nominal definition that is consistently satisfiable. Thus, Russell's implied answer to the *Riddle of Analysis* is this:

Russell's Answer: The paradox of set theory doomed all prospects of an analytic theory of sets. However, it did not preclude, but rather stimulated, a *stipulative* theory of sets, harshness, arbitrariness, and all.

We now have the means to take the measure of *Frege's Sorrow*. In Frege's reply to Russell he shows concurrent disposition to what, in effect, are Russell's notions of philosophical analysis and mathematical definition. When Frege registered his concern that "not only the foundations of my arithmetic, but also the *sole possible* foundations of arithmetic seem to vanish" (Frege, 1902/1967, pp. 127–8, emphasis added), he shows himself seized with the belief that the foundations of arithmetic are securable only by means of an analytic theory, which the paradox demonstrates is an entirely forlorn hope. But when in the very next line of his letter to Russell, Frege thinks that it may be possible to retain "the essentials of my proofs," perhaps, by reformulating his Rule V, he shows himself ready to turn a stipulative approach to sets.

Indeed this was precisely the approach that Frege did take. In the appendix to volume two of the *Grundgesetze*, Frege replaced Rule V with the more complicated Rule V'. It did not work. An inconsistency was still derivable, although it is evident that Frege was not aware of this at the time. Some twenty years later in the unpublished paper of 1924–5, Frege had given up on the set theory long since. In the end, Frege's fidelity to analytic theories won the day, and lost it, too.

Russell's technique of mathematical, as opposed to conceptual or philosophical, analysis is hardly reconcilable to his newfound affection for Moorean platonism. In all essentials the theory of types is a reversion to idealism.

Initially conceived of as a fit subject for analysis, sets were stricken by paradox, and Russell's subsequent treatment of them qualifies as dialectical. One changes the subject matter of set theory by adding qualifications. One adds these qualifications in ways that preserve, as much as is possible, a linkage with the old notion. In this, nominal definitions are resorted to and, twice over, idealist themes are sounded. First, set theory, if true at all, is less true than a *consistent* analysis would have been. Second, the theorist's knowledge of sets is in part a matter of his own creation. Russell's recovery from the inconsistency of sets in *The Principle of Mathematics* is hardly distinguishable from his recovery from the inconsistency of space in the *Foundations of Geometry*.

Frege's Sorrow proclaims the death of analytic theories for sets, and with it the methodological disablement of intuitions and counterintuitions conceived of analytically. Historically, sets went stipulative among those who, like Russell, had surrendered to the exuberant blandishments of Moorean analysis. The way of stipulation is a way straight back to idealism, which is no bad thing perhaps. Whatever our general philosophical tastes, idealism not only makes sense of stipulation, it also offers hospitality to the dialecticized conceptions of intuitiveness and counterintuitiveness sketched in an earlier section. It gives a home to all that remains of these notions when *Frege's Sorrow* holds sway. One of the enchantments of idealism is that sets of beliefs satisfying its provisions do not routinely announce that they do. It is possible for a theorist whose theory is for real. This was the moral of the generation-shift in the history of sets, from those who saw sets as stipulations to those who now see them as winking away in Plato's Heaven and Cantor's Paradise. There is a twofold moral in this. One person's stipulations may well be his granddaughter's analyses. There is little more to such transitions than the overtaking of acceptance by belief, something that befalls us like the measles or a sneeze. The other part of the moral is that the granddaughter is always mistaken if idealism is true.

It might be objected that the goings-on in 1902 hardly call the tune for century's end. Besides, perhaps Russell was a cynic. Then, too, consider once again:

Even the notion of a cat, let alone a class or number, is a human artefact, rooted in innate predisposition and cultural tradition. The very notion of an object at all, concrete or abstract, is a human contribution a feature of our inherited apparatus for organizing the amorphous welter of neural input. (Quine, 1992, p. 6)

And

there can be no evidence for one ontology as over and against another, so long anyway as we can express a one-to-one correlation between them ...; but once we have an ontology, we can change it with impunity. For abstract objects this is unsurprising.... For familiar bodies it is less intuitive. (Quine, 1992, p. 8)

This is idealism if anything is, and in 1992 a goodly part of Quine's philosophical charm and importance is his repeated effort over nearly a half-century to wriggle out of it – that is, to retain the doctrine but deny its true philosophical character. Ninety years after Russell contrived the thing he called "mathematical analysis," Quine is still at it:

This global ontological structuralism may seem abruptly at odds with realism, let alone naturalism. (Quine, 1992, p. 9)

Not to worry. Naturalism actually saves the day. It tells the story of how we conceptualize and probe the world, namely, in ways that satisfy conditions on idealism. But the story is told with realist presumptions. So, since we humans are in reality idealists, we are not idealists after all![23]

Perhaps this is a slight caricature of Quine's considered position. Even so, this does not prevent the caricatured position from being a *position*, that is, what James calls "a live option." The position, lightly sketched in the Prologue, is this:

Human beings take the realist stance. They make their intercourse with and contrive their theories of the world on the realist presumption that the world is really there and is the way we take it to be because much of what we say about the world really is so. Since we ourselves are in the world and interact with it, we ourselves are in the ambit of human inquiry under realist presumptions. But when we examine the details of that interaction we see that we negotiate with the world, and form theories of the world, in ways that satisfy the definition of idealism. This is anomalous, of course. We see ourselves as proceeding realistically, as probing the world as it is apart from our probes and with probes that really disclose how things are "out there." But when we examine what we ourselves say about our probings, what we say, if true, makes us idealists. Theories constructed under the presumptions of realism tell us – at a minimum – that realism cannot be true if the theories are.

This is the "anomalous realism" of the Prologue, the idealism that dare not speak its name. Anomalous realism is not something to fret about overmuch. It is a fact of life – and a fact of nature, if Quine's general line on ontology is true. Better that we should try to understand how we come to be anomalous in our realism. If we were to bring this off, we might also be able to redeem our pledge in regard to stipulative theories and to the stipulative conception of truth. This is a book about conflict resolution in the abstract sciences. It cannot, for all the obvious reasons, be a book about everything relevantly connected to abstract theories and the enterprise of conflict resolution. Even so, unless we have a good idea of how abstract theories operate, of what they are and where their limitations lie, we are likely to get the story of conflict resolution wrong. So I shall tarry awhile with anomalous realism and try to sketch an answer – a conjecture – about how we come to be anomalous realists.

How do we come to be takers of the realist stance? I conjecture a connection with survival. The realist stance is an economical way of paying attention. It is a way of taking the world seriously enough to negotiate it with care. We may even imagine that we are hard-wired to proceed realistically in matters when not doing so could prove an evolutionary liability. An efficient way of taking things seriously is taking them for real. I conjecture that where it matters most we find ourselves stocked with realist convictions. We deeply, and in a way unshakably, believe fire to be hot, and we had better deal with the on-rushing bus realistically, on pain of a bad outcome. A belief is *intractable* when it persists in the face of sincere assent to its negation. If we are idealists about fires, on-rushing buses, and the like, the intractability of our realist conceptions will make of our idealism a notion. Our belief in idealism is notional, a matter of *assent*, that is, the endpoints of speech acts, hence a pragmatic affair. Our contrary realist conceptions are psychological. Thus, the idealist finds himself in the same boat as a parched Buddhist who believes the material world to be a total illusion: He will thank you for a cold drink if offered. His Buddhism is notional or pragmatic. His Buddhism is not psychological. If our evolutionary conjecture has anything going for it – if it has theoretical legs, so to speak – the less a matter for survival the less we will see of conflicts between psychological and pragmatic belief. Take sets again. It is not that there are no set theorists who take the realist stance toward sets. Nature is efficient. Better that the stance be taken for all if taken for any. The set theorist will have psychological beliefs about sets if he is a realist about sets, but with a difference. His psychological beliefs are not likely to be intractable; they are not likely to concur with the pragmatic belief in its negation. I have conjectured that our disposition to take the realist stance across the board is a matter of efficiency. After all, stances are cheap and realist stances are risk-averse. This, if true, does not prevent it also being true that whether there are sets and how they are does not matter *at all*. Nothing we say about sets – nothing that is *true* of them – is in whatever way going to do us the slightest harm. It matters to a degree that thirteen really is greater than twelve, since this guides my choice of a baker's dozen the next time I am in the *patissérie*. But whether twelve and thirteen are *sets* does not even register on the scale of mattering. Such things, and theories about them, are indifferent to human welfare, hence as I shall say *indifferent*, full stop. Seen this way, two things become clearer. First, it *ought* to be the case that the very idea of intuitive set theory is just silly.[24] But, as we know, it is a fact that for lots of set theorists their theories are theories that they take *analytically*. Why do they do it? It is because our inclination toward realism is an inclination across the board. On the other hand, and second, if for whatever reason it is supposed that our interest in sets is one for which no guidance is possible by way of analytic intuitions, and given the indifference of sets, there is no other way of getting at the truth about sets than making them up, with stipulation – idealism perhaps in its purest form – playing a central role.

The phenomenon of generational-shift is now explicable. When Russell began the business of set theoretic reconstruction, he had lost, as he thought, the intuitive concept of set. Thinking so, he must also have thought that his reconstruction would have to proceed without knowing what sets are, that is, and more carefully, without any psychological beliefs about sets with regard to which he would not also assent to the statement, "you have no warrant for that belief." As Russell saw his own circumstances, set theory became a boot-strapping enterprise in 1902. No one doing set theory in this way could hold its theories as objects of psychological belief. Russell did not know what sets were, but he did know what the old sets were wanted for. Knowing this, he contrived his stipulations accordingly. Sets would be what we make of them, and what we make of them would be a matter of what we make them *for*, whether for transfinite arithmetic or for logicism. It also bears on our question that even as Russell was reinventing set theory, he wrote it up realistically. The realism of the exposition is an open invitation to psychological belief. Russell tells us that a condition of the adequacy of his "mathematical analyses" is its disposition to be believed. The theory of types is written in a way to invite belief. In the early days, in the days of its making up, pragmatic belief was possible, but psychological belief likely not. Once set theory ceases being transacted by people to whom the bootstrapping is transparent, psychological belief becomes a possibility and, under press of the attractiveness of the mode of its exposition, together with the comprehensiveness of our disposition toward realism, a likelihood.

Pragmatic belief is not a state of mind. It is what we spoke of in Chapter 3; it is differentially the speech act of *assent* and the object of it. Even so, pragmatic believing is rooted in psychological believing. People sincerely assent only to what they regard as assent-worthy, and this is a matter of psychological belief sooner or later. This rootedness of the pragmatic in the psychological reminds us of the utter dominance of psychological belief. Human beings are recurring in such states all their waking lives. Being in them is as natural as breathing, and as hard to resist. Our propensity to be in such states is nothing more or less than our across the board disposition to take the realist stance. This being so, because the attractiveness of the realist stance is comprehensive, there is a disposition to displace, over time, pragmatic belief with psychological belief to the same effect. If this is so, it explains the generational shift in set theory. A boot-strap operation in 1902, in the fullness of time it is an analytic theory that captures the ordinary concept of set, an intuitive concept that, by the very meaning of "set," excludes Russell's paradox, as far as we know.

It remains to say a little about stipulative theories. We said that stipulative theories make true what was not true before. Fiction is one way of doing this. It is harder than it at first seems, to specify truth conditions – or such other kinds of condition as may apply in those cases – on an author's creation of new truths. It is also harder than it might seem to determine what truths follow from those made true by the author, the sense in which they *are* true, and how, typically, they are contradicted or at least contraindicated by the rest of the world's truths not the authors's own.[25]

Whatever Russell was up to in 1902, he was not doing what Conan Doyle did in 1902 by way of making it true that Sir Henry Baskerville had purchased his missing boot in Toronto. Russell was making a fiction all right, but he was making it not for the purpose of making a good story or a gripping "read"; he was contriving an account that would do the job of old sets in the foundations of mathematics. Both storytelling and theoretical stipulation are serious undertakings; where they differ is in their truth conditions, or other conditions on warranted assertability such as they may be. There is little consensus about the semantics (or, come to that, pragmatics) of fiction. But nearly everyone agrees that

(1) Sherlock Holmes lived in London

is true – or if not true, all right to assert – because of a network of relations that takes (1) back to sentences penned by Doyle. It is precisely here that set theory parts company. Every sentence in the theory of types owes its truth if true, or whatever else makes it all right to assert, to a network of relations that take such sentences back to sentences forwarded by Russell. But that they were penned *by Russell* is inconsequential for their truth or assertability. A similar claim, thought it might be true, is less plausibly pressed in the case of Doyle's authorship. What matters for the difference at hand is that the penner of Russell's stipulations was a *mathematician*, whereas the penner of Doyle's fictions was a *storyteller*. In each case, the qualification here affects differently on the penner's access to conditions under which alone his enterprise succeeds. That the maker of the truth about Baskerville's boot is a writer of fiction bears on whether the lines penned meet the conditions, whatever they may be, for being a story, and even more so an interesting, engaging, well-told story. Similarly, that the maker of the truths of type theory was a mathematician is a fact that gave Russell access to conditions, whatever they are in detail, by which the theory of types is a *theory* (rather than a short story or an outright lie) and by which, too, it is theory that commands our interest, and even our assent. I do not think that we know very much about such conditions. Here, too, there is cause for conjecture. I have said that human beings have a weighty disposition toward states of psychological belief. It appears that there is another human trait equal to the first in weight and pervasiveness. Human beings are natural cooperators. In statement-making discourse, the trait manifests itself as a disposition to *truth-tell*. We are in other words *sincere* in our utterances, which is what telling the truth comes to – the truth as we see it. If Doyle had thought up type theory, it would never have seen the light of day. His having ploughed ahead while lacking the qualifications to do so would have convicted him of *prima facie* insincerity (think of the difficulties *idiot savants* and other sheer prodigies have in being taken seriously) (Kanigel, 1991). Russell's qualifications allow us to think him sincere, hence to think of his type-theoretic utterances as true to the best of Russell's knowledge. Of course, as Russell expressly tells us in his discussion of the stipulations of "mathematical analysis," this is not the understanding with which Russell forwarded the statements of type theory. But sincerity of a lesser grade trailed along and is centrally important. Russell, as we may suppose, had the psychological belief – or the hope – that his statements were worthy objects of assent. Indeed, how could they not be, if assenting to them causes no harm and also helps the cause of a philosophical program in the foundations of mathematics? Russell's stipulations were made by a cooperator to a highly regulated community of cooperators, and in a mode of discourse tailor-made for cooperative discourse, which presents its utterer as sincere; as a truth-teller. If the realistic stance is a disposition to have states of mind in which the world figures realistically, our disposition toward cooperation is a disposition to tell it "like it is," that is, objectively. Both traits are in us so deeply that the passage from a newly stipulated theory in the category of the playful, to something well worth considering, to the "real thing," can be swallow-swift.

This urge to see as true the assertions of stipulated theories we chance to make up is one thing. It is another thing entirely that we see such theories as true in the same sense as a favorite realist theory, and not in the same sense as the literary truths of fiction. On its face, this is odd to the point of implausibility. If the stipulations of type theories belong in a camp other than their own, then they belong with the truths of fiction, also stipulated. Why then do we see them in the way we see the truths of realist theories? If Quine's structuralist ontology is correct, we have an answer: *There are no realist theories*; all systematic responses to the neural whirl are a human contribution. But Quine's position implies idealism (or so I say), and idealism answers our question. We assimilate the truths of stipulated theories to the truths of nonstipulated theories because, though there is a difference between stipulation and not, stipulative theories and nonstipulative theories must satisfy the same basic conditions on theoretical adequacy if idealism is true.

In proposing in the preceding chapter a substantively rather than notionally tripartite approach to implication, inference, and argument, we set ourselves up for an obvious challenge. How does the tripartiness of this *troika* of notions assist us in the dispute over *ex falso*? We said that if the *Reconciliation strategy* is to have any chance of success, there must be antecedent recognition of the different things that paraconsistentist's and classicist's intuitions could be true *of*. Our question for the classicist, as for the relevantist, is not whether *ex falso* is true, but what, if anything, it is true of. If we supposed that Lewis was speaking for them all, we would have it that on the classical account of *ex falso* is true of all three. Hence, a contradiction *implies* any proposition; a person holding a contradiction is rationally permitted, if not obliged, to believe everything whatever; and an arguer caught in a contradiction is committed to concede everything whatever.

This is too much. The classicist might be right about implication, but no one will seriously suppose that there is any plausible sense in which he can be right about inference and argument. Logicians, as all theorists do, sometimes try to subdue the sting of their counterintuitive claims by way of technical neologisms. We might be tempted to see Lewis's boast in this spirit. A classical analysis is correct for implication, and for inference, and argument, where these latter two are understood in ways peculiar to logical theory. This, of course, is not what Lewis said. He said that systems such as S5 were correct for implication, inference, and argument, in the ordinary meaning of the terms. Perhaps he meant what *logicians* ordinarily mean by these terms, while at the office so to speak. Whether he did or did not, the important point is that the only way in which a classical analysis has any prospect of success across our threefold distinction is by making truth-preservation the core deductive virtue. Thus, where Φ is a contradiction, its implication of any other is truth-preserving. Where Φ is believed by S, then S's (contrary-to-fact) belief in everything does no damage to truth-preservation. Likewise, when an arguer subscribes to Φ, his obligations as a preserver of truth are not ignored by any subsequent conversation, whatever it may be. This is fine as far as it goes, but it does not go far enough. No one genuinely interested in a theory of inference or in a theory of argument will accept for a minute that truth-preservation is all, or nearly all, there is to the deductive niceties.

If this is so, there is reason to expect a classical account to be true of implication and false of inference and argument. Equally, there is reason to expect that a paraconsistent account will be *underwritten by intuitions* that are true of inference and argument, but that there is no need to extend, or profit in extending, to implication. We have seen that relevant systems such as *R* and dialethic systems such as **PR** fail to be good accounts of inference and argument. Aggregative and preservationist logics are better, but not "better" enough. This is because their sponsors have not taken sufficiently to heart the depth of the cleavage that we saw in the triple {implication, inference, argument}. But their *intuitions* are better off. Sensing that *ex falso* could not be a state condition on inference or a pragma-dialectical rule of argument, they get inference and argument right in that respect. Their mistake is that they take a good thing too far, and apply the same stricture to implication. That they would do *that* suggests a disposition to err in the opposite direction. This is the mistake of attributing willy-nilly what is incontrovertibly true of implication to inference and argument. Doing so puts them squarely in the ambit of Harman's complaints about the unjustified latitudinarianism of supposing that Simplification, Addition, and *modus ponens* are state conditions on belief-revision or pragma-dialectic rules of argument.

If there is to be something deserving of the names of deductive inference and deductive argument, it may be that a deductively correct inference and a deductively correct argument will have something essential to do with truth-preservation. Some of the best of the current work on the psychology of deduction (Rips, 1994) suggests that the purported connection can be seen in the relation of *restriction*. Take *modus ponens* as an example. If *modus ponens* reflects truth conditions on implication, it will also serve as a deductive constraint on inference or belief-revision if two conditions are met. The first is that *modus ponens* will be subject to constraints that have no place in the theory of implication. It will thus be a *restriction* of *modus ponens*. The other is that this restriction also be reconfigured so as to reflect the change in role from truth condition to state condition. It is the same way with argument, but with somewhat different constraints, and a reconfiguration that honors the difference between a truth condition and a social rule. The full story of how those constraints work, and of the differences appropriate to the different reconfigurations of truth conditions into state conditions and pragma-dialectical rules, has yet to be told (Rips, 1994; Sperber and Wilson, 1986; cf. Johnson-Laird and Byrne, 1991). Also see Woods (1998a, Ch. 4). Its telling lies beyond the plan of what is presently before us. Our task has been to revisit *Reconciliation* as a resolution strategy for conflict in the abstract sciences. *Reconciliation*, as we now see, is not only a benign variation on the philosopher's old favorite, ambiguation, it is also something counseled by a Principle of Charity (Wilson, 1959). In its present form, it bids disputants to maximize the impact of one another's intuitions, even (indeed especially) when they disagree. *Reconciliation* is a way of satisfying the Charity Principle. It takes the relevantist's firmly held conviction that deduction must be held to relevance conditions, and displays them in the theory of deductive inference, and of deductive argument, too. Similarly, *Reconciliation* takes the strong conviction that *ex falso* cannot be true and makes it stick for inference and argument. There is a pattern to this that emphasizes *Reconciliation's* essentially consensual character. Those who dislike *ex falso* are seized by the pretheoretical conviction that it cannot be true. Once derived in classical systems, it strikes them as a counterexample. Others are less struck. For them it is at most a surprise. *Reconciliation* is now deployed as follows:

Step one: Concerning a matter on which parties disagree, try to find another matter that converts their disagreement into agreement. Thus classicists and generic paraconsistent's disagree about whether *ex falso* is true of implication, but they both agree that it is not true of inference or of argument (except in the logician's own technical uses of these terms, which do not count).

Step two: Try to determine whether this new agreement is a "disagreement-buster." It is indeed a disagreement-buster if a) every complaint of the original objector about *ex falso* as a condition on implication is accepted by both parties in relation to inference and/or argument, and b) the original objector can think up nothing new against *ex falso* as a condition on implication.

Reconciliation is not a panacea. Its deployment does not guarantee a favorable outcome. Still, even when once played, though it does not induce the original objector to yield, it does significantly affect the burden of proof. The original objector may persist in his heartfelt conviction, his strong conviction, that *ex falso* is not true of implication, but, as we now see, it is an intuition he must consider not *pleading*. Here is the reason. There is a phenomenon to be explained. It is the original objector's intuition about *ex falso* together with the classicist's failure to share it. *This* needs to be explained. It could be that paraconsistentists are right and classicists are just blind. Perhaps they are stupid. Perhaps there is no fact of the matter about logic. Perhaps it is something else. Perhaps one party or another, or both, is *confused*. In reflecting on these possibilities, we would do well to proceed *abductively*, that is, to look for the best explanation. I shall take it without further ado that the best explanation here is also, defeasibly, the most charitable. That is to say, it is one that characterizes the parties without having to attribute to them claims independently known to be implausible or ill-evidenced. It is implausible – really reaching, so to speak – that about half the logicians presently on the planet are logically blind or just stupid. It is less implausible that they might be confused. This is what I conjecture: The *ex falso* debate has been socked with confusion, a fact that *Reconciliation* helps to get us to see. Here again is the structure of that confusion. The generic paraconsistentist's heartfelt conviction is that *ex falso* cannot be true of implication results from his conflation of implication with inference and/or argument, with regard to which all parties are agreed *ex falso* is false. On the other hand, classicists who find in *ex falso* as a condition on implication nothing more than surprise, will, if the present conjecture is correct about the conflation of implication with inference and/or argument, tend to preserve *ex falso* as a condition on these latter. That is, they will so conduct themselves in the theory of inference and argument as to indicate that their view in all strictness is that *ex falso* is true of this latter pair. It is precisely what one sees happen. One sees it in the invention of the ideal logician or the omniscient inferer. One sees it in systems of belief-revision such as AGM. As we say, AGM systems possess the two properties required by any belief revision theory that is content to repose itself on a classical base logic. One is that the theory's belief sets are consistent. The other is that beliefs (or in some variants, commitments) are closed under consequence. The latter condition implies that a reasoner fulfilling these conditions is deductively omniscient; the former implies that he is hyperlogical, and a good thing, too. For condition one cancels what would otherwise be an intolerable consequence of condition two, namely, that inferring everything is sometimes the right response to a contradiction. There would have been no need to make the ideal reasoner hyperconsistent except that AGM inference fulfills the canons of classical logic, *ex falso* and all.

Both parties, I say, are guilty of the conflation of separate things. If this is right, the confusions should have proved costly – different costs for different exploitations of the confusion. If classicists are right about *ex falso* and implication, then the cost for the paraconsistent implication theorist is predictable. His theory of implication will be awash in complexity, and it will lose both of what he antecedently regarded as virtues of a logic of implication. Similarly for the classicist, whose classicism over *ex falso* he is disposed to carry over to the theory of inference. Here, too, the cost is predictable. It will call forth AGM and other ideal logician theories of inference, in which deductive competence bears no relation to anything that beings like us could soberly begin to aspire to. In other words, they are false.

Reconciliation helps in driving the present abduction. It sets us up for inference to the best explanation. For what best explains the fact that all parties agree on what *ex falso* is false of, and yet disagree about what it is true of? (Is it logical blindness, or stupidity; or could it even be that implication is *harder to understand* than inference or argument?) Furthermore, what best explains that the complaint about *ex falso* as a condition on implication is an undisputed truism about inference and argument? My conjecture is that generically paraconsistent theorists have got inference and argument right as regards *ex falso*, and have illegitimately appropriated implication to that theoretical model. Classical inference theorists, such as those who back AGM, have made a similar mistake but in the opposite direction. Judging *ex falso* to be true of implication, they have illegitimately appropriated inference and argument to that theoretical model. In each case, the mistake and the cause of mindless-seeming disagreements about *ex falso* (and equally mindless-seeming disagreements about the extent to which *our* rationality consists in approximations to the behavior of some Big Daddy Logician) is best explained by an everyday and theoretical commonplace: Human beings have a disposition to confuse even modestly similar things.

There is a significant distinction that bears on what we have being saying about the realist stance. The distinction is that between *theory* and *heuristics*. Heuristics I understand in Quine's way. They are aids to the imagination. They help the theorist in thinking up his theories. It cannot seriously be doubted that in the business of thinking up his theories, there are some things the theorist cannot do without, including his most confident and enduring convictions as to what facts the theory must honor. Even so, not every belief required by the theorist to conceptualize and organize his theory need itself be a theorem of the theory. A case in point is any true first-order theory eligible as input to Quine proxy functions, briefly met with in the Prologue. All such theories must be extensional. Yet, for all kinds of purely extensional theories, there is not the slightest chance of our being able to think them up in a purely extensional language. For that, we need *English* or some such. In such cases, the intensionality of the thinking-up language is indispensable; but it would be a mistake to import those indispensable intensionalities into any theory governed by proxy functions. This is a make sufficiently bad and – I fear sufficiently pervasive – to qualify for a name. I call it the

Heuristic Fallacy: Let H be a body of heuristics with respect to the construction of some theory T. Then if P is a belief from H, which is indispensable to the construction of T, then the inference that T is incomplete unless it sanctions the derivation of P is a fallacy.

People who take the realist stance – or better, who, like all of us, have it thrust on them, are people disposed, just so, to commit the Heuristic Fallacy. We may well imagine that there are all sorts of things we must think of sets as being as a condition of getting anything deserving of the name of set theory up and running. Perhaps these imaginings – these insistent ideas as Peirce calls them, must make their way into the theory (some of them, anyhow). But if their entry is attended by the presumptions of the realist stance, there is a good chance that the *Heuristic Fallacy* will have been triggered, even if their admittance there is defeasible and a matter of sufferance. On the other hand, a *stipulated* entry of such things to theory all but eliminates susceptibility to the fallacy, even though such stipulations might be theoretically unfruitful or otherwise strategically ill-advised.

Given the extent of its exposure to the *Heuristic Fallacy*, it may be thought sound general strategy to *resist* the realist stance when it comes to thinking up our abstract theories.

Perhaps this is the occasion to make brief mention of a particular variation of *Reconciliation* that mention of the *Heuristic Fallacy* calls to mind. On the traditional conception, originating with Aristotle, a fallacy is an argument, or a piece of reasoning, that appears to be correct but is not, in fact, correct. More fully considered, fallacies satisfy three conditions.

1. They are attractive or seductive mistakes.
2. They are widely committed.
3. They are difficult to correct; in other words recidivism is high.

These are conditions that challenge the standard *rhetorical* role of attributions of fallaciousness, which is to make an accusation of some seriousness against one's interlocuter. According to one expert, a charge of fallaciousness is a particularly harsh form of indictment, and, for that reason, should not be lightly indulged (Walton, 1995). It would seem, however, that the truth is the reverse of this. If fallacies are mistakes we all make, and are errors that we find it easy to commit and difficult to avoid recommitting, then attributing fallaciousness is attributing an error that even the attentive and intelligent are routinely disposed to make. This being so, to characterize an opponent's reasoning as fallacious is to concede that it is reasoning that even an intelligent and informed person could succumb to. The reconciliation strategy enters the picture in the following way. The attributer of fallaciousness can be taken as saying "Judged by criteria that appear to be correct, your reasoning is sound. Judged by other criteria (which might not even seem correct), your reasoning is unsound."

5

Sets and Truth

... a sufficient, and at the same time general, criterion of truth cannot possibly be found.

Immanuel Kant, *Critique of Pure Reason*, 1781

Indeed, even at this stage, I predict a time when there will be mathematical investigations of calculi containing contradictions, and people will actually be proud of having emancipated themselves from consistence.

Ludwig Wittgenstein, *Philosophical Remarks*, 1930

THE DIALETHIC MINI-HISTORY OF SET THEORY

In the one hundred years since 1902, mathematicians have struggled to rehabilitate the theory of sets. There are plenty of versions: Zermelo-Frankel (ZF), ZF with Choice (ZFC), von Neumann-Bernays-Gödel (NBG), Kelley-Morse (KM), *Math Logic* (ML), *New Foundations* (NF), *Set Theory and Its Logic* (STL), and others. They are not equivalent, but each does roughly as well as any other in mimicking the integers. These modern theories are, so far as we can see, consistent and free of the Russell Paradox. Each is a development of the iterative conception of set and each is an articulation of the cumulative hierarchy. For all its post-Paradox centrality, the hierarchical approach to sets has attracted an ambush of its own in the dialetheist's insistence that the approach is entirely mistaken. Here, then, is the **Dialethist's Tale.**

Once upon a time, people had a lovely and intuitive concept of sets with which to do the technical business of transfinite arithmetic. But in 1902, along came that trouble-maker Bertie Russell with his infamous Paradox, and our lovely intuitive concept was destroyed. What were the people to do? They needed sets to do modern mathematics. In the end, they just made them up! The most widely used of these made-up conceptions of sets is ZF, which in turn is a construction on what is called the cumulative hierarchy.

This cost us realism in mathematics. It stuck us with a nervy postmodernism in mathematical epistemics. Such losses are philosophically painful, and they cannot

155

convincingly be compensated for by naming them "idealism." Better that we stick with *real* sets, rather than stipulations on the *word* "set" in the manner of Russell in 1903. In so doing, we enable ourselves to deny every tenet of this so-called minimal idealism, save one (or, more strictly, half of one). The half-tenet that is retained is that the concept of set is inconsistent; the half-tenet that is *not* retained is that this is a bad enough misfortune to force us to *make up* a consistent successor concept.

The very idea of sets is inconsistent, that is, the *intuitive* idea – the idea that tells us what it is to *be* a set. This being so, as long as we wish to do set theory, to say with accuracy how sets *actually are*, we will have to honor the intuitive fact of actually inconsistent sets.

And now at last we have a principled reason to reject *ex falso*. It is this: The true story of how sets actually are cannot be told if classical *ex falso* is true. An analytic theory of sets requires us to *change our logic*.

End of story.

It is an interesting story, and also a philosophical one. It emphasizes the sound-proof-of-a surprise side of the divide that is *Philosophy's Most Difficult Problem*. It offers prospects of a return to realism in logic and set theory. It offers prospects, as well, of reinstating the method of conceptual analysis and the recovery of intuitive concepts in an epistemologically fruitful way. There is a cost of course. It is the cost of the ontological embrace of actual inconsistencies and the attendant cost of moderating logic and cancelling the menace of detonation, as Schotch and Jennings call it. We shall need to weigh these costs carefully.

The realism that I am attributing to dialethists need not be seen as platonism. It need not even imply the necessity of dialethic truths. What it does imply is the rejection of any form of fictionalism in set theory, or semantics, or anywhere else save fiction itself. It permits conventionalism only of the most sober and benign sort. After all: "Convention cannot create what was not there to begin with" (Mortensen, 1995, p. 14). It is also compatible with a certain kind of what might be called "analogical decisionism," according to which "mathematics, especially pure mathematics, is more *like* a decision than a discovery of a preexisting truth" (Mortensen, 1995, p. 13). But all that this means is that the truths of pure mathematics "are internal to mathematical theories in a way *like* rules are internal to games, and quite *unlike* the way empirical claims about the physical world are true or false" (Mortensen, 1995, p. 14).

Even so, it is a point of some importance that dialethism offers the philo-sophically minded mathematician the chance to ply the method of analytical intuitions, and to insist that the intuitive axioms on sets tell us what real sets really are, whereas any acknowledgment of *Frege's Sorrow* about sets drives the mathematician, as it did Russell, beyond the ambit of analytical intuitions in ways that even a modestly conventionalistly minded dialethist would find regrettable.

The modern conception of set arises late in the nineteenth century from Cantor's work in analysis, chiefly his investigations of real functions. Cantor had sided with those who regarded functions as arbitrary, that is, as abso-lutely any correspondence between reals and reals. This gives rise to some odd-seeming functions. There are functions that pass through every interval between *a* and *b* without being continuous. There are continuous functions that are not differentiable. There was even Dirichlet's "shotgun" function, which was nowhere continuous without being either derivative or integral.

So promiscuous a conception of function was bound to meet with resistance, and did. The French were displeased, notably Baire, Borel, and Lebesque. In an effort to discipline unruly noncontinuous functions, Borel and others formulated a hierarchy of sets of real numbers. Borel sets are rather attractive. At their simplest, they are the closed sets and their complements are open sets. A closed set is a set of real numbers that contrives all its accumulation points. The Borel hierarchy of these made up of, first, the open sets, then the closed sets, then countable unions of closed sets, complements of these, then countable unions of all prior sets, complements of these, then sets that collect the preceding levels of sets; and finally the Borel sets, the union of all that goes before.

With Borel sets ready to hand, well-behaved functions are definable. Before long, it became clear that Dirichlet's function was integral. This proved to be telling. Even a well-disciplined notion of set will underwrite an extremely broad and abstract notion of function. What this suggests is that sets need to be disciplined. For, if they are, they can validate functions that seemed initially too ill-disciplined for their own good.

What is this conception of set? It is the idea that a set is a collection of antecedently existing objects, and that the principle of collection is "mental" rather than "natural"; that is, collections are not required to be natural kinds. They are any ensembles that the collector can consistently imagine or describe. Antecedence, here, is not a temporal notion. Formally, it is a relation that is irreflexive, antisymmetric, and transitive.

Two consequences of note are immediate. One is that there can be no collection of all collections. Self-collection violates the requirement that the elements of a collection exist independently of the collecting of them. The other is that if there were a collection of all collections not members of themselves, this just would be the collection of all collections, which, as we have seen, does not exist. We note in passing that the Dialethists' Story is wrong in its historical details. The present concept of set *preceded* Russell's Paradox. Russell's own conception of set, though influential, especially among philosophers, was not the dominant pre-Paradox conception.

Cantorian sets, like Borel sets, form a hierarchy. In the precise account of them given by Gödel (1947), hierarchies are made up of ranks. At rank 0 we have two possibilities, depending on whether our intended theory of sets is pure or impure. If pure, the theory restricts membership of sets to sets; if impure, sets are allowed to have sets or individuals as members. Thus, at rank 0, we have either the empty set and nothing else or we have the empty set and all sets of individuals, as well as the individuals themselves. The individuals are assumed to have been antecedently given. If there are n individuals there will be at rank 2^n sets of individuals.

Rank 1 is all sets of individuals and of sets formed at rank 0. (For ease of exposition, we are now describing the "impure" case). Rank 2 is made up of all sets of individuals, sets of rank 0 and sets of rank 1; and so on. Immediately after all of the ranks 0, 1, 2, 3,...comes rank ω. ω, a transfinite number, is the smallest limit ordinal, and is the successor of no number. It is the ordinal number of the sequences of all integers. Thus rank ω is made up of all sets of individuals and of sets formed at ranks, 0, 1, 2, 3, ... Rank $\omega + 1$ is constituted by all sets of individuals and sets formed at stages 0, 1, 2, 3, ..., ω. Rank $\omega + 2$ comprises all sets of individuals and sets formed at ages 0, 1, 2, 3, ..., $\omega, \omega + 1$. Similarly for rank $\omega + 3$, and so on. Immediately after Ranks 0, 1, 2, 3, ..., ω, $\omega + 1$, $\omega + 2$, $\omega + 3$, ..., comes ranks $\omega + \omega$ (or $\omega \times 2$), which collects all sets of individuals and sets collected at earlier ranks. Thereafter come rank $\omega \times 3$, rank $\omega \times 4$, ..., rank $\omega \times \omega$.

This hierarchy is known as the *cumulative hierarchy* of sets **CH**. It is easy to see why, since any given rank (other than rank 0) accumulates all the sets occurring at any preceding rank. Correspondingly, this conception of set is called the *iterative* conception. "To iterate" means "to repeat," and the Cantorian idea is that any rank repeats the sets occurring at all lower ranks as a member of the sets in that rank.

"The informal argument that the paradoxes are blocked by ZF is that its axioms are true in cumulative hierarchy of sets where (i) unlike the theory of types, a set may have members of various (ordinal) levels, but (ii) as in the theory of types, the level of a set is greater than that of each of its members" [Feferman, 1984, p. 238]. Feferman continues:

Thus in ZF there is no set of all sets, nor any Russell set $\{x : x \notin x\}$ (which would be universal since $\forall x(x \notin x)$ holds in ZF). Nor is there a set of all ordinal numbers (and so the Burali-Forti paradox is blocked). (1984, p. 238)

Other set theories take on some different features, for example, the theory BG proposed by Bernays and revised by Gödel. Like ZF, BG is lodged in the cumulative hierarchy, but now the hierarchy is "extended to one further top level at which we find the *proper classes*; classes of lower levels are identified with sets" (Martin, 1984, p. 238). Thus,

In BG one proves the existence of the class V of all sets, but not of all classes; further we have a class ON of all sets which are ordinal numbers. V and ON are proper classes and $V \notin V$, $ON \notin ON$. (Martin, 1984, p. 238)

Again, the received solution of the Russell paradox is that only those instances of the Abstraction axiom

$$\exists y \forall x (x \in y \rightarrow \Phi)$$

that hold in the cumulative hierarchy are sets that exist. There are *no* sets other than those in the hierarchy. Virtually every post-Russellian set theory that has been proposed – exceptions are Quine's New Foundations (NF) and Math Logic (ML) – holds in some or other initial segment of the cumulative hierarchy. So it will matter weightily if the dialethist is right in saying that this basic idea is no good.

Frege's Sorrow is of the view that there is no concept of set. *Frege's Sorrow* was not *quite* shared by Russell, as we have seen. Although the Tarski Paradox induced in the breast of Koryé the counterpart of *Frege's Sorrow* – this is our *Koryé's Sorrow* – it was not a view *quite* shared by Tarski. For Tarski, as for Russell, the paradoxes were *disappointments*. This is worth saying in a formal way. Thus

Russell's Disappointment is the position that, although the Paradox of sets makes it impossible to do the business of set theory in the manner of intuitively analytical theories, the sheer need to *have* a theory of sets requires that the business of sets be done in some other way, i.e., in the way we have been calling *formal*.

In the same spirit,

Tarski's Disappointment is the position that, although the Liar Paradox makes it impossible to do the business of semantics in the manner of intuitively analytical theories, the sheer need to have a semantical theory requires that the business of semantics be done in some other way, i.e., in the way we have been calling formal.

We see in the difference between **Sorrow** and **Disappointment** a difference of treatment options against a background of triagic disagreement. Frege, Russell, and Tarski alike saw the paradoxes as wholly destructive of intuitive set theory and intuitive semantics. Their shared triagic judgment was an answer to the question, "How bad is it, this damage done by the paradoxes?" and their shared answer was that the damage was as bad as it gets. Where they parted company was on the question of whether *treatment* was possible. After some initial uncertainty, Frege's answer was No; set theory could not recover from the damage inflicted by the Paradox. Russell, in the instance of set theory, and Tarski, in the instance of semantics, held out for *prosthetic treatment*. If set theory and intuitive semantics could not, so to speak, regrow their own shattered limbs, prosthetic devices would have to be fashioned. They would have to be fashioned because mathematics and model theory could not do without a theory of sets and a theory of truth, however rejigged and slicked up. Unless we are manic functionalists, we may think that a prosthetic arm violates our idea of what it is to be an arm, but no one will deny that, in *extremis*, the prosthetic arm is a wonderful successor to a destroyed real thing.

In contrast to *Sorrow* and *Disappointment* is the dialethist's own triagic heterodoxy. It is a significant development in the modern history of the abstract sciences, and it too deserves a name:

Routley's Serenity:[1] The diagnosis according to which the Russell and Tarski paradox show the inconsistency of the very ideas of set and of truth (or, more strictly, of statement) is entirely correct. What is not correct is the triagic judgement that this is very bad news, still less that it is as bad as it gets. In fact, strictly speaking, no *treatment* is needed or justifiable, even though some collateral adjustments are required.

For the dialethist, the recognition that there are inconsistent sets and inconsistent truths is like the discovery that in our mortality we are all prone to fatal illness. A Fregean is someone who gives up on medicine altogether; a Russellian and a Tarskian will feel the need of radical reconstruction; but the Routleyian sees the wisdom of staying out of draughts, eating well, dressing warmly, and getting enough sleep. There is no getting away from it: we come to fatal ends. There is no getting away from it: there are inconsistent sets and inconsistent truths. Human mortality is no more a bar to a satisfying and consequential life than are the inconsistencies of set theory and semantics a bar to perfectly serviceable theories of sets and of truth. All that is needed for this is a trimming of the sails of logic.

Against this is the conviction that logic is the least legitimately pruneable of the branches of theory, that logical principles are, or should be, "the last to go." In contemporary times, the point has been advanced, although in different ways, by philosophers as different as Carnap and Quine. In an influential essay Quine writes, "But there remains the fact – a fact of science itself – that science is a conceptual bridge of our own making, linking sensory stimulation to sensory stimulation; there is no extrasensory perception" (1981b, p. 2).

Contrary to the opinion of traditional empiricists, nothing whatever is knowable a priori ("there is no extrasensory perception") or by way of meanings alone. The traditional empiricist account of how knowledge of logic is possible is incorrect. Logic, like any other theory, is underdetermined by the data and so is an expression of our "conceptual sovereignty," which is far reaching enough to make of logic and other theories a "free creation" and something of a "put-up job" as Eddington said. The metaphor of "conceptual sovereignty" refers to "the domain in which [the theorist] can revise theory while saving data" (Quine 1960, p. 6). No part of any theory is immune from revision: not biology, not physics, not mathematics – and not logic. Theories are like fields of force. The closer they are to the field's sensory periphery, the more responsive they can be expected to be to sensory variation, whereas the more removed from the periphery, the less occasion there is for responsiveness to sensory variation. Far from the periphery, theories are governed by other more pragmatic constraints – simplicity, familiarity, explanatory power, and so on.

Logic is at the furthest remove from the periphery of sense.[2] So in matters of conceptual reorganization, logic will be the last to go, *in practice*. But go it can; its resistance to revision is not immunity in principle.

On the face of it, Quine's view of theories and things is tailor-made for the *Routley's Serenity*. If Quine is right, this should prove a methodological bonanza for the dialethist. It gives us an intelligent question to put to him. We can ask him what it is that he seeks to do with his nonclassical approach that he cannot do strictly, and with what benefits. That is, we can ask him to supplement his own analytical confidence with economic arguments that might appeal even to the analytically other-minded.

It might be thought that we are now moving too quickly and rather carelessly and that, under Quine's supposed auspices, with too latitudinarian a sweep. Although Quine does say that there is no bar in principle to logical revision, he himself does not recommend revision except as a last resort and certainly not before other things are reshuffled, mathematics for example. "Mathematics before logic," in the spirit of "Woman and children first," could serve as Quine's motto.

But if the dialethist wishes to have the methodological encouragement of Quine's holism, he had better have especially good reasons for tampering with logic rather than with set theory. Let us see.

The dialethist's response to this challenge can be set out as follows: (1) The reason for *not* tampering with the mathematics (i.e., intuitive, inconsistent set theory) is that the post-Russellian mathematics (e.g., ZF set theory) is bad mathematics; (2) The reason *for* tampering with the logic rather than the mathematics, apart from the desire to avert bad mathematics, is that classical logic misdescribes and misanalyzes the nature of paradox.

Consider (1): The reason for not tampering with the mathematics (intuitive set theory) is that the ensuing mathematics, that is, nonparadoxical set theory such as ZF, is bad mathematics. A primary metamathematical task of post-Russellian set theory is to evade Russell's paradox. This can be done in various ways, but the basic idea is that it can no longer be allowed that there is a single "unitary concept" of set. Here is Quine describing the point as it applies to Russell's theory of types:

> But the theory of types has unnatural and inconvenient consequence.... [T]he universal class V gives way to an infinite series of quasi-universal classes, one for each type. The negation-*x* ceases to comprise all nonmembers of *x*, and comes to comprise only those nonmembers of *x* which are the next lower in type than *x*. Even the null class Λ gives way to an infinite series of null classes. The Boolean class algebra no longer applies to classes in general, but is reproduced rather within each type. The same is true of the calculus of relations. Even arithmetic ... proves to be subject to the same reduplication. Thus the numbers cease to be unique. (1953b, pp. 91–2)

In standard post-Russellian set theories, the paradox is averted (so far as one can tell – freedom from paradox has not been proven) and, although different set theories do so differently, they all provide for the set theoretic construal of number theory, and so of the rest of mathematics, in ways that get the job done. In other words, a set theory such as ZF is not provably inconsistent, and it gets number theory right. How could this possibly qualify as bad mathematics?

Part of the dialethist's answer is, as we have seen, that such set theories are unacceptably *ad hoc*. Thus, (3)

The *ad hoc*ness of ZF (let ZF now stand in for them all) is central, deep and disabling. The technical maneuverings, having nothing to do essentially with the notion of set-hood, are counterintuitive and have little to recommend them save for the evasion of paradox. The counterintuitiveness makes the ensuing set theory unnatural and difficult to work with. In consequence, the central concept of sethood is unconvincingly described, if not out-and-out mangled. No post-Russellian theory can convincingly say that it provides a faithful account of what sets *are*.[3]

What is more, (4)

An *ad hoc* theory is not needed. Provided that one makes the requisite adjustment to the base logic, intuitive set theory can be persisted with, never mind that it is inconsistent.

And (5)

The idea of the cumulative hierarchy is itself dependent upon a prior (and different) notion of set, and the construction is dependent upon the prior notion of ordinal.[4] Moreover, until we manage to specify "how long" the construction is to go on, that is, how far the ordinals are to extend, the cumulative hierarchy is ill-defined. This is an issue that can be settled only set-theoretically. In fact, the original motivation for set theory was precisely to map the ordinals and their infinities. Given the *intuitive* notion of set we *can* specify the ordinals and thus can define the cumulative hierarchy. There seems to be no other way of adequately specifying the ordinals. ZF, for example, cannot provide the ordinals without lapsing into circularity, since ZF is pegged to the cumulative hierarchy.

Finally (6)

A further difficulty is that there are, as we have seen, instances of the Abstraction axiom that are somehow legitimate but which do not hold in the cumulative hierarchy. These include: the universal set V, the absolute complement of any set in the hierarchy, and any collection of members of arbitrarily high rank.

It is sometimes said that the exclusion of these structures by the cumulative hierarchy does not matter, since they are not required for ordinary mathematics. True enough,

... ZF and BG (when augmented by AC the Axiom of Choice, also intuitively true in the cumulative hierarchy), provide a framework in which practically all of current mathematics can be systematically worked out the exceptions are marginal; one which will be at the center of attention ... is the informal general theory of *mathematical structures*, particularly the *theory of categories*. (Feferman, 1984, p. 238, emphasis added)

The dialethist is troubled by the exclusion of category theory from the cumulative hierarchy. He might agree with Feferman that category theory is "marginal," but neither of them is prepared to say that it is unimportant.

Various attempts to remedy or mitigate such difficulties do not satisfy the dialethist. One is to identify the large categories, such as the collection of all groups, as proper classes. Proper classes are subcollections of the hierarchy that do not occur in the hierarchy as sets and cannot be members of other collections. But this, they say, is unsatisfactory. The cumulative hierarchy was proposed in the first place as the *analysis of sethood*. If we are forced to acknowledge that there are sets outside the hierarchy, "this just shows that the analysis is just wrong." Moreover, calling them by a different *name* is "a trivial evasion" (Priest, 1987, p. 42).

Finally, the requirement that proper classes not be members of other collections has no satisfactory rationale. For let K be a determinate proper class with determinate members. Then it is carelessly imperious to deny the collection {K}. So

...set theory's failure to embrace the notion of arbitrary *category* (or structure) is really just another way of expressing its failure to capture completely the notion of [set determined by an] arbitrary property. (Bell, 1981, p. 356)

In sum, the dialethic case against post-Russellian set theories pegged to the cumulative hierarchy (which is all the mathematically established ones) is this:

- The cumulative hierarchy is incompletely specified, or is circularly so.
- Its acknowledgment of sets external to the hierarchy either leaves sets inaccurately specified or involves a baptismal evasion.
- Proper classes, so-called by the baptismal evasion, are held to unreasonable, indeed incredible, conditions.

It is now clear that the dialethic charge against nonintuitive set theory runs a good deal more deeply than that it is *ad hoc*. That it surely is, but saying so secures no dialectical advantage against those who think that all theories, including all the good ones, are *ad hoc*. Even if *ad hoc*ery violates a *desideratum* on theories, that contemporary set theory fails the *desideratum* would hardly ground a charge of bad mathematics. So it is essential to the success of his program that the dialethist has deepened his objection in the ways that we have just examined.

We now turn to the dialethist's second claim (2): The reason for tampering with logic rather than the mathematics, apart from the desire to avert bad mathematics, is that classical logic *misanalyses* the nature of paradox. Thus, we again see in play the dialethist's most central *philosophical* claim:

Analytical theories are the safe way of being realists in the purely abstract sciences, and analytical theories cannot function without the guidance of basic intuitions, of the very idea of things.

In a classical system a paradox such as that $\mathbf{R} \in \mathbf{R}$ iff $\mathbf{R} \notin \mathbf{R}$ is characterized as something that "leads to contradiction" and that the sentence proclaiming that $\mathbf{R} \in \mathbf{R} \wedge \mathbf{R} \notin \mathbf{R}$ is "a self-contradictory theorem"

(Quine, 1953a, p. 90). Now it is true that "$\mathbf{R} \in \mathbf{R}$ iff $\mathbf{R} \notin \mathbf{R}$" together with Excluded Middle implies "$\mathbf{R} \in \mathbf{R} \wedge \mathbf{R} \notin \mathbf{R}$" and that this is a contradiction, correctly described in the semantic metalanguage as unsatisfied by every valuation v. However, it does not suffice as a characterization of the paradox merely to say that it implies a contradiction. To see that this is so, let us consider the semantics for our strongly paraconsistent system **PR** (Priest, Routley, and Norman, 1989, p. 168). Let $V = \{\{T\}, \{F\}, \{T, F\}\}$. $\{T\}$ is the classical *true only*; $\{F\}$ the classical *false only*; $\{T, F\}$ is the paradoxical *true and false*.

A valuation is a map v from the set of the sentences of **PR** to V such that

(Ia) $T \in v(\neg \Phi)$ iff $F \in v(\Phi)$
(Ib) $F \in v(\neg \Phi)$ iff $T \in v(\Phi)$
(IIa) $T \in v(\Phi \wedge \psi)$ iff $T \in v(\Phi)$ and $T \in v(\psi)$
(IIb) $F \in v(\Phi \wedge \psi)$ iff $F \in v(\Phi)$ or $F \in v(\psi)$
(IIIa) $T \in v(\Phi \vee \psi)$ iff $T \in v(\Phi)$ or $T \in v(\psi)$
(IIIb) $F \in v(\Phi \vee \psi)$ iff $F \in v(\Phi)$ and $F \in v(\psi)$

Logical consequence: $\Sigma \Vdash \Phi$ iff for all valuations v either $T \in (\Phi)$ or some
    $\psi \in \Sigma, T \notin v(\psi)$.
Logical truth: $\Vdash \Phi$ iff for all valuations v, $T \in v(\Phi)$.

The semantics hook up with a De Morgan *lattice* in the following way (Priest, Routley, and Norman, 1989, p. 189). We define a map from the sentences of **PR** to $\{T,F\}$: $T \in v(\Phi)$ iff $\Phi \in T$, and $F \in v(\Phi)$ iff $\Phi \in F$. Thus v is a (zero degree) valuation of the sort defined in the paragraph above. More precisely, the semantics above make v a map to $V \cup \{\Lambda\}$, and thus allow for truth value gaps. The additional condition that $a \in T$ or $\alpha^* \in T$ suffices to make v a map to V.

On the dialethist's view, this is the *correct* characterization of paradoxes, a characterization that challenges the received idea that any valid argument whose conclusion is a statement in the form $\ulcorner \Phi \wedge \neg \Phi \urcorner$ is automatically a *reductio* of its premise set. By dialethic lights, *Philosophy's Most Difficult Problem* is even more difficult than it first appeared. For not only do we need a principled distinction between a sound proof of a surprise and a valid *reductio* of the implying premises, we also need a principled subdistinction of that same distinction for whose conclusions are contradictions. The semantics of **PR** provide that both "$\mathbf{R} \in \mathbf{R}$" and "$\mathbf{R} \notin \mathbf{R}$" come out as true *and* false, and not merely as false for every valuation, as in classical systems. So the dialethist will say that classical logic gives the wrong analysis of paradox, and that dialethic logic gets the concept of paradox right.

Dialethic logics in the manner of **PR** are of striking philosophical interest, needless to say. When adjoined to ordinary set theory they permit the comprehension schema full existential sway. Detonation is averted by a modest adjustment of the classical apparatus (I shall return to this point below), so why not rehabilitate intuitive set theory, paradox and all? Notice that if we say that the Russell set simply *cannot* exist we give the dialethist room to charge us with prejudice (and a rebuke from Robert Meyer), which is an invitation to stalemate. If, so as to repel the charge, we say that the Russell set fails to comply with the Law of Noncontradiction, the dialethist will reply, correctly, that it does not, since LNC holds in **PR**. This will not give the critic much comfort, since LNC is also false dialethically. But if we say that we are simply not prepared to admit paradoxical objects – to give them the benefit of our ontological commitment, so to speak – doing so can hardly serve as a basis on which to complain to a dialethist about his different ontic tastes. A point by Meyer already briefly noted in Chapter 3 comes again to mind. What, he asks, is our reason for thinking that everything whatever, everything of every kind, without the remotest chance of an exception, must be consistent? Why could not the susceptibility that sets have to inconsistency be regarded as a rather natural, though also vivid, mark of their extraordinariness?

Perhaps we can now see, if we did not earlier, that Meyer is posing questions that are not entirely easy to turn aside. That, it might be said, is part of the charm of dialethic versions of paraconsistent logic. It sticks the classicist with a question that he is almost guaranteed to mismanage (one thinks of the rather frantic exactions of Aristotle in Book Γ of the *Metaphysics*).

Dialethic logic offers itself as the liberator of intuitive set theory, as an analytical theory of how sets actually are, no thanks to us. The cost of the liberation appears to be slight – some paradoxical objects are admitted into the ontology of sets. At this juncture the necessity to reengage the analysis of entailment presses sharply. *Ex falso quodlibet* is intolerable in a system such as **PR**. The question is, by what means does it avoid its generation? It is an interesting technical (and historical) aspect of the affair that a small adjustment to a truth functional component of **PR** does the trick. Disjunctive Syllogism is not allowed to be universally valid.

DISJUNCTIVE SYLLOGISM LOST?

In so saying, are we not back where we have no good reason being, namely, in making hopeless runs at Disjunctive Syllogism? No. In **PR** Disjunctive Syllogism not only fails, but fails in a perfectly natural (one might even say, predictable) way, unlike its alleged failure in relevant systems such as R. It fails in **PR** in a natural and intuitive way if we take at all seriously the most distinctive and philosophically interesting feature of **PR**'s semantics; the paradoxical truth value {T,F}. In the proof of absolute inconsistency, let Φ be some *paradoxical* sentence. Then Φ gives its own negation, $\ulcorner \neg \Phi \urcorner$. So, with regards to

    1. Φ
    2. $\neg \Phi$
    3. $\neg \Phi \vee \psi$
    4. ψ

our valuation v takes Φ into {T,F} and $\ulcorner\neg\Phi\urcorner$ also into {T,F}. Disjunctive Syllogism is thereby disarmed. Φ in line 1 cannot semantically exclude $\ulcorner\neg\Phi\urcorner$ in line 3, and the late-medieval objection of the Cologne School is *now* given a context that exonerates it. What is more, DS is not truth-preserving for paradoxical inputs. With Φ and $\ulcorner\neg\Phi\urcorner$ is true (because true and false) and $\ulcorner\neg\Phi\urcorner$ is false (because true and false), but ψ is just false.

Except for the contraposition principle, this is **PR**'s only major departure from the classical transformation rules. It is said to be a small cost, a cost justified by the larger benefit of the repatriation of intuitive set theory. The departure cannot be complained of on any such grounds as that DS is self-evidently correct or that, as a law of logic, it cannot be trifled with. For we are now transacting the debate from the methodological perch of holism, the doctrine that no logical principle is immune from revision. What is more, the deviation from DS is itself slight.[5] DS is valid in **PR** except where its premises take the paradoxical truth value. Where DS fails in **PR** it fails for just the right cases and is left otherwise untouched, which is again to say that DS fails in **PR** *intuitively*.

The value of the rehabilitation of intuitive set theory, Russell set and all, will vary proportionately with the difficulties in working with a successor theory. For some idea of this, let us return to Quine. Quine's writings on set theory show no sympathy for the dialethist's complaint that standard systems are not natural to work with and are counterintuitive in a number of respects. For Quine, "[i]ntuition is not to be trusted here [i.e., in set theory]" and, in fact, "[c]ommon sense is bankrupt [in set theory], for it wound up in contradiction" (Quine, 1966/1995, p. 27). In his review in *The Journal of Symbolic Logic*, Henson suggests that the "surprising results concerning NF serve to emphasize the fact that there is no informal or intuitive idea of 'set' which leads to the axioms of NF" (Henson, 1975, p. 242).

Quine's main contributions to the philosophical investigation of sets apart from NF are presented in *Mathematical Logic* (ML) [1951] and *Set Theory and Its Logic* (STL) [1969]. I shall here confine my remarks to NF and STL. Henson sets our stage rather nicely, although it must be emphasized that he was not criticizing Quine at this point in his review. Between them, Quine and Henson are saying: (1) Intuitions in set theory have no legitimate place. Intuitions let us down in the past, for they led to contradiction. (2) The job of set theory is not to satisfy intuitions or even to handle things in a natural way; it is "to grapple with the paradoxes, whether by von Neumann's method of non-elements or by Russell's hierarchy of types, or by some other probably equally artificial device" (Quine, 1951a, p. 138). (3) The job of set theory is to furnish what beyond quantification theory is needed for the foundations of mathematics.

In doing set theory *blindly*, Quine is inviting disaster. And he gets it. The Burali-Forti paradox is derivable in ML (which is a natural enlargement of NF), an indication that something is wrong with NF (Rosser, 1942). There is. NF has no standard model; "no integration of the membership predicate (with the right identity relation) compatible with the axioms of NF could make well-orderings of both the less-than relation among ordinals and that among finite cardinals" (Wang, 1986, p. 640). The class N_n of natural numbers in ML is the intersection of all classes (not sets) containing 0 and closed under successorhood. This being so, if NF is consistent then the class N_n of ML is not a set. This is not a disaster, since the sethood of N_n is securable by further postulation; but it is disappointing nevertheless, since the postulate is *ad hoc* (not that the point would cut much ice with Quine).

I have not come close to proving that Quine's anti-intuitivism *led* him to these difficulties. Even so, the dialethist may insist that it had some role in them, and in this he agrees with many of Quine's critics. Nor have I said that Quine's anti-intuitivism is a dominant opinion among working set theorists. In fact, it is not, as we saw in the preceding chapter.

Doing set theory nowadays all but means working in ZF or some extension of it. The subject has become a sophisticated and beautiful part of mathematics. For many it is underlain by what is regarded as a highly intuitive concept, the iterative notion of set. This issues in the "generation" of sets rank by rank up into the transfinite. [D.M.] Martin calls it "the standard concept of set." (Ullian, 1986, p. 584)

The iterative theory is basically the simple theory of types extended to infinities of every increasing type or rank. To complicate matters, Quine has recently plumped for the iterated concept, pointing out that it makes the transition to ZF smoother, and remarking "[t]hus I do not see Russell's theory of types as dormant *commonsense* awakened. Still I do not see it as somewhat akin to that" (Quine, 1973, p. 121).

What then is to be made of the paraconsistentist's insistence that standard set theories are sufficiently unnatural and *ad hoc* to confer on the rehabil-itation of naive set theory a benefit sufficient to offset (and more) the cost of crimping Disjunctive Syllogism? As I have said, it depends on how badly off the standard accounts really are. On the one hand, Quine's views about the lack of intuitive concepts and of commonsense and of naturalness in set theory coheres with dialethic sentiments about ZF, never mind that Quine himself does not regret these things or find his work troubled by them. On the other hand, many contemporary logicians find that ZF *is* a natural set theory to work with and that it answers to an intuitive, in fact the settled, concept of set. Who is to say? The dialethist's intuitions not only tell him what it is to be a set but also that, of the two alternatives, only the second is thinkable. Like all analytically minded theorists, the dialethist reserves more of a role for counterintuitiveness than seems appropriate for it. He sees less of a gap between counterintuitiveness and counterexample than I think there actually is, as was the burden of the previous chapter to show. On the other hand, minimizing the difference is precisely what the method of analytic intutions requires. For if you do not count on your intuition that such and such consequence is unacceptable, that is, a counterexample, what, pray, is the point in the method of analytical intuitions? And it is a fundamental dialethic intuition that the very idea of set is not unintelligible, and therefore that things of this very conception are real. Beyond that, the dialethic theorist is free to make the accommodations that his intuitions require. He can be a holist about logic in a form that qualifies him as a fallibilist, and he can be an economist if he is ready to see as real those accommodations required to safeguard his fundamental intuitions.[6]

It seems that the dialethist has managed to make enough of a case to justify the following claims: Paraconsistent logicians, operating intuitively, holistically, and from a supplementary cost-benefit perspective, make a defense of dialethic systems such as **PR**, which has a perfectly cogent structure and that it would be folly to dismiss out of hand. Nevertheless, in as much as their argument turns on the cost that must be borne by someone working in ZF (say), the question of how much weight to give the argument is obscured by uncertainties about how much of a cost ZF exacts and of how fruitful it is to work in a rival paraconsistent set theory.

There is something about the dialethist's approach that makes possible some headway with the adjudication problem in the theory of entailment. A primary objective of classical systems was to offer an account analysis of the entailment relation. This, anyhow, is how a great many philosophers understood such systems. The principal objective of relevant systems such as R was to offer a better *analysis* of entailment, a theory that corrected what relevantists took to be conceptual deficiencies in the strict account. We have seen that this sort of head-to-head confrontation between the two accounts is extremely difficult to adjudicate. This was a rivalry pursued from the stalemate-prone precincts of analytic theories. Even when the adjudication attempted to find analytically adequate common ground on which a consensus might be constructed, the actual history of this contention reveals a chronic repositioning of give-and-take under the provenance of intractable intuitions. After a while it becomes hard to resist the thought that adjudication is inescapably mired in stalemate.

The present approach breathes new life into these contentions. Although **PR** absorbs most of the distinctive features of R, and thus gives a relevant theory of entailment, the entailment theory that it provides is a byproduct secured in the pursuit of a different, and larger, objective. **PR** has different and, some might say, more interesting fish to catch, and relevant entailment just happens to get caught in the net.

An important purpose of **PR** is to give what it takes to be a more faithful account of paradoxicality than is provided by rival systems; and it seeks to do this because of its aspiration to be the base logic in the rehabilitation of mathematics. **PR** thus places itself substantively in the service of intuitive set theory, and in so doing aims to give specific and concrete purport to the classical notion that logic is and ought to be the handmaiden of all the sciences, including mathematics. The service it offers to mathematics will carry an adjudicative edge only if mathematics is better off thus served, and if the service it promises to mathematics it actually delivers. These two conditions are interconnected in an interesting way. If dialethic set theory were to turn out to be the better set theory, easier to work with and fully equipped to discharge its standing obligations to number theory and the other mathematical sciences, this would indicate that both conditions had indeed been met. The adjudicational nod could non-question-beggingly be given to **PR**, including that part of **PR** that provides for relevant entailment; for as a *byproduct* of the larger accomplishment, relevant entailment might be considered a cost worth bearing.

We meet with dialethic set theory in its comparatively early days, and there is one, not by any means decisive, test that it cannot at present be expected to have met. Is it a set theory that the community of practitioners might be prepared to embrace as their own? As of now, it is too early to say. There are legions of set theorists for whom the cumulative hierarchy and ZF are just fine to work with. To the extent that ZF is held by practitioners to be the standard and intuitive contemporary treatment of sets, and so not in need of replacement, the mathematical jury will have to stay out on dialethic set theory and the logical jury will have to stay out on **PR**, the logical theory that sustains it.

The adjudication problem for **PR** and its rivals cannot yet pretend to a definitive resolution. Resolution need not be despaired of, since there are intelligent steps to take now. We should send dialethic set theory and its supporting logic to the ongoing mathematical research program and see how they fare there. There is no doubt that the onus falls more weightily on the paraconsistent mathematical logician, such are the dialethical burdens of his antiestablishmentarianism, but it is a burden he should be ready to assume and eager to discharge.

THE CURRENT STATE OF DIALETHIC SET THEORY

Much of the work on dialethic sets begins with a first-order language, supplemented by the membership predicate ε. There are axiom schemata for abstraction and extension. Priest calls such a system **N** (Priest et al., 1989, pp. 178–86). **N** is good for lots of things. The empty set and the universal sets are definable there, as are the usual operations on sets. **N** also yields characterizations of relations, functions, and injections. This is pretty much all the set theory that the ordinary mathematician needs, including the category theorist. One can also get a definition of infinite sets, von Neumann ordinals, limit ordinals, and cardinals, but it may be that Cantor's theorem cannot be proved. The status of inaccessible cardinals is also unclear in **N**.

The Russell set, **R**, can be proved, of course. However, it is not yet known whether or not this makes **N** trivial, that is, whether every sentence of **N** is provable. This is an embarrassment. Still, Brady has shown that if the logic of **N** were altered to include relevance conditions, then **N** would be provably non-trivial but, even so, it would remain open just how far contradictions spread in **N** (Brady, 1989). All that is proved so far is that they do not spread *everywhere*.

More can be said about **N** (see Priest et al., 1989; Brady, 1989 in the places cited. See also Mortensen, 1995, pp. 141–6). But this much is already clear. **N** presents the theorist with some interesting open questions. An essential question that it closes, as it must, is the derivability of **R**. Another question it leaves open – and it would have been better had it been possible to close it – is the absolute inconsistency of **N**. For all its interest, **N** is not trouble-free.

INTUITIVE SEMANTICS

The existence of paradox serves as the fulcrum for **PR**'s alleged strategic advantage. Paradox is not limited to set theory; it visits naive semantics as well. It is natural to ask whether the dialethist's strategic maneuver can be made to generalize for other theories in which paradox is unitarily derivable, and whether it will do better there than it has heretofore done in the theory of sets. If it turned out that it did generalize, its attractiveness would be considerably enhanced. We would then have some reason to forward **PR** as a general instrument for the recovery of theories in which paradox is provable and whose nonparadoxical reconstructions are less than satisfactory. An obvious test of the conjecture is whether a naive yet inconsistent semantics embedded in a dialethic logic is, all things considered, the better theory.

The Tarski sentence spells classical trouble for semantics. The paradox is rooted in semantic closure, that is, in the fulfillment of a set of conditions that, for ease of exposition, and following Priest, I shall now state for formulas of one free variable (Priest, 1987, pp. 31–2 and Ch. 5). We will say that a theory T fulfills the Tarski conditions if:

(1) For every formula Φ of T there exists a term of T, ϕ, such that ϕ names Φ.
(2) There is a satisfaction predicate, "Sat," and a formula, Sat(xy), such that each is an instantiation of the schema.
(3) Sat $(t, \phi) \rightarrow \Phi(vt)$ is a theorem, where t is any term, Φ is any formula in which the variable v is free, and $\Phi(v/t)$ is the formula in which each free occurrence of "v" is interchanged with "t." (We assume the "usual precautions concerning the binding of variables free in t") (Priest, 1987, p. 13).
(4) The transformation rule $\{\Phi \rightarrow \neg\Phi\} \vdash \Phi \wedge \neg\Phi$ is a valid rule in the base logic of T.

Any theory of fulfilling these Tarski conditions is inconsistent. To show this, put for Φ the formula *Sat(vv)*. Then schema (3) yields

$$Sat(\neg Sat(vv)\ \neg Sat(vv))\ \leftrightarrow\ \neg Sat(\neg Sat(vv)\ \neg Sat(vv)).$$

By application of transformation rule (4), the contradiction is expressly derivable. Here, too, the contradiction is got from an antecedent equivalence of a formula of T and its negation. The contradiction thus is got from an analysis of the unitary concept of Sat.

Although the Tarski Paradox is frequently derived from the analysis of the concept of truth, or more strictly, bivalent sentence, for technical reasons it is simpler to proceed from the schema for satisfaction. It is not difficult, with the aid of supplementary formal procedures, to define denotation and truth, and to prove for them variations of the present paradox. It is more to our purposes here to take note of the fact that language of theory T is a formal language. Tarski believed that natural languages, such as English or Dutch, also honor the Tarski conditions. Not everyone agrees. Whereas it has been said that Tarski has *shown* that "any natural language containing semantic terms and the possibility of self-reference may be expected to be inconsistent, and to produce . . . paradoxes" (Bartley, 1984, p. 219), some philosophers are of the view that "Tarski showed no such thing, whatever he may have thought. . . . [W]hether natural languages satisfy [the Tarski] conditions is another matter; most analysts conclude that they do not, and I agree" (Post, 1987, p. 265).

It matters deeply whether semantically closed natural languages satisfy the Tarski conditions. If their satisfaction were achieved only by semantically closed artificial languages, the offending paradox could be considered a nuisance entirely eligible for artificial repair. In contrast, one might hope that the semantics of a natural would not dance to purely artificial manipulation, that the semantics of a natural language would be as natural as possible. So it matters whether English and Dutch satisfy those conditions.

Let us say that a natural language L fulfills the Tarski conditions if and only if

(5) For every string of L, Φ, there is a noun phrase, φ, its name.
(6) There is a binary predicate of L, Sat, such that every sentence, instantiating the schema,

   Sat (t, Φ) iff Φ (t)

is true, where Φ is any string of L that becomes a sentence of L on insertion of the noun phrase t, and where the parentheses indicate such insertion.
(7) There exists a valid and truth-preserving argument schema:

   Φ iff it is not the case that Φ.

(8) Therefore, Φ and it is not the case that Φ (Priest, 1987, pp. 14–15).

It is easy to see that any natural language fulfilling these three new conditions will contain contradictions. But we have not yet shown that English, for
example, *does* satisfy these conditions.

Let us be clear about the issue at hand. If English does not fulfill the Tarski
conditions, then the semantical theory of English can quite reasonably be held
to a consistency requirement. If only artificial languages of a certain particular
design satisfy the Tarski conditions, this could be said to be a design flaw that
one could be expected to remedy as artificially as needs be. But if English does
satisfy these conditions, then a true contradiction is generated by the rules of
English itself, concerning which there is an enormous antecedent presumption
of *correctness*. True, a semanticist of classical bent may be drawn to the idea that
the semantic inconsistency of, as it were, regularly and naturally constituted
English shows that the theory of English is defective. If he means to act on such
a notion, the classical semanticist must do two things. He must non-question-
beggingly *identify* the defective components; that is, he must try to find some
theorems that he is able to discredit quite independently of their contribution
to the paradox. He must *replace* those theorems with paradox-resistant counterparts that are still recognizably rules of English. As long as the classical
semanticist undertakes to comply with a condition of empirical adequacy on
his theory of English, he does not have a free hand in blocking the paradox just
as he likes. Of course, the dialethist is persuaded that the classical semanticist
is bound to fail these requirements. If he is right, and if these seem sensible requirements, the dialethist again has a fulcrum for his dialethic maneuver: The
correct description of English includes the assertion that English contains *true*
contradictions and, that being so, there is no realistic alternative but to under-
take the semantics of English intuitively and to embed the inconsistency in a
base logic that enables the theorist to accept the contradiction and to contain it.

It is important to take note of the ways in which semanticists of classical
bent have tried to show either that English does not fulfill the Tarski conditions
or, in case it does, how its paradox-averting reform might be achieved. It is
not at all clear that the regulation of English in such ways as keep paradox at
bay will manage to leave key questions unbegged, or to avoid the mangling
of English itself or, finally, to avoid the embrace of theoretical constraints of
such counterintuitiveness that their employment would lay not a jot of a claim
on acceptance, save for their discouragement of paradox. At this juncture, the
paraconsistentist has an invitation worth considering. Why, he says, not try
semantics intuitively but with a dialethic base logic? We might get to like it.
After all, he will add, the *classical* remedies are not very impressive.[7] The main
classical remedies are these.

(1) Suspend the law of Bivalence and declare that the Tarski sentence is
    neither true nor false. This will not take care of the Strengthened Liar,
    got from the Liar by the systematic replacements of occurrences of
    "false" with occurrences of "not true."

(2) Banish sentential self-reference. By this remedy "This sentence is true," no less than "This sentence is not true," would be declared grammatically illicit, and so, too, "What I am now saying is not difficult for a native speaker of English to pronounce." Also ruled out is the celebrated intervention, several minutes into the program of, "This news is coming to you in the World Service of the BBC," to say nothing of Gödel's incompleteness theorems. It is difficult to see that blanket banishment of self-reference can be justified.

(3) Banish *semantic* self-reference, that is, self-reference involving ascription of the truth-predicate, and replace with a hierarchical rule that permits semantic attribution to sentences of next lowest rank. Thus the predicate "true" turns out to be typically ambiguous. There is *no* empirical evidence that the English predicate "true" is typically ambiguous.[8] Moreover, the Strengthened Liar is extendable to sentences endorsed by the hierarchical rule. Sentences such as "This sentence is not true at its rank," will follow directly from the hierarchical prohibition itself (Priest, 1987, p. 24).

Let us understand by the term "intuitive semantics" a theory of truth having at least the following three features. First, the theory preserves deep-seated intuitive beliefs about truth. Second, the theory is part of the very language whose truth predicate it seeks to characterize. Since fulfillment of the first two conditions yields the Tarski paradox, a third condition proposes that the paradox be dissolved or disarmed by the theory in ways compatible with adherence to conditions one and two.

It is commonly believed – by Tarski and legions of those who came after – that no theory could fulfill the three conditions simultaneously. If this is right, intuitive semantics is impossible. It can only be expected, then, that post-Tarskian semantics will be nonstandard in one way or another. The principal difference between recent theories and Tarski's own is that, whereas Tarski was prepared to abandon intuitiveness quite radically, other theories have tried to stay closer to home. They have sought to make their treatment of the paradox square as much as possible with the spirit of intuitive semantics. In particular, they have tried to keep the logic of their accounts as classical as possible.

Intuitive semantics turns on this fundamental principle:

FP: For all Φ, if Φ is a well-formed formula of a language L, then so too is $\psi(\Phi)$, where ψ is a (one-place) semantic expression of L.

Thus, *FP* provides for the well-formedness of the following expressions of English:

"The cat is on the mat" is true.

"The dog is in the manger" is false.

"Peter Strawson admires the present king of France" is not true.

"I can't utter a sentence of English" is false.

TYPICAL AMBIGUITY

At the heart of Tarksi's hierarchical approach to semantics (which mimics the type theoretic approach to sets) is the rejection of *FP*. In fact, the similarities with type theory suggest that a natural way of gaining *entré* to a Tarskian hierarchical semantics is by way of a discussion of set theory. In so doing, it is helpful to have at hand two of our earlier quotations from Quine and a further remark from Ullian. Here again is Quine:

> But the theory of types has unnatural and inconvenient consequences. ... The universal class – (as in '5') gives way to an infinite series of quasi-universal classes, one for each type. The negation -*x* ceases to comprise all nonmembers of *x*, and comes to comprise only those nonmembers of *x* which are the next lower in type than *x*. Even the null class Λ gives way to an infinite series of null classes. The Boolean class algebra no longer applies to classes in general, but is reproduced rather within each type. The same is true of the calculus of relations. Even arithmetic ... proves to be subject to the same reduplication: Thus the numbers cease to be unique. (Quine, 1953b, pp. 91–2)

and

> When we ... pursue general set theory, we must grapple with the paradoxes, whether by von Neumann's method of non-elements or by Russell's hierarchy of types or by some other equally artificial device. (Quine, 1951a, p. 138)

And now Ullian:

> Quine's purpose of STL [*Set Theory and Its Logic*] is [not] to ride any intuitive concept. It was his purpose to stand back and see what, beyond quantificational logic, is needed for a foundation of mathematics. For Quine this means what formal apparatus is needed, not what "concepts" are needed. (1986, p. 584)

In Russell's type-theoretic approach, the elements of the domain of discourse are stratified within an infinite hierarchy of disjoint types. Sets are constructed only of elements of next lower type. If we took individuals to be of lowest type then the set of individuals would be of next higher type, sets of sets of individuals of still next higher type, and so on.

It is easy to see that Russell's set is banished for its violation of type constraints. Unavoidable self-membership fails the principle that a set contain as members only entities of the type next lower than it. It is also easy to see that on this account the predicate "is a set" is typically ambiguous, that is, ambiguous as between "*x* is a set of type 1," "*x* is a set of type 2," and so on without end. Sethood, then, is infinitely ambiguous in type theory. Does it matter that type theory runs roughshod over our intuition that "*x* is a set" is *not* infinitely ambiguous in English? Not if, following Quine, we think that it is not the business of set theory to answer to intuitions and that the overarching adequacy condition on a theory of set is that it give a consistent foundation for mathematics.

How does this situation compare with Tarski's hierarchical semantics for L? Tarski proposes that the sentences of a language L be arrayed over an infinity of *ranks* (cf. Russell's types). Sentences of rank 1 contain no semantic-predicates, sentences of rank 2 ascribe semantic predicates to sentences of rank 1, sentences of rank 3 ascribe semantic-predicates to those of rank 2, and so on. On this approach, the intuitive principle FP is suspended and the Tarski sentence is disallowed.

If type theory makes the predicate "*x* is a set" infinitely ambiguous, then a hierarchical semantics of the Tarski sort makes the English predicate "is true" also infinitely ambiguous. And just as the radical ambiguity of sethood offended our intuitions, the radical ambiguity of truth does so as well, perhaps more so. Does this matter?

A distinction between intrinsic and instrumental functions bears on our questions. Set theory, if it has an intrinsic function at all, is required to tell us faithfully and accurately what *sets* are and how they behave. But it is also clear that set theory has an instrumental function, namely, to provide for mathematics. The distinction allows reexpression of Quine's conception of a good theory of sets.

I   The principal job of set theory is to perform its instrumental functions in the foundations of mathematics.

II Thanks to the debacle of the Russell paradox, it is not obvious that set theory has *any* intrinsic function.

Is this also the way with semantics, with the theory of truth for a natural language? It hardly seems so. A theory of truth seems not to possess an instrumental function in anything like the way that set theory does. Its job is to tell a story – intrinsically – about truth, and to do so in a way that is not ruined by the Tarski inconsistency. One way in which semantics avoids ruination is Tarski's way. But now the truth predicate is infinitely ambiguous and, it could be said, truth has been made unrecognizable. We have made it *nonexistent*. If so, the theory may be thought to have failed in the exercise of intrinsic function. This is precisely what the dialethist thinks *has* happened.

KGH THEORIES

Whatever we make of the Tarski hierarchal approach, lots of people do not like it. It is less intuitive, they say, than it needs to be. So let us turn to what earlier I called the KGH approach. "K" is for Kripke (1975 or 1984 references here are to reprints), "H" is for Herzberger (1982), and "G" is for Gupta (1982). Since the papers of Herzberger and Gupta share more of a common position about naive semantics, it is perhaps best that we detach the "K" and deal with it separately.[9] We begin with the *Fixed Point Method*.

Let L be a language and I an interpretation of it with domain D. Predicates of L are interpreted as subsets of the Cartesian product of D, D × D; that is, predicates denote n-ary relations on D. Suppose that we wish to enrich L by adding the n-ary predicate "F^n." Let the interpretation of "F^n" be the ordered pair $\langle R_1, R_2 \rangle$ of n-ary relations on D. R_1 is the extension of "F^n" and R_2 its counterextension.

Axioms are now added to regulate the behavior of "F^n" in the enlargement of L. This involves taking an existing model and extending it with such axioms. The axioms must be in conditional or biconditional form. They admit of, but do not require, quantifiers. The model for the new axioms is specified by transfinite induction. In the present example, axioms are biconditionals $\lceil \Phi$ iff $\psi \rceil$ where Φ is a sentence "$F^n(a_1, a_2, \ldots, a_n)$" and ψ is a formula in which some of the "a_i" occur. The model enlarges incrementally. This is done by adjusting the interpretation of "F" in such a way that instances of Φ take the same truth value, T or F, as ψ took at the prior stage. At each succeeding stage new sentences emerge as true instantiations of ψ, and this requires a corresponding expansion of instances of Φ to preserve the truth of the axiom.

In due course, "through the magic of infinity" (Mortensen, 1995, p. 136), a stage will be reached at which, for each instance of ψ true or false on the interpretation, Φ will already have been "fixed" in the same truth value. These junctures are *fixed points*.

Let $I_o = I$, the interpretation with which the stage-process begins. The method of fixed points spawns a series of interpretations $I_o, I_1, I_2, \ldots$. In successive interpretations of I_o, the ordered pair $\langle R_1, R_2 \rangle$ which is the interpretation of "F^n" is altered by extension of R_1 and of R_2. This succession of alterations can be thought of as governed by a rule, $X, I_{n+1} = X(I_n)$. Then a fixed point of X is an interpretation $I_\lambda : X(I_\lambda) = I_\lambda$ which gives a model of our enriched language.

As applied to Kripke's theory[10] of truth, "Tr" in our new predicate and its governing axiom is the biconditional

$$T_r(\Phi) \equiv \Phi.$$

"T_r" is now interpreted by the ordered pair $\langle R_1, R_2 \rangle$ of one-place relations on D. Under I_o "T_r" is given the empty interpretation $\langle \theta, \theta \rangle$. Successive interpretations are constructed from a function ψ until a fixed point is achieved. X adds new sentences to R_1 and R_2, the extension and counterextension of "T_r." Then $X(\langle R_1, R_2 \rangle) = \langle R1', R_2' \rangle$. Informally, if R_1 contains the sentence "Sally is wise," then R_1' contains "'Sally is wise' is true." X permits the acceptance or rejection of "'Sally is wise' is true" just when it is permitted to accept or reject "Sally is wise." At any given stage save one, "T_r" is only partially interpreted by its corresponding couple $\langle R_1*, R_2* \rangle$. The exception is I_λ, where λ is a limit ordinal. In that case, $R_{i,\lambda} = V_\alpha < \lambda R_i, \propto (i = 1,2)$. Thus, as those who know her will attest, "'Sally is wise' is true" will on the first application of X come out true whereas "'Buffalo is the capital of New York' is true" will come out false. At that stage neither "'Sally is wise' is true" is true' nor "'Buffalo is the capital of New York' is true" is true' will have an interpretation. However, on its next application, X provides that the former of these is true and the latter is false. Hence, at each stage prior to the fixed point, X gives only a partial understanding of sentences predicating the word "true." However, at a fixed point, application of X tells us no more about the extension and counterextension of "T_r" than was specified at the prior application. At this point, there is no more about "T_r" that can or need be told. "At such a point we will say that the model is saturated and the language contains its own truth predicate" (Gilmore, 1967/1974, p. 139).

As we can see, X is a *monotonic* operator. If it evaluates a "true"-containing sentence as true at a stage, all successive applications are likewise true. This requires that some sentences never require a truth value, notably the Liar sentence. Let α be the smallest ordinal for which I_α makes that sentence true (or false). Application of X gives $I_{\alpha+1}$, on which the sentence is false (or true), and monotonicity is violated at the cost of incompleteness.[11]

In the Gupta-Herzberger treatment we find a new characterization of paradox, and a naive semantic reaction to it. Jointly, they are intended to yield a characterization of the paradox, which makes possible a treatment of it that honors *FP* and as many as possible of our deep-seated intuitive beliefs about truth (Herzberger calls these our "primordial beliefs about the concept of truth" [1984, p. 133]). In particular, the GH theory attempts to keep nonstandard maneuvers to a minimum.

In some theories, paradoxical statements are supplied with a many-valued, or a supervaluational, or a fixed point semantics, in which the paradoxes emerge as neither true nor false – which is precisely what they are not intuitively speaking. Or they are given valuations under which they are both true and false without further ado and, thus "glutted," are surrendered to the tender mercies of dialethic logic.

In the GH treatment, such nonstandardness is avoided. Its models are ordinary ones and its valuations are classically two-valued. Paradoxical statements come out neither as true and false *tout court*, nor as truth-valueless.[12] Instead they are represented as statements that are *systematically unstable* in their valuations. A paradox is thus a statement that is true on one valuational level, false on a higher level, true again on a level still higher, and so on. We might imagine a semantic digital scale, displaying "1s" for truth and "0s" for falsehood. Now if a paradox were weighed on such a scale, the instrument would display "1s" and "0s" alternately without stopping. The scale would never come to semantic rest, and this would mean that the paradoxical sentence is "unstable."

Such a semantics is said to be *naturally* hierarchical or semihierarchical, so it can "accommodate this kind of semantic instability without contradiction or incoherence. It makes available a novel reconstruction of what some people have felt to be the "inconsistency of natural languages" (Herzberger, 1982, p. 132). Technically what makes this approach work is a modification of Kripke's theory of truth so as to allow classical valuations. As we saw, Kripke's treatment is monotonic and inductive; the GH approach is nonmonotonic and semiinductive.

Let us suppose that in taking things in the GH way we start out with a classical model, thus with a truth predicate whose extension and counterextension close all gaps. What would our semantic construction look like? At finite levels, valuations would be inconsistent from one level to another. The Simple Liar (e.g., "I am false") would fetch truth at one level, falsity at the next, truth at the still next, and so on. Ordinary statements of finite rank (these are said to be "grounded," that is, they have a truth value in the smallest fixed point) would sooner or later sort themselves out and fetch fixed and stable values.[13] In Kripke's adaptation, $\{X_1, X_2\}$ are the extension and counterextension respectively of a truth predicate T_x. Now where we want to interpret T_x, we must have it that for any sentence A, "A satisfies (falsifies) T_x iff A is true (false) by the evaluation rules for L. When this happens the pair $\{X_1, X_2\}$ is called a fixed point" (Kripke, 1975, p. 67). What is particularly interesting about a GH theory is that it seriously puts to the test the hypothesis that on a sufficiently high level *all* grounded statements would be stable, thus "leaving only pockets of instability around paradoxical statements."

Since Kripke's original theory is monotonic, it cannot carry out its levels-constructions even to the first infinite level. Again, montonicity here means that "if the interpretation of the [truth predicate] T_x is extended by giving it a definite truth value for cases that were previously undefined, no truth value previously established changes or becomes undefined; at most, certain previously undefined truth values become defined" (Kripke, 1975, p. 68). Kripke's construction places unions of all previous stages at each limit stage, "so that at stage ω it would put the Liar statement in the extension of T and it would also put that same statement in the counter extension of T. But then the Liar statement would get evaluated as true and false at stage $\omega\ldots$, which is absurd (Herzberger, 1982, p. 136; cf. Kripke, 1975, p. 64). Herzberger thinks that things can be put right by changing the "union rule," which determines limit stages.

A useful summary of Kripke's and also of Martin's and Woodruff's theories is provided by Martin himself in the Introduction to his collection of essays. As we have said, this is not quite GH, which employs "variants of the inductive construction just described" and which is "more classical in spirit...Herzberger...proposes a "semi-inductive" characterization of the truth-predicate to replace the inductive characterization...[so as to] study the behavior of the paradoxes within exclusively classical, two-valued models. Gupta's investigation rejects the inductive characterization of the truth-predicate also, and suggests...a "revision" rule of truth, which builds up successively better *approximations* to represent truth" (Martin, 1984, pp. 7–8).

The part of the idea that is common to both papers can be put roughly as follows. At the beginning of the construction, the extension and counterextension of the truth predicate are considered empty so that all attributions and denials of truth are without truth value. Those sentences that do not contain the truth-predicate can still be evaluated *intuitively*, some as true, some as false. These constitute the extension and counterextension, respectively, of the truth-predicate at the next higher level, from which it follows that more sentences are evaluated as true (false) at this level. The process thus begun is continued into the transfinite. The resulting extensions (counterextensions) are swept together into a "fixed point" – a point at which all sentences that will belong to the extension (counterextension) of "true" already do.

Virtually everything the dialethist has said against classical post-Paradox set theory he will repeat against both the Tarskian and KGH approaches to semantics (although he will also appreciate the attempt by the latter three to preserve a concept of truth that more closely resembles the intuitive notion abandoned by Tarski). Against Tarski there is a twofold resistance. The Liar does *not* destroy the very idea of truth, that is, truth in its intuitive conception. Correspondingly, to visit transfinite ambiguity upon a contrived successor notion is too much *ad hoc*ery for comfort. The same reservations apply to GH, though moderated in recognition of their (modest) relaxation of Tarskian austerities. To these could be added the objection that the GH analysis, in which a sentence is paradoxical when it is true and false but only in different "respects," is an evasion, and the further complaint that approaches such as Kripke's, in which consistency is purchased at the cost of completeness, are the duals of accounts just as interesting and worthwhile in which this cost-benefit profile is reversed. And once again the dialethist has an invitation to send out:

If you would rather avoid the *ad hoc*ness and counterintuitiveness of the GH theory, why not give a dialethically intuitive semantics a try?

The argument that dialethic logic is vindicated by its role in the rehabilitation of intuitive set theory is, let me say again, far from conclusive. It may turn out, for example, that dialethic set theory, in the manner of Brady or Mortensen, is too complex and difficult to work with, and that on balance, the benefits of staying with ZF outweigh the advantages of dialethically intuitive theories.[14]

It is the same way with semantics. Dialethically intuitive semantics is still in its infancy and has yet to establish a track record of sustained achievement. There is little doubt that it will have difficulty in establishing its *bona fides* (see below). One of those difficulties pertains to the proof that natural languages fulfill the Tarski conditions, by virtue of which they are made inconsistent. That proof turns on the acceptability of Tarski's Convention T that, though in the passages above is represented as a condition on satisfaction, can also without technical loss be construed as a condition on truth. Thus, by Convention T we have it that

$$\Phi \text{ is true iff } \psi$$

where ψ is a placeholder for a sentence and Φ a placeholder for a quotation-name or structural description of the same sentence.

It might be thought that Convention T gives nothing but comfort, since what could be more obviously sound than sentences like "'Snow is white' is true iff snow is white"? Nevertheless, some philosophers have found Convention T to be troublesome. James Cargile proposes that since the argument for the inconsistency of English pivots on Convention T, we should take the conclusion that establishes the inconsistency as a *reductio ad absurdum* of Convention T (shades of the Great Delta Debate discussed later in this chapter) (Cargile, 1979, p. 252). Hintikka, on the other hand, thinks that Convention T is falsified by a counterexample (Hintikka, 1975, p. 208). Hintikka offers the following case (I here retain his numbering):

(3) "Any corporal can become a general" is true if any corporal can become a general which can be rewritten as
(3*) If any corporal can become a general, the sentence "Any corporal can become a general" is true.

In (3), "The first 'any' (the "any" in the quotation), obviously has the force of a universal quantifier, whereas the second "any" (the "any" outside the quotes) operates as an existential quantifier" (Hintikka, 1975, p. 208). It might seem to the reader that the proffered counterexample could be undone by pleading that quantificational ambiguity suffices to show that (3) does not instantiate (one half of) Convention T. Hintikka sees this coming and deftly resists the complaint and others as well.

It is not to our present purpose to attempt to settle the disagreement between Hintikka and his critics. Both Cargile and Hintikka are disposed to think that there is something defective in proofs of the inconsistency of truth in natural languages. But our present task is to investigate the dialethic option in semantics, an option that presupposes the soundness of those proofs. Even so, it is far from obvious that a dialethically naive semantics is trouble-free. I repeat here what was earlier said about dialethic set theory. Why not give them both a chance to win a cost-benefit contest with their rivals? And why not place a moratorium on wrangles about entailment, pending the outcome of these larger cost-benefit determinations? If it is thought that this is not much progress with the adjudication problem in the theory of entailment, I can only say that it is some progress, and that some is better than none. At a minimum, we now have some idea of what could count as a solution to the adjudication problem; and that, anyhow, is something.

THE CURRENT STATE OF DIALETHIC SEMANTICS

Priest has produced a formal semantics, call it **Sem**, which honors the claim that English is a semantically closed language that fulfills Tarski's closure conditions (Priest, 1987, pp. 157–87). **Sem** also respects Convention **T**:

For every closed formula Φ, $\mathbf{T}\,\phi \leftrightarrow \Phi$ is provable.

Sem is rather complex. For one thing, it requires a fair number of axioms, twenty-six in all. But complexity is not really a primary concern, since it can always be said that this is just how complex semantics *is*. It is striking, however, that it is not known whether **Sem** is inconsistent: What makes this especially surprising is the proof noted earlier, that any theory containing its own satisfaction predicate *is* inconsistent. Is not the agnosticism of **Sem** with regard to its own inconsistency inconsistent with this fact? It is not, as it happens, since the earlier inconsistency result used a satisfaction relation restricted to one free variable, whereas satisfaction in **Sem** is more general and, moreover, is needed to make **Sem** work. Perhaps, then, **Sem** is consistent. Priest suggests that if so, it is an accident having to do with technical constraints on how to formulate the satisfaction predicate that are nonetheless inessential to the story that **Sem** tells about its primary semantic targets. In any event, there are also ways – technical ways – to make **Sem** inconsistent, say by feeding into the axioms of **Sem** some additional Peano arithmetic (Priest, 1987, p. 164).

Nevertheless, on the point in question, **Sem** is asymmetrical with **N**, for there is no doubt that **N** is inconsistent and that it can be shown to be so without merely technical supplementation. On the other central paraconsistent question, **Sem** and **N** are at parity. Even when **Sem** is made inconsistent by technical artifice, it is not known that it is not trivial. Of course, *ex falso* fails in **Sem**, as does the Curry paradox:

$$\{\Phi \rightarrow (\Phi \rightarrow \psi)\} \vdash \Phi \rightarrow \psi.$$

But it is not known that other trivializers are not derivable in **Sem**. "The techniques for proving non-triviality are still in their infancy," as Priest says, "and I do not yet know of any that can be shown to solve the problem for [theories such as **Sem**]" (Priest, 1987, p. 165).

Something rather interesting does come out of **Sem**, all the same. In a celebrated paper of 1965, "Theories of Meaning in Learnable Languages," Donald Davidson argued that a natural language is unlearnable by beings like us unless we can understand an open and potentially infinite set of sentences on the basis of finite information (Davidson, 1965). Theories of meaning must account for this ability. Davidson thinks that this *is* explained by the fact that the semantics of such languages are recursively enumerable. But Priest defends a claim that for languages of appropriate richness, rich enough (approximately) for arithmetic, consistency is incompatible with recursive enumerability (Priest, 1987, pp. 170–1). Such languages thus cannot be learned, if Davidson is right. This is a setback. Unless **Sem** is inconsistent.

It is evident that, like **N**, **Sem** has its work cut out for it, and then some. As things stand, neither theory has been able to make good on the fundamental dialethic insight that there are provably inconsistent theories that are provably nontrivial. Even so, Priest has furnished a complete and sound semantics for the extensional connectives (Priest, 1987, Ch. 5). It is a strikingly familiar semantics. "But for one change, dropping the assumption that truth-value is

unique, [it] is exactly classical. Even the sets of logical truths are the same."
This is because "classical logic is just the special case where no parameter
(and hence no formula) takes the dialethic value $\{0,1\}$" (Priest, 1987, p. 96).
Dialethic logic thus subsumes classical logic.

The system **PR** is a paraconsistent system that is also relevant, and the object
language of which is intensional. Priest's basic system in *In Contradiction* is a
good deal less so. If we dub this latter account **PP**, then unlike **PR**, **PP** is not a
relevant logic and its object language is extensional. The difference between
PR and **PP** lies in their respective semantics. The desirability of **PP** is obvious.
It conflicts with classical logic on fewer issues than does **PR**. It is a logic less
easily dismissed by a classicist than **PR**, and this matters for the question of
stalemate.[15]

Quantifiers are handled in **PP** somewhat straightforwardly, although with
an occasional bump en route. A restricted universal quantifier can no longer
be construed as a quantifier plus a truth function, this because of "the loss of
inferential force" of material implication. If we turned to an intensional connective with a view to restoring "inferential force," some of the dependencies
between restricted universal and restricted existential quantification are lost.
So, says Priest, a new approach to restricted quantification may be needed.
This alone may give the wavering classicist pause, but on no account can he
merely say that the dissident does not know what he is talking about.

GAPS AND GLUTS

It might fairly be said that theories K and GH disqualify themselves from any
serious claim upon intuitiveness except in the minimal sense of fulfilling condition *FP*. If Tarski's hierarchy has made truth unrecognizable, so, too, have
the "quasi-hierarchies" of Kripke, Gupta, and Herzberger. Primordial beliefs
about the concept of truth have been trifled with too much. It is perhaps ironic
that a GH-theorist would find fault with Kripke's non-truth-valued treatment,
nonstandard though it may be. The idea of truth value gaps (now detached
from the rest of Kripke's theory) is certainly not the offense to primordial
semantic beliefs that the infinitely wise typical ambiguity of "true" is, as wit-
nesses the alleged truth valuelessness of "John's children have the measles"
when John has no children. So why not take the gappy approach seriously; and
why not ask of the dialethist whether he can demonstrate the superiority of
gluttony?

Parsons has suggested that a truth valueless semantics fares no worse than
paradoxical semantics but, also no better. In hockey parlance, they are tied at
the end of the third period, and there is no provision for overtime. This, if true,
is already a major victory for dialethism, for it holds a seriously contending
theory to a *tie*. It is not necessary to review Parson's argument in detail, though
it will richly repay serious study for someone interested in following it up. I
am more concerned with the metatheoretical question, "What if Parsons is

right?" Before turning to this, I shall reproduce one of Parsons' arguments in order to give something of the flavor of his overall thesis.

Suppose, on meeting a gapper, we were to ask him to state his position on the Liar sentence **T**. He says

(1) **T** is not true and **T** is not false.

But, by his own lights, (1) itself is not true and not false, and so he has not managed to state his own position truly. How so? Suppose that **T** is "¬T*a*" where "*a*" denotes that very sentence. If we assert that the liar sentence is not true, what we say has the form of

(2) ¬T*b*

where "*b*" is the liar sentence, which is also named by "a."
Therefore,

(3) *a* = *b*

But (2) and (3) jointly imply

¬Ta

that is the Liar itself *and* that has been derived from two *true* premises. So the "standard gapper solution is to consider [(2)] as on par with the liar sentence itself, and to deny that either of these have [*sic*] a truth-value" (Parsons, 1990, p. 345).

Priest notices this and presses the gapper rather hard. He is right to do so, but Priest has the same problem. He cannot express his disagreement with the gapper, and so cannot express a disagreement that lies at the heart of the formulation of his own positive position. Suppose that someone says that A and a paraconsistentist within earshot replies "¬A." This alone does not constitute disagreement, since the paraconsistentist might think that A is true as well. How can he express disagreement? He might say "A is not true" but, by his own lights, this may be consistent with A's truth. Perhaps he could assert "¬A" and add that A is not paradoxical, that is, that A is not both true and false. But if A is the Liar it is paradoxical and this statement about it is true. So Priest lacks the "means within his symbolism of expressing not being a dialethist" (Parsons, 1990, pp. 345–6, n. 10; cf. Priest, 1987, p. 139).

"In summary," says Parsons, "both gappers and glutters are apparently unable to assert certain of their own views correctly if their theories are correct" (Parsons, 1990, p. 346). Parsons thinks that relief from this problem can be got by carefully *drawing* a sentence and asserting its negation. But if correct, this is a remedy no less for the glutter than for the gapper. So, on the point at issue, the gapper and the glutter remain at parity. The tie has not been broken.[16]

Parsons' parity-thesis is extremely interesting, and a boost for the dialethist. But since we cannot win on the strength of a tie, two related questions require attention. One is whether the tie can be broken on cost-benefit grounds or whether we are landed back in a stalemate. It also raises the question of whether the stalemate might yield to the following cost-benefit argument and bestow the laurel on the gapper. Suppose that it is true that, underdetermined by the data, resorting to gaps or to gluts are equally good but incompatible semantic reactions. Of course, gluttony requires a nonclassical logic, and gappiness, too; but the logic needed for the one is a good deal more nonstandard than the logic needed for the other. Gluttony demands the rejection of *ex falso*, but thanks to supervaluations, gappiness does not.[17] The non-truth-valued semantics can make do with classical entailments and dialethic semantics cannot. This does not settle the matter; to think that it did would beg the question against the dialethist. It is bad strategic news for the dialethist all the same, for on the point at issue he must now repose his confidence in nonclassical entailment on its indispensability for naive *set theory*. This means that his case for the superiority of dialethic semantics defers to his case for the superiority of dialethic set theory and that, it may be thought, is too much justification-weight for a set theory to bear.

ONE TRUTH VALUE ONLY?

There is a recently published argument designed to show that strong paraconsistency is almost certainly hopelessly wrong. In its simplest terms, it is the following:

  i Dialethism collapses if expressible only in the form, "For some propositions Φ, Φ is true and Φ is not true" (as opposed to "For some propositions Φ, Φ is true and Φ is false").
  ii There is no good reason to think that the second characterization in (i) is ever available to the dialethist.
  iii Therefore, dialethism collapses.

Such is the view of R. M. Sainsbury. It is not readily apparent why Sainsbury thinks that the dialethist cannot allow his position to be described as the acknowledgment, for some Φ, that Φ is both true and false. Much of the charm of the glutter's enterprise is that it is precisely contradictions that he does affirm. If Sainsbury's point were simply that the glutters should not do that, he would beg the question rather flagrantly. In any event, Priest embraces a rule that takes him from untruth to falsity; and if there were something to be said for proclaiming dialethism in the manner of (i), he could always proclaim it in the manner of the derived

  (ii) Φ is true and false.

Sainsbury implies that this will not do. Falsehood is a theoretical epicycle, not needed for semantics or for theories of rational belief (Sainsbury, 1988, pp. 143–4).

If this is right, and there is no decisive or even adequate reason for introducing the truth value falsity into the vocabularies of our theories, then the dialethist must revert to (i) to say what he thinks as a dialethist. Yet, Sainsbury says that this is just what no one *should* say, not even a would-be glutter. Why does he think this? The answer lies in the reasons given for the rejection of falsity as theoretically superfluous. "All our semantics" can make do with just one truth-value or semantic property, and that is the property of truth.

Assent to a proposition is a means of recognizing the presence of this property, dissent from a proposition is recognition of its absence, and abstention from a proposition recognizes neither its presence nor its absence. Negation takes truth into its *absence* and *vice versa*; conjunction takes the absence of truth into the absence of truth; and so on. What we cannot have, says Sainsbury, is the joint presence and absence of truth. Thus is precluded the intended semantic role for (i), which is the attempted recognition of the fusion, so to speak, of the presence and absence of truth. Supposing for now that this is impossible, how would the reinstatement of falsity change things? Its difference lies precisely in the difference between the absence of the property and the presence of a contradictory property.

In the language of presence and absence, if truth and falsity are two genuine semantic properties, then they are susceptible to presence, to absence, and to neither. The glutter also will say that they are susceptible to joint presence. Sainsbury must allow for this possibility, since to deny it is to beg the question against the dialethist, who can now be seen as arguing as follows. Given that truth and falsity are two semantic properties each eligible for instantiation by propositions, then the derivations of the Russell and Tarski paradoxes count as some reason for saying that these properties *can* co-present themselves.

But, if *falsity* does not exist, there is no question of its co-presence with truth, and neither the Russell derivation nor the Tarski derivation provide any support for the idea that truth itself might be entirely present and entirely absent. If we can rid ourselves of falsity, we can disarm dialethism in one fell swoop.

I demur from this. The Tarski and Russell derivations go through swimmingly with "not true" substituting for "false." This, among other things, is the moral of the Strengthened Liar. Sainsbury has an answer for this. The Tarski sentence **T** cannot recognize the presence of the absence of its own truth. Absences are not, as with the absence of Pierre noticed by Sartre's waiter, also as presences. Absences do not present themselves; they are not presences.

If so, then Sainsbury can say that "there is no room for dialethism, since there is no room for a distinction between nontruth and falsity in the domain of propositions." His challenge to the dialethist is to justify the distinction on grounds other than the dialethists' own need for it. He believes that the challenge is unlikely to be met.

I think that it can be met. Consider

(3) The present King of France is bald

concerning which we might wish to say

(4) Proposition (3) is neither true nor false.

Falsity being unavailable, we might try instead

(5) Proposition (3) is neither true nor not true

for how else are we going to express our abstention with regard to (3), our recognition of neither the presence or absence of truth in it? Now (5) yields

(6) Truth is both present and absent in proposition (3)

which is disallowed by Sainsbury's own theory. That is, the abstention-elucidating

(7) Truth is neither present nor absent in (3)

cannot be asserted truly. But (7), and (6) too, is a statement of precisely what we *did* recognize in (3) when we abstained from it.

Perhaps repairs can be made. We might change the conditions on abstention. Instead of saying what cannot be said truly if the theory is correct, that in abstaining from Φ we recognize neither the presence nor absence of truth, why not say

(8) Φ is true and $\ulcorner \neg\Phi \urcorner$ is not true

thus mimicking "The present King of France is not bald and he is not not-bald either"? But (8) does not do this well enough; in fact, it falls foul of the negation rule which takes nontruth into truth.

What is wanted, of course, is a distinction between external and internal negation. Thus, proposition (3) is not true but then neither is

(3*) The present King of France is not not bald either.

The distinction cannot be made to work without a suitably orthogonal second truth value. The untruth of (3*) coupled with the untruth of (3) requires a verdict of

(9) Proposition (3) is neither true nor false

So much the worse for the distinction between external and internal negation, so Sainsbury might say. He would do well to ponder, for his own theory sinks him without it.

RATIONAL CRITICIZABILITY

In Chapter 3, we considered, and rejected, the possibility that a paraconsistent logic might be justified as a theoretical description of how cognitive agents get on with the business of belief-revision and preference-management under conditions of inconsistency. Notwithstanding, it is fundamental to the para-consistent program that *sometimes* it is rational to believe a (noticed) contra-diction – thus the paraconsistent's proposal to rehabilitate naive set theory. It is here that an objection presses hard for attention. It can be argued, as Popper and others have done, that the admissibility of contradiction makes nonsense of the idea of rational criticizability. If someone believes some body of propositions Σ, what rational grounds could there be for abandoning or amending Σ? In the standard case, this is done if a further proposition Φ is put forward that, although accepted by the person in question, is inconsistent with Σ. Given that inconsistencies are now rationally allowable, it would appear that there is nothing to prevent our friend from simply adding Φ to Σ and living with the ensuing inconsistency. Thus does *ad hominem* argumentation lose its purchase, and conflict in the abstract sciences threatens to drift off into an anarchic rhetoric.

The paraconsistentist carries no brief for promiscuous inconsistency. His own toleration of inconsistent propositions is highly selective. Even so, it is crucial that he try to produce some reasonably effective general criterion for determining when it is all right to embrace an inconsistency and when not. True, he could offer the rule: Embrace only those inconsistencies for which there exists compelling evidence of truth. But this can hardly be said to be an effective rule and, in any event, critics will take it to be question-begging.

We have been suggesting that the paraconsistentist might well benefit by taking into account Elster's distinction in *Ulysses and the Sirens* (Elster, 1984, pp. 65–77). Paradox, says Elster, is more or less interesting depending on whether it involves "contradictions of the mind" or "inconsistency stemming from a *single* belief," and that therefore follows from "a single unitary project." In speaking of contradictions of the mind, Elster is thinking of situations in which "contradictory attitudes coexist passively, and it is possible for the indi-vidual to abandon one of them when the inconsistency is pointed out to him." Contradictions of the mind are expressible as beliefs in the form $\ulcorner A \wedge \neg A \urcorner$ got by Adjunction. The other kind of inconsistency is more interesting because it is somehow deeper and more intractable. If an attitude, the lurking incon-sistency may be "constitutive of the personality as a whole"; if a belief, the inconsistency may lurk in a unitary concept.

Sentences acquire the paradoxical value when they are equivalent to their own negations. To this we can now add: An inconsistency is *unitary* if it arises by way of the *conceptual content* of the paradox-generating object. In the case of set theory, the paradox-generating object is the Russell set, and in the case of semantic theory it is the Tarski sentence. The distinction that Elster is after, adapted here for our purposes, is that between a contradiction got by Adjunction and a contradiction implied by an antecedent equivalence got by the *analysis of a unitary concept*.

The dialethist may now have an answer to the question, "Which inconsis-tencies is it all right to hold?" He might say that it is a necessary condition

on rational commitment to an inconsistency that it have been derived from
an equivalence got by unitary conceptual analysis. He cannot offer this as a
sufficient condition, since that would make it all right to embrace the paradox
of the village barber who shaves all and only those villagers who do not shave
themselves. Whether the paraconsistentist *can* offer up a sufficient condition
will turn on his finding a way of excluding the Barber which does not also exclude the Russell set or the Tarski sentence. This he might try to do as follows.
He might say that holding the inconsistency is all right when there is independent reason to affirm the existence of the paradox-generating object. Thus, the
Russell construction exists, and is a set. The Tarski construction exists and is
a statement. But there exists no such barber as that described in the Barber
Paradox.

I do not for a moment believe that the proposed conditions on rationally em-
braceable inconsistencies are decisive or problem-free. For one thing, a critic
has plenty of room to complain that the difference between the Barber, the
Russell set, and the Tarski sentence is psychological, and so a matter of the degree of antecedent obviousness of the paradoxical specifications. Even so, the
paraconsistentist in turn has room to counter that at a minimum his proffered
conditions serve to demarcate *prima facie candidates* for rational acceptance,
and then to submit those candidates to the further test of costs and benefits
that we have lately been examining. By such a test, there would presumably
be nothing gained by tonsorial enthusiasms for the Barber.[18] But if the Russell
set were embraced, intuitive set theory could be given a dialethic rehabilitation; and if the Tarski sentence were admitted, a paraconsistent restoration
of naive semantics could be produced. At least these things can be tried; they
underwrite a coherent research program. It is well to note, however, that this is
less of a defense than the realistically minded dialethist would wish to have. It
is one thing to have proofs that certain statements actually take the paradoxical value and that the state of affairs that it inconsistently describes actually
obtains as described. It is another thing to propose that true inconsistencies
be *tolerated* for the good that they do in the resurrection of intuitive set theory
or intuitive semantics. Short of such *proofs*, it is hard to see how the dialethist
could overcome, in a way that does not beg questions, the classicist's insistence
that if true inconsistencies are the price of intuitive theories, of theories transacted in the manner of conceptual analysis, then so much the worse for them.
If the dialethist has an answer to press it can only be along the following lines:
By the method of analytic intuitions, the intuitive conception of X reveals what
it is to be an X. Frege and Russell tried to take this approach to sets. There was
a setback. The Russell Paradox showed that inconsistencies were in what it is
to be a set. Both Frege and Russell misjudged what the analysis of the intuitive
concept of set told them. Frege was drawn to the view that there is nothing
to the idea of set, hence, nothing for set theory to be a theory of. Russell shared
the first part of this reaction, and then struggled on stipulatively, or formally.
The success or failure of the ensuing stipulations on the word "set" would be

a matter of what mathematicians were willing to accept, of how they fared in the relevant dialectical community. But the theory could not pretend to be an account of what it is to be a set. We are trying to see the realistically minded dialethist in a different light. His is a different understanding of the Paradox. What the Paradox shows is that it lies in what it is to *be* a set that these really are inconsistent sets. And the suggestion of lines above that inconsistencies are acceptable as true when they inhere in an Elsterian "unitary concept" is a suggestion to this same effect. If the analysis of an intuitive concept, of what it is to be an X, if the very idea of Xhood yields an inconsistency, then it lies in the very nature of Xhood that there are inconsistent Xs.

A test, of sorts, has been proposed by Parsons (1990). It is a "rough guideline" for telling that contradictions are both true and false, and so is a guideline only for the dialethically inclined. It easily adapts to our present interest in contradictions that arise from the analysis of unitary concepts. It is this: Wherever a nondialethic response would be to assign such a contradiction *no* truth value, the dialethist should assign it both. In the case of the Tarski contradiction, the truth-value "gapper" infers that the Tarski statement from which it arises does not exist; and this denies to the contradiction any truth value at all. In the case of the Russell contradiction, the gapper infers that the Russell set does not exist and, again, the contradiction is engapped and disarmed.

Gaps are decreed precisely where the dialethist imposes gluts. Gappers and glutters react differently to presuppositions. The gapper reasons as follows: For propositions Φ and ψ, if Φ is provisionally evaluated as both T and F, and if Φ carries a *presupposition* ψ whose falsity would exchange glut for gap, then *deny* the presupposition and make the exchange. The glutter, on the other hand, affirms the presupposition; so there is no exchange to make.

It will not work. Either the principle is overstrong or it provides us no guidance. Consider the celebrated case in which Larry has no children. Then the principle would seem to provide that where the gapper withholds both T and F from "Larry's children have the measles and Larry's children do not have the measles," the glutter should assign it both. But no glutter will see it so. The principle is too strong. It cannot be allowed to command the exchange of gaps for gluts in all cases.

What makes the principle workable for the Tarski case and the Russell case, but not for Larry's children? In the cases for which it works, there is no prior reason to suspect presupposition-failure. It is for gappers the provisional and paradoxical evaluation that places the presupposition into question. With Larry's children, unless we have independent reason to think that the presupposition fails there is no chance that the conjunction "Larry's children have the measles and Larry's children do not have the measles" will attract, however provisionally, the paradoxical value.

The test will not work without the understanding that Φ bears on the provisional evaluation of paradoxicality *and* that it does so independently of any assumption of the failure of ψ, its presupposition. With Larry's children, the failure of the presupposition is a premise in the derivation of the conclusion that the conjunction is glutty (or the different conclusion that it is gappy). In the other cases, the failure of presupposition ψ is inferred (by gappers) from the premise that Φ is paradoxical.

In the end, gappers and glutters part company on whether to deny or to affirm a presupposition. But Parson's principle gives no guidance to the dialethist about when he is to do this. In particular, it does not tell him whether or not to affirm the existence of the Barber. The Barber is important. We return to it in Chapter 7.

DIALETHISM AND REDUCTIO ARGUMENTS

In the earlier section on Rational Criticizability, we wondered whether the dialethist could offer us a principled basis on which to distinguish those inconsistencies that are the rightful targets of criticism from those that are not. We have been reminded that one of the attractions of dialethic approaches to set theory and semantics is recovery of the kind of realism that attends the method of analytical intuitions. If we have learned something from the chapter preceding this one, we would do well to make the present point more softly: *If* there is something to be said in general for the method of analytic intuitions, then perhaps there is something to be said for the method of analytic intuitions in set theory and semantics. But at what price? Are true inconsistencies not too much for an intuitive set theory and an intuitive semantics to ask for? And how in any event are the true inconsistencies to be distinguished from the false? In his more realistic moments, the dialethist is drawn to an idea that proffers assurance on both points. We ask which inconsistencies is it all right to pledge to, and the answer is, those whose truth we have a demonstration of. And we ask, what, pray, could justify the expense of pledging to inconsistencies, and the answer is that it is no expense to speak of when the inconsistencies are *true*.

At the heart of this give-and-take, and of the main issue in the section on Rational Criticizability, is whether there exists a principled method for distinguishing correct dialethic proofs from correct *reductio* arguments. It is a vital question, as we can now see. That inconsistencies are acceptable when they rise from the analysis of a unitary concept is a suggestion that places a large burden on the concept of a unitary concept. I will say just below why there is little comfort for the dialethist in the concept of a unitary concept, but let us for now try to give the idea free rein. *Reductio* arguments are a sharp discouragement. For if any argument really does proceed from the analysis of a unitary concept, *reductios* surely do. But *reductios* do not establish the acceptability of the inconsistencies they prove, but rather the unacceptability of that which implies them. When implied by a unitary concept, *reductios* constitute a wholly adequate basis for the rejection of the concept; for insisting that nothing could be an instance of it. So then, let Δ be a valid argument purporting to prove the truth of a contradiction. Schematically Δ has the form

1. Φ_1
2. Φ_2
   .

   .

   .

$$\frac{\text{n. } \Phi_n}{\psi \wedge \neg \psi}$$

Why is not this a *reductio* argument? It is true that not everyone is happy with *reductio* arguments. Intuitionists allege their overdetermination when it comes to mathematical epistemics, especially in their use in existence-proofs (which offend against the intuitionist's constructivist sensibilities). But no intuitionist is ready to assert, unless he is also a dialethist, that *reductio* arguments *always* "fail" in the manner in which, for the dialethist, they sometimes do. So we may safely leave to one side the intuitionist's epistemologically peculiar disaffection with the *reductio* form of argument.

The classicist believes that any valid argument of the Δ sort is a good *reductio*. The dialethist believes that some of these are not, and that those that are not are proofs of the truth of their inconsistent conclusions. How does he tell which is which? The dialethist sees his proof of a dialethic truth as a valid argument whose contradictory conclusion does not amount to a *reductio*. Thus dialethic proofs are *reductio*-types of argument that fail in a certain way. It is helpful to remind ourselves that a natural, although not a necessary, setting for *reductio* argumentation is one of dialectical give and take between parties who find themselves deadlocked over some issue. The dialectical context suggests a way of distinguishing good *reductio*s from good dialethic proofs: Δ is a dialethic demonstration for parties P and A (or dialectical communities **dc'** and **dc''**) when the premises from which the inconsistency was derived were not initially in doubt by P (or **dc''**) and A (or **dc''**) and not subsequently put in doubt for them by the contradiction they imply. But the actual record of such matters reveals no general disposition to crimp the generality of the *reductio's* central presumption. The widely received wisdom is that the *reductio's* power is such that there is no matter of whatever degree of antecedent certainty and dialectical accord against which it could not, if valid, be a successful ambush argument. The dialethist has an uphill struggle against his opponent's classical intractability; and there is a structural reason for it. For the dialethist to succeed in convincing his opponent that a Δ-argument is sometimes a correct dialethic demonstration, he must succeed in argument-contexts of the Δ sort in pinning on his classical opponent a *Converse Begging the Question* violation, in which he refuses to accept his own virtual premises, that is, premises that he should accept. But nothing in the nature of CBQ, nothing of what might be called its communal epistemics, supports a finding of guilty when a party refuses to accept the truth of a premise set or an assumption set, for the very reason that the sets are inconsistent.

We find ourselves in the midst of what we might call the Great Delta Debate. If there is anything to be said for the dialectical structure we have called CBQ, then the Great Delta Debate is not one that a dialethist can win against an attentive classicist. The CBQ-structure encompasses sets of statements that a disputant can be supposed to have accepted because, although not expressly conceded, they are of a sort that he would or should concede, if he called them to mind. Built into the structure are factors that make it both conservative and populist. One is guilty of a *CBQ* violation when one refuses to concede what "everyone knows" or what is "too obvious for words," or, more carefully, does so without the support of supplementary argument. The Great Delta Debate presents a discouraging backdrop for anyone intent on marshaling supplementary arguments in vindication of his refusal to concede that premises implying contradictions are collectively false (i.e., false *only*). Any such argument must be an argument that demonstrates the existence of true inconsistencies. It will therefore be an argument of the Δ sort, of a sort that already is landed in the impasse of the Great Delta Debate.

Perhaps we have been too ready to throw in the towel on behalf of the dialethist, who may not be much inclined to thank us for our trouble. Graham Priest has written a whole book devoted to the cause of Δ-arguments which succeed as dialethic demonstrations. *Beyond the Limits of Thought* is a work of considerable ingenuity and endless fascination. But if what I have been saying in the past few pages is correct, all of Priest's arguments that purport to establish dialethic limits are

    (a) construable as *reductios*

and are so

    (b) in such a way that the dialethist lacks the dialectical resources to gainsay.

Priest's arguments are interpretations of positions that have dogged the tracks of philosophy from the very beginning. In the modern period alone are arguments ranging from Berkeley's Paradox (which Priest [but not Berkeley] thinks demonstrates that something both is and is not in the cause of conceivable things) (Priest, 1995, pp. 65–77); to Hegel's proofs that the limits of thought are marked by true contradictions (Priest, 1995, pp. 113–21); to interpretations of Cantor's arguments that purport to dissolve the paradoxes of the infinite but that, says Priest, "merely relocated them" (Priest, 1995, pp. 140); to dialethic construals of the Russell and Tarski paradoxes; to Frege's concept *horse*, and Quine's unavoidable but self-canceling indeterminacy of translation; and more besides. All of these arguments have a common structure.

Let *T* be a totality. Then there exists an object α and an operation f for which there is a demonstration Δ that $f(\alpha) \in T$ and $f(\alpha) \notin T$. It can be conceded that if three things can be demonstrated independently, this will serve as a basis for saying that Δ is a dialethic demonstration rather than a *reductio*. They are (i) the existence of α, (ii) the existence of f, and (iii) the existence of *T*. In all the cases that Priest considers, the contradiction flows from the premises asserting the truth of (i), (ii) and (iii). So much, then, for the Elsterian idea that the paradoxes in question are inconsistencies that flow from a *unitary concept*. Of course, as we have seen, the validity of Δ itself is enough to move a classicist to see the establishment of (i), (ii) and (iii) as unrealizable, since, if realized, a contradiction would be true. Even the dialethist will grant that any such Δ is construable as a *reductio*, indeed that this is a wholly natural way of seeing such arguments. But suppose in a particular case that a dialethist wants to resist the construal. He must offer independent proofs for the existence of α, f, and *T*. Suppose he gives arguments Γ in which he appears to have succeeded in this attempt. Now Γ will be consistent with Δ or not. If not, either Γ is mistaken, or Δ is, or neither is; that is, dialethism is true with respect to *them*. Since dialethism is itself the issue, neither the third alternative, nor its negation, is available to disputants; and both the first and the second deny the dialethist something his defense cannot do without. On the other hand, if Γ is consistent with Δ, Γ is destroyed unless Δ is not a *reductio*. But this is indeed what the dialethist is supposed to be in process of proving rather than assuming. We may therefore conclude that the dialethist is unable to meet the burden of his own case. He must be able to show that although all the Δ's of present interest are construable as *reductios*, he has the dialethical resources to discourage such construal. Question-begging denies him those resources.

Is not sauce for the goose also sauce for the gander? The classicist wants to see all Δ's as *reductios*. The dialethist will press with a qualification: Every Δ is indeed a *reductio* (*red*) unless it is a dialethic demonstration (*dd*). This nothing the classicist cannot accept, since his own position is captured by the equivalence of "All Δ's are *reds* unless they are *dds*."

In Locke's original conception, an *ad ignorantiam* is a dialectical maneuver in the following form (see Chapter 1):

If you, my opponent, cannot give a better proof of your position than I have given for the opposite view, then you must submit or be silent.

Ad ignoratium challenges distribute the burden of proof. If the challenge is correct, the person challenged has the greater burden. His proof must be "better" than that of his opponent. Aside from heavily conventionalized contexts such as criminal and civil trials, it is notoriously difficult to apportion proof-onera in non-question-begging ways.[19] In real-life dialectical practice, the greater burden – oneric primacy – is often reserved for the party who challenges received opinion. It is easy to misunderstand the dialectical role of received opinion in contexts of oneric assignment. Some see it as a kind of doxastic privileging by powerful elites; others see it as favoritism toward popular views; others as a kind of doxastic conservatism, and so on.[20] In its use here, received opinion is an assignor of oneric primacy. Let us say that

Def. Burden Proof In a dialectical exchange D between parties over an issue I, with one party espousing I and the other party opposing it, the *burden of proof* at a juncture of D with respect to I falls on the party for whom the challenge "Prove I" (or "Prove not-I") would be a *misplaced ad ignorantium* move at that juncture.

One of the attractions of *Def Burden Proof* is that it leaves it open that an *ad ignoratium* challenge would be a mistake for both parties, either at a given juncture or even at all. This acknowledges the real-life fact that burdens come and go as the dialectic unfolds, and it leaves it open that there might be some points of disagreement for which no burden of proof is assignable (e.g., "Bananas taste lovely"). Still, the definition is no more helpful in specifying the idea of burden of proof than is the heretofore unexplained idea of *ad ignorantium* moves that are *misplaced*.

A better way of approaching oneric primacy is by reflecting on the conditions that initiate a disagreement. A disagreement is always a disagreement between parties, whether individuals or dialectical communities. Disputes always have initiation conditions. There are two modes of initiation, one-step and two-step. In a one-step initiation, one party A challenges a position held by the other in ways that constitute A's disagreement with it. In a two-step initiation, one party P forwards a proposition, represents it as true to his opposite number A, and A challenges the proposition in ways that constitutes his disagreement with it. In the one-step initiation, the initiator is the challenger A, and in the two-step initiation it is the protagonist P. It is economically attractive to suppose that the burden of proof (if there is one) rests with the initiator. By these lights, in one-step initiations P is not deemed a participant, and certainly not an initiator, even though it is a view that he holds that gets the disagreement started. This offers the idea of "received opinion" some useful work to do. In a dispute D between parties A and P, received opinions are propositions effectively open to challenge in D by one-step initiation.

There is more to be said about burden of proof, its differential twists and turns, than needs to be said here. We may suppose that we have captured the twin notions of *initial* and *overarching* burden of proof, leaving undeveloped ancillary shifts and initiations. We may agree, then, that in the matter of the Great Delta Debate, the burden of proof rests with the party who insists that not all Δ's are *reductios* and that some are dialethic demonstrations, since the contrary view is the received opinion, not only in the ordinary sense of the term, but in the more technical meaning we have here given it. Thus, the dialethist lies exposed to a powerful and nonmisplaced *ad ignorantiam*. It is he who has the burden to show that some Δs are *dds*. But, as we have seen, it is not a burden that he has the dialectical resources to discharge in a non-question-begging way. This leaves his opponent with no case to answer; there is no burden of proof that rests with him. He did not have such a burden at the opening stage of the dispute and did not acquire it when it became clear that his attacker could not go on without begging the question. The *ad ignorantium* charge requires that the person to whom it is addressed must make a better argument than any argument his opponent is obliged to make. Since in the circumstances of the case the classicist is required to make no argument, the burden that falls on his attacker (and is left undischarged by him) is a burden that satisfies the *ad ignorantiam* charge. In the matter of showing that the classicist is wrong in holding that no Δ's are *dds*, the dialethist must either yield or leave the field.

I have been assuming that the dialethist has been wanting to *prove* that there are true contradictions. We have been considering two ways in which this could be tried. One is to deploy the method of analytic intuitions and to brazen it out in the spirit of *Routley's Serenity*. But Chapter 4 took pains to show the dialectical impotence of the method of analytical intuitions, a fault it has irrespective of the nature or dramatic value of the issue in dispute. If that case was well made, there is nothing good to say for the method of analytic intuitions in support of or in discouragement of dialethic objectives, when they are under dispute. Analytic intuitions are dialectically impotent when directed at the analytically other-minded.

The second approach to the establishment of genuinely dialethic truths was by way of Priest's arguments. But Priest's arguments all have the form of Δ arguments, and it has been a burden of the past several pages to show that it cannot be determined in a non-question-begging way that since Δs are not refutations but, rather, are dialectical demonstrations. And if this is right, the dialethical turn is powerless against a wholly appropriate *ad ignorantium* challenge.

Even if these criticisms of the dialethist's prospects are well founded, they do not show that the dialethist lacks resources to advance his position. After all, my *ad ignorantium* complaint turned on the point whether the dialethist could prove that some Δs are dialethic demonstrations. I said that on this question he must yield or quit the field. But nobody thinks, or should think, that this was the only means of dialethic advancement.[21]

In the chapter to come, I want to reflect on a way in which the dialethist might wish to show that contradictions are sometimes true, but that do nothing to attract the idea that proofs of this sort are *reductios*. As we will see, an attraction of this approach is that it has no need of tying the idea of true contradiction to the suggestion that they are contradictions that flow from the analysis of a "unitary concept."

6

Fiction

Do I contradict myself? Very well then I contradict myself (I am large, I contain multitudes).

Walt Whitman, *Leaves of Grass*, 1855

I am clear that [the philosophy of mathematics] would have to be a fictionalist account, legitimizing the use of mathematics and all its intratheoretic distinctions in the course of that use, unaffected by disbelief in the entities mathematics purports to be about.

Bas van Fraassen, *The Scientific Image*, 1980

In this chapter, I want to try to redeem the pledge of stipulationism in (especially) the abstract sciences. At the same time, I wish to seek for consideration what might fairly be said to motivate the fundamental principle of dialethism, the existence, namely, of true contradictions. On the face of it, the place to pursue their objectives is in the theory of fiction, or in that branch of it known as the *logic of fiction* (Woods, 1974). Fiction is, or appears to be, a paradigm of truth-making stipulations, in the sentences made true by the author's sayso and those that follow from them under suitably constrained closure conditions. And fiction endorses, and thereby makes true, sentences having the character of contradictions – or so it would appear. A third reason for exploring the logic of fiction is the mere fact of *fictionalism* in the foundation of mathematics (Woods, 1974). It is a fact that summons up an initially attractive principle:

Fictionalism: Since fictionalism with regard to a theory T proposes that T's truths are secured and validated by analogy with how the truths of fiction themselves are secured and validated, fictionalism with respect to T cannot be an adequate semantic theory of T unless the theory of fiction borrowed by T's fictionalism is in its own right a semantically adequate theory of *fiction*.[1]

A fourth reason for exploring the logic of fiction is that it gives us another go at rehabilitating the method of intuitions. I shall show below that doing so is also the last chance for this beleaguered methodology. So we must say something about theories of fiction.

Apart from what was said in Chapter 4, we have done little to vindicate our notion of truth by stipulation. Also needed, if one is of dialethic bent, is a defense of true contradictions that does not fall foul of the *ad ignorantiam* and the Converse Begging the Question fallacy.

We saw in the previous chapter that the appeal of dialethism is bound up with the appeal of theories such as **N** and **Sem**, and I have counseled the dialethist to get on with the job of developing these theories and demonstrating their superiority to rivals. It might be asked for whose imagined benefit is this advice so freely offered. Some will say that it will do the dialethist no good unless he has some chance of persuading the orthodox that some things are genuinely true and false. But the orthodox can be expected to say, "Do not bother." If anything is obvious, it is surely the falsity of any such claim. The recalcitrant can help themselves to our cost-benefit strategy for *ex falso*. True, they will say, we get a verdict against *ex falso* in **PP** and **N** and **Sem**, but only at the higher cost still of a central claim that is obviously false. So perhaps it was a stupid strategy. Some people, for example Parsons, are prepared to concede that it is a victory of sorts for dialethist that they do some things as well, but no better than, other (gappy) heterodoxies. But impressing Parsons may not impress the orthodox, for Parsons is certainly not one of them.

A PRIMORDIAL INTUITION

Dialethists lack the comfort of what Herzberger calls a "primordial belief." With naive set theory, primordial intuitions reposed in the axioms. When the axioms proved an embarrassment, analytical intuitiveness gave way to something rather more formal (hence something less about sets). With semantics, the disquotationality of truth was a primordial intuition. After the paradox the intuition held fast, even though it lost all chance of straightforward formulation. With relevant entailment theory, primordial intuitions were the falsity of *ex falso* and the diagnosis that found it irrelevant. Even though they lie open to a rival's contradiction or to paradoxical eventuation, primordial intuitions are what get the theorist going. They give him something for his theory to make something of.

With dialethic logic it would appear that there is nothing to be getting on with, no moorage for its most basic truth. It wants for a primordial intuition. Priest tells us that he had a devil of a time getting himself to believe that **T** is true and false. For the rest of us, Priest's "devil of a time" is an unmotivated impossibility. It would be better for the paraconsistentist to seek for a "primary datum" elsewhere. It is strategically *maladroit* to seek the datum in set theory or semantics because the established theories of such things already extend no hospitality to it there, and the disestablished theories presuppose the datum rather than produce it.

It would not be hopeless if a primary datum could not be found elsewhere (only difficult). **N** and **Sem** might evolve in ways that do eventually proclaim their superiority and, as a byproduct so to speak, license the primary datum by cost-benefit inference. But the evolution would be met with considerable and perhaps impeding hostility. The theories would evolve, if at all, in a climate of stalemate.

Paraconsistentists do look elsewhere, as we have seen (Priest and Routley, 1989, pp. 367–393; Priest, 1995, Ch. XI, XII, and XIII). I propose to look elsewhere, too, but in a different place.

FICTIONAL DISCOURSE

It has rightly been said that "literary phenomena rather convincingly show the inadequacy of most going formal semantics" (Routley, 1979, p. 3). The semantic analysis of fictional sentences, even simple ones such as "Sherlock Holmes solved the case of the speckled band," or "Sherlock Holmes lived in London," is notoriously difficult, and some will say that it is not worth the trouble. Such sentences are not needed for physics. But, whether for the fun of it or with graver objectives in view, we could approach such matters not in Quine's way, but more in Davidson's. We would ask what are the truth conditions of such sentences. Should they have none, we would ask why this is, and what, then, is their semantic status in English?

If you overhear someone say that Sherlock Holmes was headquartered in Barrie, Ontario, you might protest, "No, it was London, surely." It seems that you would be right. You would have corrected a *mistake*. It is perhaps rather remarkable that we manage such things as reference and ascription to fictional people, and descriptions of and inferences from their goings on, many of which are correct. Theories have arisen to account for such things. Some of them borrowed their insights from the theory of descriptions. On the face of it, doing so was a spectacular failure. It assimilates Sherlock Holmes to the present king of France. But this makes a nonsense of our conviction that we know who Holmes is (or was) and something, at least, about what he did.

It seems then that we need theories that are tailor-made for fiction. Various proposals have been made, most of which – at least initially – try to honor the following pretheoretic intuitions:

[A] Reference is possible to fictional beings even though they do not exist.
[B] Some sentences about fictional beings are true.
[C] Some inferences about fictional beings and events are correct.
[D] These three facts are made possible, in a central way, by virtue of the creative authority of the authors of fiction. Indeed, the primary and basic criterion of truth for fictional sentences is the author's sayso.
[E] It is possible in a fictional truth to make references to real things. For example, "Sherlock Holmes lived in London" is true. If we wished to

speaking tendentiously we could, by analogy with the axioms of naive set theory, speak of these as axioms of naive theories of fictionality.

It is worth noting that the author's creative license mimics that of God himself according to the Genesis account. For it is there apparent that "in contrast to sexual creation myths in which [the world] is brought into being (frequently) by an earth-mother... [i]n Genesis... the forms of life are *spoken* into existence" (Frye, 1981, p. 106). Thus, in the beginning was the Word. Just as God made provision for Adam linguistically, Conan Doyle made provision for Holmes in the same way, a lesser accomplishment to be sure, but similar. It is lesser because although Adam was made to *exist*, Sherlock Holmes was not.[2]

Among the leading contemporary theories of reference, inference, and truth with respect to fictional beings and events, we find the following:

- Quasi-actualist Meinongean accounts of fictional objects (Parsons, 1980, Ch. 3 and 7; Routely, 1979; cf. Castañeda, 1979; Zalta, 1983).
- Accounts in which fictional objects are nonactual but well-individuated objects existing in (metaphysically) possible worlds.[3]
- Nonreferential, substitutional-quantificational accounts of fictional entities (Woods, 1974).
- *De dicto* modal analyses (Plantinga, 1974, pp. 159 ff; Kaplan, 1973; Gabriel, 1979; Lewis, 1983).
- The approach in which fictional entities are nonactual, well-individuated objects inhabiting fictional worlds, some of which may not be (metaphysically) possible (Howell, 1979).
- Make-believe approaches to fiction (Walton, 1990).[4]

Robert Howell has subjected all but the last of these approaches to trenchant scrutiny, and has found them wanting. He favors the last-mentioned account, his own, but he claims for it no more than plausibility enough to warrant "the attention of all philosophers who are now concerned with fiction" (Howell, 1979, p. 174). I am of a mind to do something different — to start all over again, although in doing so I shall need to return to Howell and to a point central to his approach.

If we take seriously the intuitions – the naive axioms – recorded in [A] to [E], it becomes important to notice that truth and falsity sort themselves differently through sets of fictional sentences. This needs to be explained. For example, it appears that sentences in

Category A:

- Holmes was a detective
- Holmes lived in London
- Holmes solved the case of the speckled band
- Holmes patronized Watson
- Holmes was a marvelous deducer

are all true of the stories, and false of the world, to speak loosely. Whereas sentences in

Category B:

- Holmes was admired by Agatha Christie
- Holmes was appropriated by the later novel, *The Seven Per Cent Solution*
- Basil Rathbone caught nuances of Holmes's personality better than any other actor to date

are all true of the world, yet false of the stories, still speaking loosely. Perhaps the principle of differentiation is this. Sentences of the first sort all involve ascription to Holmes of nuclear properties only, whereas sentences of the second kind all involve attribution of nonnuclear properties only. The distinction is that of Parsons. Roughly speaking, a nuclear property is a property that helps to fix an object's identity (e.g., being a woman); nonnuclear properties do not do this (e.g., being thought about by Terence Parsons). This suggests an obvious generalization. The fictional sentences that are true of the story and false of the world are exactly those making nuclear attributions only and are endorsed by the story. Sentences that are false of the story and true of the world are exactly those that make nonnuclear attributions only and are endorsed by the way the world is.

It will not work.

- "Holmes exists" makes a nonnuclear attribution, yet is true of the stories and false of the world.
- "Holmes is an *impossibile*" makes a nonnuclear attribution, and it is false of the stories yet true of the world (see below).
- "Keith was elected President in 2055" makes a nuclear attribution that is true of the story and false of the world, but that by the reasoning just above is also false of the story and true of the world.[5] For we also have it from Bradbury: "Keith was not elected President in 2055." The contradiction is not inadvertent. The story turns on it rather centrally.

The broad difference between our two categories is this. For Category A: If S takes the truth value T, then that it does so is completely determined by the stories or by what Parsons calls their "maximal accounts."[6] For Category B: If S takes the truth value T, then, though not completely determined by the maximal account, S takes as presuppositions some true sentences from Category A.

Fixing the maximal account, that is, determining how to fill a story in, raises four hard problems: (i) What implication should be drawn from a text? (ii) What about the completeness of fictional entities? Are there properties P which a fictional entity *could* have, but such that it neither has P, nor does not have P, in fact? (iii) What is it that individuates the objects of fiction? (iv) What are we to make of inconsistent fiction? In particular, can the dialethist find his primary datum in inconsistent fiction?

Concerning (i): Do we put into the maximal account all the logical consequences of everything in the text? No. That would be overabundant. There are too many – infinitely many – consequences, and many of them, most in fact, are irrelevant to the story. What then will serve as a differentiating principle? There is a certain attractiveness about "the principle of history," which bids us to include just those logical consequences needed for the specification of Holmes's history (Woods, 1969). This is an open-ended task, no doubt, but it is perhaps no more so that of producing a history of, say Herod's life about which there is so much that is unknown.

Concerning (ii): We should strive to make the object of fiction complete except when by explicit say-so the author proves otherwise. Of course, there is a certain theoretical tidiness in not going this way. One gets a fairly nice definition of fictional objecthood and an elucidation of the claim that fictional objects do not exist. Fictional objects could be said to be a proper subset of the set of objects failing the completeness condition, yet fulfilling the author's say-so criterion of truth.

Even so, this should be resisted. It leaves it objectively the case that Holmes neither had nor did not have a mole. It confuses what we do not know with what is not so – *ordo cognescendi* with *ordo essendi*. Most of all it ignores the significant fact that we construct maximal accounts with the aid of *default logics*, as they are somewhat misleadingly called. If you tell someone over the phone from Bombay that you are looking at a tiger from your study window, he will attribute four legs and some stripes unless you tell him differently, that is, in default of information to the contrary. When we assign "default values" to variables whose actual values are unknown, we are guided by the appropriate protoypes. If we did not do this in general we would disarrange communication and all but paralyze routine inference. It is a merit of the solution that I am proposing that we handle communications and inferences about fictional goings-on in as natural a way as the genre will permit and in light of the author's pronouncements. Holmes's mole is misleading. Whether he had one or not does not matter much. But it cannot be allowed that it follows from the fact that it does not matter much that he neither had one nor did not. If we do say this, there is no nonarbitrary way of resisting the obvious generalizations, and we will find ourselves having to endure it that:

- Holmes neither did nor did not have a spine.
- Holmes neither did nor did not have a brain.
- Holmes neither did nor did not have a persisting identity over time (unlike the Countess of Snowden, apparently).

Such inferences are defeated by genre-considerations; by the fact, as with Holmes, that Doyle's writings are representational and narrative fiction, hence a form of discourse wholly, if defeasibly, amenable to prototypical reasoning. Holmes could not have been a deducer without a brain; he could not have stridden the moors without a spine; and he could not have been a human being, an embodied person of a certain sort, without a lingering *corpus*. True, the author can defeat any or all of these inferences, as Doyle did in having killed off Holmes at Reichenbach Falls yet providing for his return in later stories as the self-same person. But we are not here talking about what the author should or could make true, but what the reader should take to be true in default of contrary information from the text and prior stages of the evolving maximal account. In doing this, the reader avails himself of what any sensible person knows about the world. He helps himself to theories – to neuropsychology, to kinesiology, and even perhaps to some old-fashioned metaphysics, Aristotelean essences and all. It is the same way with Holmes's mole. Although it might not settle the question of whether he had one or did not, the dermatology of the human skin decrees that it was one way or the other.

The position that we should take on the issue of completeness is fixed by the fact that short of textual, prototypical, and other evidence internal to the emerging maximal account, there is no nonarbitrary way of deciding when to be default reasoners, and when not.

Concerning (iii): To what do we owe the ability to individuate fictional objects? If we wish to talk about the fictional in ways that comply with the axioms [A] to [E], we must make sense of how we refer to them. There is a celebrated argument about this matter. It is an argument to be found or adumbrated in writings of Plantiga and Kaplan (Kaplan, 1973) and, most emphatically, Kripke (Kaplan, 1980). It goes as follows. The name "Sherlock Holmes" cannot be said to refer to anything. If it could refer to something, there would be some way of uniquely specifying what it refers to. But with fictional names, there exists no historical connection, or possibility of it between our employment of it and any appropriate object. So the putative referent of "Sherlock Holmes" cannot be individuated.

Howell, for one, finds the objection close to devastating, and would himself accept it but for an assumption that he challenges. The assumption is that if individuation cannot be got by way of appropriate historical connections between referring devices and the right sorts of objects, then individuation cannot be got at all (1979, p. 172). Howell thinks that this does not follow, and he proposes a method of individuation that involves fixed places inside the field of Doyle's imagination, and ours, too, presumably. This is the most distinctive feature of Howell's own positive account of fictional objects, but I shall not follow it up here (1979, pp. 172–5). There is a more direct answer to the objection.

The Kripke-Kaplan-Plantinga problem can be seen as a case of a more general puzzle, often credited to Benacerraf, although Benacerraf's problem is about numbers and other theoretical abstracta (Benacerraf, 1973). On the one hand, the truth conditions for statements about abstract entities require that we hold them to abstract condition. On the other hand, their epistemic conditions – conditions in virtue of which we can be said to know them – depend on our having the appropriate causal access to them. But the two conditions are not jointly realizable; there seems to be no way for us to engage nonconcrete objects causally. So, too, it may be said, with fictional objects. The truth conditions for fictional sentences require us to give to fictional objects an ontological status (viz., nonactual), which makes them causally inaccessible to us. This places us in a dilemma according to which there are fictional truths only if we cannot know them. The Benecerraf problem is not, as such, a problem in the cognitive psychology of discourse about fictional entities. It is not concerned with the question of by what cognitive mechanism do I get to know to whom "Sherlock Holmes" refers. Rather, it is this question at a limit: Are there *any* cognitive devices that give us causal or historical access to Sherlock Holmes so long as it is held that Sherlock Homes is nonactual?

If this were simply a question of considerations in virtue of which Holmes is individuated, it could not for long resist a satisfactory answer. Holmes's completeness (see the discussion above) now provides that he is individuated by his properties, just like us. A subset of those properties will be this essence. Part of that essence will have to do with who created him (cf. Noah's origins). Of course, saying this places a fair theoretical weight on the extent to which the problem of the maximal account enjoys, in detail, a satisfactory solution. In any event, puzzles do arise. Here is one of them.

What about the arithmetic of fictional beings? If author X creates fictional object f and Y creates f′ with the same name and with most properties in common, deviating only in detail – or not at all; it does not matter for the point at hand – how do we *count* these objects? Are there two or one? The answer is: Primacy goes to the earlier author, or more carefully, to the original author. If the author of f is the original author, then if f = f′ f is a migrant in the work of the other. Even this is not entirely problem-free, but it is basically right, and we shall let it be (Woods, 1974, pp. 45–7). The main issue is whether thus individuated we have the kind of access to Holmes that Kripke-Kaplan-Plantinga, and Benecerraf, too, presumably, say that we must have but do not.

Most of us know a good deal more about Holmes than we do about Doyle. For all most of us know, Doyle might have been Jack the Ripper. And Doyle is long dead. Does this mean that he is cognitively inaccessible to us? No. Doyle and we are members of the appropriate causal (or anyhow historical) sequences, with Doyle closer to *then*, and we fugitively at home with *now*. The same is true of Holmes. We are causally (or historically) linked to Doyle who is causally linked to Holmes. Causally? Yes. Doyle brought Holmes into being; he caused him to be.

If this is so, then there *are* some nonactual things. This is what our axioms [A] and [E] demand. They demand no less than the truth of "Some things do not exist."[7]

Some commentators (Howell, 1983, p. 168), are worried about the supposed distinction between "is" and "exists." Parsons thinks that the distinction is vindicated by the strength of the theory in which he embeds it. This overlooks a possible difficulty. Parsons sees the strength of his theory in the success it has in explaining the data. Alternative theories, which Parsons does not like for other reasons, seem to some people to explain the data equally well. In such theories there is no distinction between what is and what exists (Lewis, 1978/1983; Van Inwagen, 1997, pp. 299–308). So with regard to what makes a theory of fiction good – namely, its explanatory potency – the distinction would seem to be insufficiently motivated. Perhaps this is so,[8] but I will not here take the matter further. The real trouble lies elsewhere, as we shall soon see.

INCONSISTENT FICTION

We are now at the central question of the present chapter, "What are we to make of inconsistent fictions?" Contradictions arise in three main ways.

I. There are internal contradictions, that is, contradictions expressly in the story and underwritten by the author's say-so (Heintz, 1979, p. 93, n. 11).

II. There also are external contradictions arising from the semantic tension between the fictional being's world and the real world. Thus, "Sherlock Holmes lived in London" is true by the author's say-so criterion, but it is also false, or anyhow not true, by what the world makes of London (Lewis, 1978/1983, pp. 261–2). It seems plain that whereas fiction gives rise to internal contradictions only occasionally, external contradictions are liberally and systematically produced.

III. It also appears, on reflection, that fictional objects are not possible objects, not residents of possible worlds. For suppose we said that something is a possible object just in case it belongs to a possible world, and further that something is a possible world just in case it is described by a maximal consistent set of propositions. If an object x were not a member of the actual world, but rather of some (merely) possible world W, then if W were realized, x would be actualized (Woods, 1974, pp. 75–8). But now consider: Holmes and the other objects of fiction satisfy the predicate "is fictional." Thus, given the creative authority of the author, anything satisfying the fiction-predicate could also satisfy self-contradictorily descriptions. So we may say that for any fictional object x, there is a property F such that it is possible that $Fx \wedge \neg Fx$. Furthermore, thanks to standard reduction axioms for modalities, if $M\Phi$ then $NM\Phi$. So, if "$M(Fx \wedge \neg Fx)$" is satisfied by some object, so too is "$NM(Fx \wedge \neg Fx)$." Fictional objects are necessarily possibly contradictory. But of no *possible* object is it possible that it behaves self-contradictorily. Hence, fictional objects do not belong to possible worlds. They are *impossibilia*.

In the matter of internal inconsistencies, let us assume for the sake of simplicity that Holmes squared the circle and that any accurate representation of this event implies a contradiction. So, for some predicate Φ, Holmes satisfies ⌜Φ ∧ ¬Φ⌝. It is the same way with all fictional objects, considered now as *impossibilia*.

The intuitive axioms are a bonanza. Perhaps they are too good-looking to be true. To the dialethist they offer a primordial intuition that he can make the most of. They offer him genuinely true contradictions along with an explanation of how they came to be. To the stipulationist, stipulative truths are what no merely formal theorist could aspire to. They are the results of auctorial stipulation; they are true; and they are objects that accommodate themselves to psychological belief as robust as it comes. Correspondingly, the intuitive axioms show a hospitality to realism of a caliber only slightly less than full-bore. For the truth about Sherlock is as much truth as any truth about Sherlock's creator or about the objects of number theory. They are real truths even if they are not about what is real. The axioms of fictionality converge thus upon an impressive set of philosophical desiderata. Will they stand up to scrutiny?

Among the leading accounts of inconsistent fiction, it is interesting that so little play has been given to dialethism. The first use of a paraconsistent solution appears to have been that of John Heintz in "Reference and Inference in Fiction." Here Heintz embeds discourse about fictional objects in an early system of dialectical logic due to Meyer and Routley.[9] This looks like the right thing to do. A dialectical [= dialethic] logic of the sort that Heintz recommends using is one that will admit of up to an infinite number of contradictions as theorems and yet is absolutely consistent.

Heintz is also sympathetic to Meyer's and Routley's suggestions that belief in the consistency of the world is only a metaphysical prejudice, subject neither to proof nor disproof. This, they think, legitimates the opposite belief as a conjecture, and leaves open the possibility that for all we know the world, too, is simply but not absolutely inconsistent. If so, a dialectical logic might serve as a wholly general logic, a logic of nonfictional and fictional discourse equally. There would be no need for a radically special logic for fiction.

If we accepted Heintz's suggestion, it might be that the problem of internal contradictions in fiction is solved. It would also seem that by addition of some modest modal supplementation of our logic the nature of fictional objects as *impossibilia* could be honored without disaster.

But there remains the matter of external contradictions of truths, occasioned by the author's say-so against the grain of how the world actually is. Even if the world were simply consistent, it remains true in the general case that for each sentence made true simply by the author's say-so the world will make it false. Each such sentence Φ will come out as both true and false, and so will take the paradoxical value {T,F} of a logic such as **PP**. This suggests a revision of the above approach. Now that we have numberless cases of admissible paradoxical sentences, it would be well to shift our logical base from the earlier system of Meyer and Routley to a system such as **PP**. If we did not, we could not persist with the claim that here is a wholly general logic for fiction and nonfiction alike. So why not do it?

TRIVIALITY

If we were to follow the present suggestions, we would trivialize the semantics of the world, never mind the world of fiction. As we have stated it, there is no limitation on what or how much the auctorial say-so condition can make true. Moreover, we can think of no means of constraining the say-so condition except in arbitrary and unconvincing ways. We seem to be left with just two options. One is to retain the say-so condition and give it free sway. The other is to drop it altogether. I do not say that this latter is not a possible maneuver, but rather that it would be a seriously complicating maneuver, making it impossible to say that (some of) what we say about fictional objects and events is true. So I do not want to drop it, save as a last resort.

How, then, does a say-so condition given free sway trivialize the world? It allows for the possibility, for example, that an author would make true every proposition made false by the world. And so every proposition would take the paradoxical truth value. Every sentence would be both true and false. This is ironic, since one of the virtues of **PP** was that from the admittance of *some* paradoxical sentences, not every sentence need be admitted, since *ex falso* fails in **PP**. But given an unconstrained say-so condition, this does not matter. Say-so itself trivializes the world, *directly* as it were, without the necessity of supplementary derivations. The say-so criterion is troubling in the manner of the Curry Paradox rather than the Tarski Paradox. It threatens absolute inconsistency without the mediation of an antecedently established negation-inconsistency.

It might be protested that although unrestricted say-so could trivialize the world, it has not, in fact. The class of sentences whose paradoxical value is occasioned by say-so, although interestingly rich and large, leaves most of the real world undisturbed. My view is that even if this were true, it would be cold comfort. It would still make the world trivial in principle. But it is, in fact, *not* true that the class of sentences made paradoxical by auctorial say-so leaves most of the world undisturbed. In a story by Djaich da Bloo, which is printed as an appendix to this chapter, precisely the falsehoods of this world are made true by say-so. All of them. And they are wholly resistant to the constraining mechanisms of a dialethic logic.

The world is paradoxical and completely trivial. Perhaps we should just brazen it out and say that, if so, this reinforces – in fact proves – various claims that have seriously been proposed from antiquity onward: that the world is unreal, that all is illusion, that reality is *thoroughgoingly* inconsistent, that there is no reality. It may be objected that the paradoxical triviality of the world is the stuff of tricksters and mystics, or of our ancient forebears who did not know any better. For my part, I should be reluctant to dismiss the paradoxers of yore as tricksters, mystics, or ignoramuses, but let that pass.

This is an awful result. It calls to mind an argument that, if correct, sustains a thesis every bit as startling as the triviality of the world. In "Quantum Mysteries for Everyone" (Mermin, 1981), the physicist N. B. Mermin makes the following claim. Most people believe that there is such a thing as "deep" physical reality. Nearly everyone thinks that this reality is *local*, that is that it does not, perhaps cannot, support unmediated action at a distance – a "great absurdity," according to Newton. But Bell's Theorem (Bell, 1964) demonstrates the falsity of the locality assumption. So although reality exists, nonlocality considerations require that we radically reconstrue its nature. Mermin, however, draws the opposite conclusion. What Bell's Theorem shows is that *if* reality exists it is nonlocal. But since nonlocality is unthinkable, we must conclude, *modus tolendo tollens*, that reality does not exist.

By the author's say-so criterion, the word is paradoxical and trivial, and, under present assumptions no dialethist can prevent its being so. Like most apocalyptic theses, it is descriptive of a state of affairs that is not reflected in our experience. Triviality's failure to be reflected in our experience is seen in two basic ways: (1) Most people do not know that the world is trivial. (2) Even when they do, they do not *act* on this fact. It is for them an abstract fact and not available as a premise in practical reasoning. Seen in the second way, the triviality of the world is a further instance of a challenge to the realist stance that the realist stance cannot yield to, and consequent occasion for the *mauvaise foi* discussed briefly in the Prologue.

It is clear that as a practical matter (that is, in taking the realist stance), we "edit out" the triviality of the world, much as we sometimes edit out (or suppress or ignore) inconsistencies among our beliefs or our preferences. At this juncture, semantics (now trivialized) defers to pragmatics and to the ways in which human agents edit and use information in the course of belief-management, practical reasoning, and action. Although all propositions now come with the paradoxical value {T,F}, in practice the human agent will make propositions usable by dropping T or F. In this way, as a matter of practicality, he will impose a classical working-semantics on his experience. So, contrary to what is still widely believed, any nontrivial working semantics will be parasitic on a pragmatic theory of how to make sense of experience in a trivial world. How in general does a rational agent know what truth value to retain from the set {T, F}? The answer might be that he selects the truth value that endows the proposition bearing it with the best chance of integration into his web of working beliefs, which is a "going concern" as Quine says. It may be that, experience being what it is (and we, too, being what we are), the classical working semantics is, in its particular valuations, something that is naturally selected for, or perhaps, in part, a matter of broadly cultural endowment acquired early rather than late. Whatever the details, it is a fact that, in general, the valuations are selected as the world (as we experience it) calls for them.

In special kinds of cases – perhaps the Russell set and Tarski sentence are such cases, and so, too, the internal contradictions of fiction – our experience will misfire in semantically mystifying ways, that is, in ways that would make the agent's working semantical apparatus mimic the triviality of the world if something were not done about it. The inconsistency would threaten anew to trivialize our experience. Such, as I say, might be the discovery of the paradoxes or of Keith's election and nonelection in the year 2055. At this juncture our rational agent, if he were of paraconsistent bent, might be persuaded to modify his working semantics and embed his selectively recalcitrant experiences in a dialethic logic. This he would surely wish to do were he uncertain about how to rid himself of the semantic noise. He might yield to the orthodox in the case of Russell's set and Tarski's sentence and put his money on a restructured ZF or on a hierarchical formal semantics. He also might wish to do the same for the true contradictions internal to fiction. Even so, he would need to tell us how he intended to proceed. Details of this *perestroika* would need to be laid down. Which of the [A]-[E] axioms would go, and which semantic analysis would serve, as faithfully as could be, the facts of Keith's interesting electoral life? The paraconsistentist offers contrary encouragement. Reconstruction of this kind is needed across the board. Not for set theory, not for truth theory, and not for fictional discourse. A paraconsistent optimist might propose that his own remedy possess three main attractions for any serious-minded literary semanticist. It fulfills our five intuitive axioms, so far as we can tell. In particular, it gives free sway to the author's say-so. It hints at an account of the dependency of nontrivial semantics on pragmatics. That is, the semantic provisions that we make for our experiences are captured in a good pragmatic theory about how we edit the trivial output of the semantics of the world, namely, in ways that make for the most useful kinds of belief-management (including, at a minimum, ways of belief-fixation that keep us alive). It recognizes the *partial* paradoxicality of our experience, in the light of which it motivates a paraconsistent revision of what is still basically a classical semantics for experience.

There are problems. We said that the demonstrable triviality of the world is not reflected in experience, that is, we do not experience the world as trivial. This suggests a kind of Kantianism in which practical knowledge, knowledge on which we act, knowledge that we pay the bills with, is confined to the world of experience. Notice, however, that we have given free sway to auctorial say-so. This is disastrous. It is the undoing of our theory. For, in da Bloo's story, every proposition is experienced as both true and false. Not only is the world thoroughgoingly paradoxical, but so, too, is the world as experienced.

It is, if anything is, obvious that the world is not experienced as trivial. This presents us with options. One is to reinvoke the original solution of the problem of the demonstrable triviality of the world, by pleading now a distinction between the world of experience and the world as experienced. We would now hazard that although the world of experience is demonstrably trivial, it is not experienced so. It could fairly be said that there is a distinction without a difference and that, beyond the original distinction, such discriminations are blatantly fraudulent. Even if this should be wrong, the unbridled entitlements of say-so can overwhelm every such distinction, however it is drawn. This would require invocation of infinitely many such distinctions to make experience proof against triviality.

It is clear then that we cannot give free sway to say-so. Doing so overwhelms us with a nasty semantic virulence. Let us agree that say-so might fairly be constrained by the condition that its provisions do not offend against self-evidence. It is obvious that experience is not trivial. And that is that. The semantic omnipotence of authors is now brought into line with omnipotence of God himself. God cannot create a rock too heavy for divine heft. Djaitch da Bloo cannot semantically subdue this kind of self-evidence.

Self-evidence bears all the weight of the strategy. It is doubtful that it can do so entirely successfully. Pending a convincing and clear exposition of it, could it not be protested that a like refutation will overturn the triviality of the world itself? For is it not equally self-evident that the world itself is nontrivial? The answer, as I say, awaits on an account of self-evidence. But if we are sufficiently Kantian, we could retain the original claim, arguing that since the noumenal realm is unknowable; it cannot be self-evidently known not to be trivial. This leaves us with a question: Even if it is not self-evident that the world is nontrivial, what good reasons are there for holding that it is? It will not do merely to invoke auctorial say-so, since this already has been denied free sway. What point is there to giving it free sway about the world but not about the world as experienced? It could only be the fruits of a theory that turns on and is driven by the decision to warrant the provisions of say-so differentially. We will need to produce the theory.

The dialethist has a stake in making such a theory work. It furnishes him with his primary datum. Although the world is trivial, the world of experience is not. The world of experience contains, as true inconsistencies, the internal contradictions of fiction. It would have been clearer had the primary datum revealed itself as the set of ordered pairs $\langle \Phi, \ulcorner \neg \Phi \urcorner \rangle$ with Φ true by the author's say-so and $\ulcorner \neg \Phi \urcorner$ true by the ways of the world. But the semantic virulence of unconstrained say-so denies him this datum, thanks to da Bloo's mischief. The Kantian option, although more complex and less free-flowing, will give him what he wants. Never mind about the world, the world of experience contains true inconsistencies and the world of experience is nontrivial. Is this not a proof that the logic of experience is dialethic?

Indeed it does, provided that the Kantian option is coherently sustainable independently of its production of the primary datum. Sustaining the Kantian option is largely a matter of getting clear about the provenance of say-so. Whether we contain it outright or merely in application to the world of experience, auctorial say-so is now significantly disabled as a serious theoretical tool in the semantics of fiction. Various courses are open to the theorist, none of them congenial to our original axioms. We could put it that with regard to the world of experience, say-so may be allowed to make some falsehoods true, but not all. The need to mark a boundary between some and all is a clear invitation to arbitrariness. We might take special note of the fact that the trouble caused by da Bloo's story – its semantic virulence – was occasioned by the author's employment of semantic predicates and their attachment to propositions. Once noted, we could try to mark the boundary of say-so's legitimacy precisely at the point at which virulence has its origin, and thus restrain say-so from making *semantic* fictions about propositions (shades of Tarskian metalanguages). It will not work. It is egregiously *ad hoc* on its face and, in any event, the total damage of virulence can be got by making all events both obtain and not obtain and all states of affairs both realized and not realized. Given the standard conventions about such things, the author's output would still be semantically significant. For although we would not longer permit the author's vocabulary to contain semantic predicates, by the provision of say-so itself it would still be *true* that every state of affairs is both realized and not.

Better, then, to disable the author's say-so outright. Once done, we could say that though the author proves that certain things are so, it cannot be said that they are *true*. This is not by any means a measure of desperation, for there is an antecedent intuition that favors it, quite apart from the distinction of semantic of virulence. Those attracted by the intuition find it natural to say that although Holmes did live in Baker Street and did not live in Berczy Street, it cannot be said that it is strictly true that he did. Whereupon a suggestion: Since more things are correctly assertable than are true, why not reconstrue auctorial say-so as a condition on assertibility, not truth?

Although the present proposal is a retreat from axiom [D], it tries to leave the others intact. In particular, we are still permitted to retain the conviction that "Sherlock Holmes is clever" should not be assimilated to "The present king of France is bald." Assertable fictional sentences are thus not neither true nor false; they are false (whereupon another waiting intuition is cashed). The alternative gappy strategy is an interloper from on high intent on preempting "true" (which is now unavailable anyway) or to enforce a disagreeable orthodoxy – that existence is a condition on reference.

I resist this. I do not yield on "There are things that do not exist," or on the clear-eyed notion that Holmes is individuated by the writings of Doyle extended to the maximal account. Doyle individuates Holmes by way not of truths but falsehoods. The falsehoods that carry the day are precisely the falsehoods made correctly assertable by say-so plus the maximal account. That is the trick and the charm of fiction. This is not far off the mark of physics. Its frictionless surfaces are individuated by falsehoods, for there are no frictionless surfaces. These falsehoods are knitted together mathematically, and the lot are held assertable by conventions concerning idealizations of nature, whereas the Holmes's falsehoods are knitted together by say-so and the conditions of default logic.

The present suggestion has, as I say, other virtues. It allows us to preserve the conviction that we know who Holmes is, for he is the individual individuated by those falsehoods. It also honors a somewhat weaker, but still attractive, variant of the conviction that we know what Holmes did. Knowledge implying truth, we must pull in our horns, but only sightly. We know what can rightly be asserted about what he did.

Assertability is not here a condition on truth. We are not forwarding a theory of truth in the manner of Dummett. We are not forwarding a theory of truth at all.

Anyone of Davidsonian bent will be interested to know how the sentences of fiction, presently construed, are learned. It is a conspicuous feature of what answers this question that fictional sentences do not oblige Tarski's Convention T. Let Σ be any assortment of assertable fictional sentences. That is, let Σ's sentences be those made assertable by the author's say-so plus the maximal account, and let S be any member of Σ. Then Convention T is honored in one direction,

(1) If "S" is true, then S

but not in the other,

(2) If S then "S" is true.

In fact,

(3) S but "S" is false,

is what our present hypothesis provides. Accordingly,

(4) "S" is not true.

But by a rule that is not in doubt,

(5) "Not-S" is true.

From (1), this gives

(6) Not-S.

But from the construction of Σ,

(7) S.

The orthodox will say that such is the price of trifling with Convention T. But it is worth noticing that although from (6) and (7) we have it that

(8) $S \wedge \neg S$

we do not have it that

(9) "S" is true $\wedge$ "$\neg$ S" is true

given the unavailability of (2), one half of Convention T. Still, we are left, if not with a semantic contradiction, then an inconsistency pertaining to assertibility. It is bad news all the same, since if we flesh out (8) in the manner intended we get something like

(10) It is true that "S" is assertable and "¬ S" is assertable

which, if it is not a semantic contradiction, leaves the relevant structure of its conjuncts somewhat obscure.

Worse still, since Σ can, by da Bloo-like connivance, be made to contain all sentences negating any sentence true of the world, we must say of any sentence true of the world that its negation is assertable and not. No paraconsistentist intervention will stanch this hefty multiplicity of unwanted assertibles. It comes straight from say-so itself, abetted by paraconsistently innocent principles of derivation. Both the world and the maximal accounts of fiction are awash in it, bad news in each case.

This is crucial. If we forbid an author to fill up Σ in the manner of da Bloo, we seek a perch from which to inoculate against triviality. Still, our old problem recurs: How are we to constrain say-so?

GAPPER REDUX

A further retreat is necessary. Perhaps we need to concede that though rightly assertable in some sense, the sentences of Σ are so in a way that falls short of honoring (3). If so, the detachment of:

(11) S

from

(12) "S" is correctly assertable

is banned outright. This is close to throwing in the towel entirely, but no other recourse seems possible. So the assertable sentences of Σ require some modest buildup of structure, say, by attachment of the adverbial construction "in-fiction" or "in-the-Doyle-maximal-account," or some such. If we imagine the adjustment to have been made, the sentences of Σ would now come forward reconstructed as "f(S)."

Questions press. What is the structure of these artifacts? What are their closure conditions? How are they assimilated to the routines of defaulting reasoning? Do they have truth conditions and, if so, what are they? Do they merit the restoration of Convention T? (Do we have it that "f(S)" is true if and only if f(S)?)

One thing is clear. It still cannot be allowed of any pair of "inconsistent" members of our restructured Σ, "f(S)" and "f(¬S)," that the occurrences therein of "S" and "¬S" are false. Otherwise, the following quartet would emerge:

(13) f(S)

(14) "S" is false

(15) f(¬S)

(16) "¬S" is false.

From (14) we have

(17) ¬ S

and from (16)

(18) "S" is true.

By the usual conjunction properties,

(19) "¬S" is true and "S" is true

and so

(20) "¬S and S" is true.

The gapper lurks nearby. It suffices, he says, that "S" and "¬S" are *not* false (and not true either). Given that the gapper rejects Priest's rule

(21) Φ is not true ⊢ Φ is false

the abundance of inconsistencies in the manner of (20) is evaded.

I have already said why I do not fancy a gappy approach to the theory of fiction. It makes of "The present king of France is bald" an analytical paradigm. Present-day French monarchs seem massively wrong for fiction. The gapper could concede the counterintuitiveness and plead his case strategically. Taking the gappy approach avoids triviality.

The author's say-so, now a condition of the truth (or assertability) only of sentences of the form ⌜f(Φ)⌝ and ⌜f(¬Φ)⌝, supply the paraconsistentist with a goodly supply of inconsistencies, as many as an author may please. It is still a problem as to how much it is appropriate to let semantic invaders such as da Bloo get away with. If he can get away with his invasions unconstrained, then the world is indeed trivial, and the theorist will have occasion to beat a hasty retreat to our rather desperate "Kantian" fallback. If da Bloo is allowed to get away with less, then the world is still inconsistent, and the dialethist alone can spare it the embarrassment of triviality.

It is rather quaint that the apparently harmless device of auctorial say-so, even with all the protections of the f-operator, should be so devastating for semantics. The say-so criterion, even when fitted for the production of truths only in the form ⌜f(Φ)⌝, threatens to destroy the world, such is the import of triviality. Structural affinities with the innocent-seeming axioms of set theory are apparent, as with the modest devices of semantic self-reference in the raw. The orthodox are led to the conviction that axioms for sets cannot be allowed to produce inconsistencies, and devices for semantic self-reference, too. This seems dead wrong for say-so, but I know of no way of proving this to be so.

We have resisted the idea that the tie between the glutter and the gapper should be declared broken in favor of the gapper. The gapper offers some attractive economies. Doing things his way spares us the nuisance of having to put the bridle on say-so, all the more appealing since we have not figured out a way of doing it anyhow. Whether we say that the $\ulcorner f(S_i) \urcorner$ are true or assertable but not true (provided that we hold the embedded Φ_i to condition of truth-valuelessness), the author can all but say what he likes. The gapper can propose further economies as well. It is not just that opening the gap solves a nasty problem in the naive theory of fictional discourse; it also solves or helps solve bad problems in naive set theory and semantics. Opening the gap thus saves, or brings under control, not one problem, but three, a noticeable methodological economy. A three-problems-for-the-price-of-one maneuver has a nice whiff of expectant generality about it.

The gapper's remedy exacts costs of its own, of course, as is the way with hard problems. One of the costs is that the following will no longer be true:

(22) David Lewis has done a lot of thinking about Holmes over the years.

(23) Holmes has had a large cult following in several cities of North America.

Neither will bear construal *de re* on the gappy approach, since there is nothing truth-valued to say about the value of the bound variable, as in "There is something x of whom David Lewis has done a lot of thinking about in the past few years." Neither is our f-construction here available. It is *not* true-in-fiction that David Lewis has done a lot of thinking about Holmes in the past few years.

Plainly (22) and (23) will require ruthless regimentation, the less the truth the higher the cost. Parsons's suggestion that gappers and glutters are natural rivals in some quarters – in the theory of truth, for example – should not obscure the fact that in other contexts rivalry gives way to out and out chumminess. **PR**, let us recall, permitted truth value gaps, and the early system of Meyer and Routley is thoroughly gappy, though not without the corrective of supervaluations. Perhaps a further rapprochement might be achieved in a logic of fiction of the $\ulcorner f(\Phi) \urcorner$-sort we have been examining. Some might say that there is no "perhaps" about it, and that it is obviously so inasmuch as Heintz's own dialethic apparatus is the very thing. It is not. Heintz's device works for internal contradictions only, unsupplemented by modal considerations in the manner of the f-operator. In the Bradbury story – that is, in the maximal account of it – it emerges that:

(23) Keith was elected president in 2055 $\wedge\neg$ Keith was elected president in 2055

and the indiscriminate prosperity of consequences is blocked paraconsistently. Heintz makes no provision here for semantic virulence of the da Bloo sort or for the plenitude of sentences that, made true by an author and only thus, are likewise made untrue by the world.

We have seen the utility of the f-operator. It spares a sentence false in the world from being true by the author's say-so. As long as we have no satisfactory and credible means of muffling da Bloo-like maneuvers, this is nothing but desirable, quite apart from its preserving the insight that sentences true in fiction are not strictly true. We must posit either ambiguity or tacit structural differences if we are to spare sentences the nuisance of unwanted inconsistency. Each is a hard sell. It is far from obvious that "Sherlock Holmes solved the case of the speckled band" *is* ambiguous in English, and yet calling down "f(Sherlock Holmes solved the case of the speckled band)" creates the necessity to say something about its internal structure. If we take the gappy approach that hovers in the past few pages, some guidance is given: For any sentence Φ, its occurrence in $\lceil f(\Phi) \rceil$ is non-truth-valued. This is shaky guidance indeed, for now it must be explained how it is that f manages to take the truth-valuedly robust "London has a population of one hundred million souls" into semantic oblivion in "f(London has a population of one hundred million souls)," the latter in turn being true in any story making it so. The problem recurs with truths antecedently true by the ways of the world that an author decides to borrow for a guest appearance in an f-context.

Care must be taken not to empower the f-operator carelessly. It is ludicrous to see it as a truth-value annihilator. It will work well enough only for sentences attracting antecedent allegations of truth-valuelessness, for sentences in which, for example, there are suspicions of reference-failure. "Sherlock Holmes lived in London" is, so gappers say, non-truth-valued just as it stands. It causes no harm, or no further harm, to suppose it thus in f-contexts. But, again, "London has a population of one hundred million souls" and "Paris is the capital of France" are, just as they come, free of any hint of truth-valuelessness. It cannot seriously be entertained that our modest device of adverbial qualification will send *them* into semantic oblivion.

INCOMMENSURABILITY

The f-operator is unsatisfactory in other ways. Among those who have been attracted to it, there has been a tendency to assign it multiple tasks that it cannot jointly perform. One of those tasks is to mark off the sentences of fiction from the others in ways that allow for the blockage of inconsistency's omnideducibility, while leaving those sentences as untrifled with as possible so that their maximal accounts can be entrusted to the routines of default reasoning. It is a balancing act, and hard to bring off. If the sentences of fiction are permitted to behave very like the others – if, for example, they are assigned similar closure conditions – it is difficult to keep inconsistency at bay.[10] If we stint on closure conditions, we may be able to handle problems of inconsistency, but we will have trouble providing for construction of maximal accounts in a natural way. Still, virtually all of the problems that we have here been examining will go away if we are especially hard on closure. This is an approach favored by David Lewis and one can now see why. Lewis will not permit any deduction containing mixtures of f-sentences and non-f-sentences (Lewis, 1978/1983, p. 264). Moreover,

> We should not close under the most obvious and uncontroversial sort of implication: the inference from conjuncts to conjunctions. (Lewis, 1978/1983, p. 278)

Such austerities are invasive therapy of high potency. On the other hand, some will find that the remedy has little to recommend it but the success born of avoidance. It is Groucho Marx's doctor's remedy.

> *Groucho:* Doctor, my shoulder hurts when I lift my arm like this.
> *Doctor:* Do not lift it like that.

An interim measure suggests itself. It is counterintuitive and *ad hoc* by my lights, but it may be the best that we can do for now. Perhaps we should call down for fictional sentences a strong incommensurability thesis and refuse intercourse with any but their own kind in any given maximal account, by whatever devices keep them trouble-free. It is a contrived isolationism, to be sure, one in which trouble is not defeated, but only kept at bay. Our isolationism makes illegitimate some of our more interesting questions – hardly illegitimate on their face – questions such as "If $\ulcorner f(\Phi) \urcorner$ is true and Φ false, what is the semantic status of Φ in $\ulcorner f(\Phi) \urcorner$?" Hard cases make for bad law. As in life, so at times in formal semantics.

Thus constrained, isolated sets Σ_i of fictional sentences can themselves be eased up on. Almost certainly the Σ_i will be able to tolerate more closure than is permitted by Lewis. Here more is better, for the Σ_i-maximal accounts will be easier to produce and with greater default logic-naturalness. Given that our interest is now entirely internal and story-specific, worries about the semantic-structure of f-sentences largely evaporate. The f-operator now imputes no significant structure, save that of membership in the appropriate Σ_i. And with our gaze now inward, we can let internal contradictions arise as they may. If they arise in Bradbury's way, as with Keith, nothing more is needed than Heintz's dialethic device to spare the story a trivial maximal account. If the inconsistency arises in da Bloo's way, then that story is trivial, and its triviality is proof against dialethic safety-measures, which is precisely what its author intended it to be. This is no rebuke of the dialethic literary semanticist, since his rightful interest is to keep stories from inadvertent triviality; and this, we may now say, he can manage to do.

Can the dialethist claim a primary datum? It would appear not, on reflection. The benefit of the datum is canceled by the high cost of losing all prospects for naive literary semantics.

There exists, in any event, an important asymmetry between, on the one hand, dialethic set theory and truth theory, and dialethic literary semantics on the other. If the dialethist is right, then the appeal of his approach is that it enables the recovery of naive theories of sets and truth. But, if I am right, no dialethic maneuver will rescue the naive theory of fictionality.

We have seen that far and away the most creditable justification of dialethic logic turns on a reckoning of costs and benefits. The benefits are set theory and semantics done in better ways than they are now, and the costs are slight and manageable adjustments of the classical apparatus. It is of undoubted importance that these projects are far from *faits accomplis*. The dialethic semanticist has issued interesting promissory notes but has not yet been able to redeem them; and it is not entirely foolish for the classicist to be openly skeptical about future prospects for redemption. It follows, then, that the best justification that currently exists is really a justification of the dialethic research program, not of results yet to be realized. The justification would have been stronger, would have been a justification of strongly dialethic logic itself, had we been able to find datum for the primary dialethic. But, as we have lately seen, that seems a forlorn hope. The open-minded skeptic may wish the dialethic research program well, but he has no rational basis for any greater optimism than is conveyed by "We shall have to wait and see."

The lack of a primary datum for dialethism matters for entailment, too. A primary datum, if such there were, would afford the adjudication problem or, more exactly, of that part of it that turns on the status of *ex falso*, a dialethical footing. The datum would by definition furnish a sufficient reason for saying that the world is negation-inconsistent; and since it is transparent that the world is not absolutely inconsistent, *ex falso*, affirming the opposite, would fail. If the reflections of this chapter are right, no such datum has yet been produced. This alone guarantees a similar fate for the disagreement about *ex falso*. In the end, *ex falso* and dialethism itself languish in the same methodological boat. As things are now, only an economic solution seems possible, a solution that turns on a judicious reckoning of costs and benefits. In the case of fictional semantics, the returns are in, and they seem not to attach a sufficiency of benefits over costs to give the nod to dialethism. In the case of **N**, dialethically naive set theory, and **Sem**, dialethically naive semantics, not enough of the returns are in to justify a verdict for the paraconsistentist. But, as we have said, there *is* a research program underway. It should be persisted with.

Of the prominent theories of fictionality, it remains to examine the make-believe approach. This need not detain us long, given that a fundamental tenet of make-believism is that sentences about the fictional are in no sense true. Even so, it will prove instructive to see how badly a theory of fiction serves the requirements of a fictionalist account of mathematics. We shall confine our

remarks to Kendall Walton's well-regarded book, *Mimesis as Make-Believe:
On the Foundations of the Representational Arts* (Walton, 1990).[11]

Here is a bare-bones summary of Walton's theory:

A work is fictional if it is used as a *prop* in the *game* of *make-believe*. A proposition
is fictional if there is a *prescription* to *imagine* it.

There are two observations we might make. One is that every one of the itali-
cized words is used in a nonstandard (and generally speaking, unexplained
way). The other is that the analysis of what it is to utter, for example, "Sherlock
Holmes lived on Baker Street" is psychologically unreal. Let's take these in
order. What is a prop? It is an article used in staging a play. If x is a prop, it
functions in the play as a thing of type K, though it is not a thing of type K in
fact. (Think of a cap pistol serving as a prop for a. 45 automatic.)

What is make-believe? In its primary sense, it's what actors do. Christopher
Plummer *plays* John Barrymore in ways that get the audience to see the action,
with appropriate and utterly necessary circumspection, as Barrymore's. In the
loosest possible sense, Plummer *get us* to believe that this is happening to
Barrymore. Hence *make*-believe. In a secondary sense, make-belief is pretense.
The con man pretends that he is a bank investigator. But he is not; it's all make-
believe. Make-belief and props are made for one another. When Plummer
uses the prop (e.g., the tea in the whisky bottle) in the appropriate way he
is pretending to drink whisky. But, fortunately for Plummer's liver, it is only
make-believe.

The flexible use of *game* we remember from Wittgenstein. Greater flexibility
still is encouraged by the growing popularity of game theory (it seems as if
game theorists are ready to make a game out of anything whatever). Still,
there seems to be a sound and undisturbable distinction between constitutive
conditions and rules. If so, the *conditions* under which my bursts of acoustic
disturbance constitute a stretch of discourse should not be called *rules*. Of
course, if I do not meet the conditions then I am not speaking English. And
I might fail such a condition deliberately. But it is overfigurative to say that I
have broken a rule. So *War and Peace* is not a game, and doing so harbors no
prescription to do anything. True, if I do not handle the sentences of that novel
in the right way, then it cannot be the case that I am reading a novel. If I wish
what I'm doing to be novel-reading, I must meet the constitutive connections
on novel-reading, but it is a *façon de parler* (and not a load-bearing one) that,
in these circumstances, I am obeying prescriptions.

What, finally, is it to *imagine* the proposition that Sherlock Holmes lived at
221B Baker Street? It is nothing, if "imagine" is intended as in "Imagine Paris
on D-Day!" What is meant here presumably is "Imagine that Sherlock Holmes
lived at 221B Baker Street" or "Take it for true that Sherlock Holmes." But
doing this is not pretending that this is where Holmes lived, or make-believing

it. I can pretend to be going to visit the bank that currently occupies that address, when in fact I am going to the races. If I did in fact visit 221B Baker Street, I suppose that I might (privately) make-believe that I am Sherlock Holmes, returning home from the country; or I might make-believe that I am a client, slightly late for my meeting with the great man.

There is also – as I said – the related point that the Waltonian account of what it is to read *War and Peace* is psychologically unreal.

It bears on our question that Walton's approach is shot through with a nonstandard – and in effect a technical – vocabulary. What is missing is a suitably precise explanation of it. This is not to say that it cannot be explained, but rather that it needs to be. And it needs to have an interpretation that lends the theory some plausibility.

Apart from this, there are problems that afflict the core of Walton's positive account:

To utter "Sherlock Holmes lived on Baker St.," is to assert "The Sherlock Holmes stories are such that one who engages in pretense of kind K in a game authorized for it makes it fictional of himself that he speak truly" (where pretense of kind K is exemplified by the utterance of "Sherlock Holmes lived on Baker St.").

There is a question about the second occurrence of "Sherlock Holmes lived on Baker St." Is it an occurrence of that of which the quoted sentence on lines 1–3 is also an occurrence? If so, is this sentence in its second occurrence open to the substitution that the quoted passage authorizes? If so, how do you stop an infinite regress? If not, of what sentence is the "second" occurrence an occurrence? If, on the other hand, they are both occurrences of the same sentence, and the second occurrence is not open to the substitution authorized by the identity sanctioned by the quoted passage, why not?

Assuming this problem to have been dealt with, assume that what in this passage is said to be "asserted" is true. Then, since to utter "Sherlock Holmes lived on Baker St." is to make a true assertion, it would seem to follow that when Walton's conditions are met a fictional sentence is *true*. But the heart and soul of the Walton approach is that fictional sentences are not true, and that this does not matter.

Perhaps Walton's is not a convincing account of *fiction*. But this need not preclude its affording some insight to the foundations of mathematics, however improbably that might be. Applied to mathematics, Walton's theory provides as follows: To utter "Every number has a successor," is to assert "Arithmetic is such that one who engages in pretense of kind K is a game authorized for it makes it fictional of himself that he speaks truly" (where pretense of kind K is exemplified by the utterance of "Every number has a successor"). (Of course, the prior questions apply here, as well.)

Let X, Y, and Z be arithmetic sentences satisfying the conditions of this passage. And let

(1) X
(2) Y
(3) Z

be proof of Z in Peano arithmetic. In this proof X, Y, and Z are uttered; and by the provisions of the quoted passage, in each case this is to assert what we might call "a Waltonian." So we have three Waltonians, X_w, Y_w, and Z_w. When we arrange these as

X_w
Y_w
Z_w

the last thing that we get is a proof of Z_w (to say nothing of Z). So is it not the case that Waltonization destroys all the proofs of number theory? (A fate similar to a quite general type of problem for paraphrase theories of fiction.)

We began this chapter in hopes of finding a dialethic primary datum. We also thought we could lend encouragement to theories of stipulative truth. The attraction of such a theory is the finish it lends to theories we have been calling formal. A formal theory is studded with statements that commend themselves to the theorist in various ways, some of which are economic. But they need not appeal to the point of psychological belief; indeed, we have seen that statements accepted by the formal theorist can be objects of his disbelief. Still less does the appeal of such statements lie in their objective truth; indeed even their stoutest boosters will concede that if true at all, their truth is "internal" to the theory or to the "game" in which the theory is a systematic set of moves.

How, then, could a stipulationist account of truth "finish" a formal theory? It could finish it by removing its anomalies. Instead of the embarrassment of accepting claims that he actually disbelieves, the formal theorist can now believe them while continuing to believe that they are not so independent of the theorist's intervention. He can also explain his difficulty in finding them robustly true, for indeed, while they are robustly true, they are not so independent of the theorist's creative involvement. Had we found a relatively unproblematic primary datum for dialethism, it also could have deputized as a primary datum for stipulative truth. We would have had it that authors stipulate in ways that make things true, that these are real truths albeit, *modulo* their stipulative status, about unreal things and states of affairs. In their place, stipulative theories could produce all the truths of which they are capable, without the silliness of having to pretend that sets and integers and the like have the reality of a punch in the eye. Truths about sets would then be real truths about unreal objects; objects, in other words, having no status prior to the theorist's creative intervention.

The difficulties that plague the dialethist's primary datum also discourage a like motivation of stipulative truth. While discouraging, the difficulties are not decisive. We may note that stipulative truth is well catered for even if my minimalist idealism is true. Many people think that idealism *is* true; and a good many more go about their theoretical business as if it were true, apparently innocent of the fact. Even so, it behooves us to say more about what stipulation is before offering it up for serious theoretical consideration.

STIPULATIONISM

With a little imagination, we could think up a set theory as stupid as da Bloo's story about Richardo Bosque. In a way, it would not matter, even if we gave truth-by-stipulation as free a sway as in the beginning we gave to the author's say-so. At the very heart of any account of fictionality is the truth-making potency of auctorial say-so. It is no less so that any serious stab at a stipulationist theory of truth must acknowledge the primacy of the insight that any stipulation produces a *truth*.

Left unfettered, these are vastly overgenerous entitlements. It is fortunate that, in fiction and abstract theorizing alike, there is in fact some evidence of resistance. It is a *quid pro quo* of da Bloo's semantic promiscuity that the truths that he created are true-in-the story, true-in-"The Mischief of Ricardo Bosque." They are truths all right, but not truths that have outgrown their hyphenation. It is the same way with stupid set theories (and some that are not stupid at all). The fruits of their constituting stipulations are hyphenated truths only. They are true-in-system S, untrue-in-system S*, and so on. They are *playful* theories in the manner of Chapter 4, in which a playful theorist is someone who is content that the output of his stipulations be hyphenated truths at best. Stupidity is not the central condition, even though some playful theories are stupid.

A like recharacterization awaits our idea of a formal theory. Here, too, a theory's formality is parasitic on what can be said about its theorist. A theory is formal to the extent that its theorist seeks and realizes that he may not succeed in finding truths that stand some chance of shedding their hyphens.

Russell saw that stipulations produced hyphenated truths, and that communal acceptance of them eliminates the hyphens. When this happens, as appears to have been the case with ZF, the theory in question is, or approximates to being a *received theory*. A received theory is a set of stipulated truths, and others proved from them, whose acceptance by the appropriate disciplinary community or disciplinary culture induces in such *groups* hyphenation-blindness. Integral to this blindness is the sheer psychological heft of the realist stance, abetted by the conditions under which later generations are *introduced* to such truths (they are introduced to them as settled facts). It is a powerful dynamic. Stipulated truths are hyphenated truths. If, as sometimes happens, these become received truths, those who see them so fail to see their hyphens; in fact, fail to see their hypens even if they are made aware of the received theory's stipulative origins. In this some may detect grave defections from rationality and honesty. I see the psychological dominance of the realist stance.[12]

If, as we have tried to show, there is nothing to be said for the method of intuitions, it will matter considerably what we have to say about received opinions as presently conceived. But before taking the question up, it may be of interest to see some parallels with fiction. A stupid story, such as "The Mischief of Ricardo Bosque," is constituted by closures on auctorial stipulations. The truths of that story are hyphenated. Because it is a stupid story, it will never attain a story's counterpart of the status of a received theory. I will say that such a story has no chance of becoming *canonical*. "Canonical" here is a technical term. A story, or a literary corpus, is canonical in a community of readers or in a literary culture to the extent to which its readers or the members of that culture have become blind to the hyphenation of its truths. Stupid stories are unlikely to become canonical, but stupidity is not the central point. Most works of fiction are not canonical; or, if canonical, are so for extremely small readerships and for short periods of (cultural) time.

Philosophers such as van Fraassen see the need to reconcile what I have been calling abstract sciences to the model of fiction. I see it the other way round. Fiction needs to be reconciled to the paradigm of abstract truth. Here is why. All known theories of fiction, save those that make no allowance for hyphenation-blindness, run into an extremely bad problem. Imagine a Russellian inventory of the furniture of the world, a complete ledger of how the world is. Then, except for real-world truths borrowed for fiction, no fictional truth will be on that list. In a great many instances, truths that are on the list will give every sign of being incompatible with those that are not, the very propositions of which we *want* to say that they are made true by an author's say-so. In the Russellian inventory there is a serious discouragement of unhyphenated fictional truths. This is why, toward the end of Chapter 5, we came close to throwing in the towel. I propose that we do so now. The semantic *phenomenology* of fictional truth is undisturbed by this decision. The semantic *reality* is that all truths of fiction are hyphenated, and that this being so, there is no discouragement to be found in a Russellian inventory of the ways of the world.

If, however, we take this to be our model for the semantics of abstract theories, then there is no abstract truth that is not hyphenated. This would be as it should be except for a point of major difference between fiction and set theory, let us say. The difference lies in the Russellian inventories record of the existence and career of every individual human being up to now. Sherlock Holmes will not – and cannot – be found in that number. Much of what Holmes did in life involves the world in ways contradicted by how the world has been in fact. And yet it is extremely difficult to see how a Russellian inventory of the world is like discouragement for set theory. If this matter is incompatible with the existence and behavior of sets, then that a Russellian inventory at a time fails to record sets and their behavior is nothing more than reflection of a fact of irreducible importance for stipulationism. It is that stipulated truths cannot predate their originating conditions. In this, Russellian inventories disclose their fundamental dynamics. Facts are inventoried, but not before they become facts. This is so independently of what brought those facts about, whether a destructive cyclone on the Canadian prairie or an insightful piece of mathematical innovation in Cambridge.

I conclude that of the pair {the logic of fiction, the stipulationist theory of truth}, if either has lessons to learn from the other, it is not in a manner that satisfies our principle **Fictionalism**. It is the other way round. *Fictionalism* bids the stipulationst to do his basic training in the logic of fiction. In truth, it is the logician of fiction who seeks guidance from the stipulationist. Stipulationism is a stable theory precisely where the logic of fiction lacks purchase. Fiction has a quite general rival in the way the world is. Stipulationists' rivals are other stipulationists. The rivalry between fiction and the world is settled conclusively in favor of the world. Rival stipulationisms retain similarities with fiction, but with a crucial difference. When contending stipulationist theories lack the resolution device of received opinion, they must reinvolve their variable and compatible origins. But the standing of the received opinion is nevertheless available in principle to one or the other. We said that canonical fiction resembles in certain respects the received opinion in an abstract science; and so it does. But here, too, there is a difference. Even canonical fiction is contradicted by the world.

Some will see in the contrast between fiction and stipulationistically abstract theories an even more serious disadvantage for fiction. If what we have been saying in these pages is correct, fiction lacks a coherent policy for closure, whereas set theory – to persist with this example – enjoys closure procedures that are by contrast plausible. Bearing on this is the primary deranger of smooth closure conditions for fiction. It is the inordinate difficulty in integrating what we know of Holmes with what we know of other things. Set theory, on the other hand, fits all of mathematics, and most of what mathematics is needed for, like a glove. This is not to say that stories lack maximal accounts at the level of readerly practice. Even the untutored reader is adept in discerning what follows from what among the sentences penned by Doyle. The difficulty, and the failure, prevents itself at the level of theory. With stories, we lack a coherent theory of maximal account construction; whereas for sets the maximal account is, to all extents and purposes, set theory itself.

A last word is owed the issue of realism. In its classical forms, realism requires objectivity, and objectivity implies both antecedence and independence. If, for example, we are classical realists about some true proposition p forwarded by a theorist S, then p is made true by how the world already was (or is concurrently with p's initial utterance) and independently of S's role in thinking and uttering p. Stipulationism fails classical realism at its origins. Stipulated truths arise from the stipulator's action and depend for their truth on a fit with no antecedent state of the world. But origins aside, stipulated truths behave as realist truths do. The world may produce a truth at a time t, and a stipulator may produce another truth at time t. If, in the fullness of time, the stipulator's truth attains the status of received opinion, it functions in its theoretical deployments, in its integrations with other truths, indiscernibly from realist truths. With the appropriate diffidence, there is little in the way of ontological assay that is not open to the stipulationist. If p is a truth of ZF, then p expresses a fact about ZF sets. If a stipulationist theory T is unrivaled at t, then the facts T expresses are likewise unrivaled; and this is all but unqualified recognition of T's ontology.

Perhaps there is realism enough in stipulationism to warrant reconsideration of something proposed in the Prologue. "Collectively," we said, "the cost of the idealist strategy is the abandonment of realism." Provided that even our minimal idealism produces sentences that are less than true, the assessment would seem to hold. But stipulationism is not idealism, not even in the version we have been calling "minimal." Even in the early days there are idealist *dicta* to which no stipulating type theorist or ZFer had to plight his fidelity. He need not have conceded – and did not – that his set theoretic truths were either partial or comparative – a matter of degree; he need not have conceded – and did not – that ordinary concepts are inherently (or even routinely) inconsistent; and he need not have conceded – and did not – that all thinking except "metaphysical" thinking is defective. So described, the type theorist or ZFer makes almost a blanket rejection of idealism. What he retains is the absence of antecedent objectivity for his truths at the point of their origination. This cannot do realism any ultimate damage unless it can be shown that the truths of ZF, for example, the truths in the mathematical closure of originating and enabling stipulations must, on account of those origins, wholly exclude any possibility of their being really true. But saying so takes some nerve. It is nerve enough to suggest concession of a genetic fallacy.

If there is anything good to be said about these speculations, two consequences of note stand out. One is that the anomalous realism of the Prologue is perhaps not all that anomalous. The second is that pluralism in logic and the foundations of mathematics, and the other abstract sciences, is no more embarrassing, no more unbearable, than the pluralism implied by theoretical rivalries in the empirical sciences. In each case, resolution devices exist, in principle; and in each case they turn on estimates of costs and benefits (always in the one case, and frequently in the other). Let it not be forgotten that hyphenation and the other forms of system-relativization are applications of the resolution device we have been calling *Reconciliation* or ambiguity-on-purpose.

Before we quit the issue of realism, it is worthwhile to note in passing the existence of what we could call **"Can't Help It Realism,"** which, nearly enough, is my realist stance. The *Can't Help It Realist* stands to realism proper as the weak AI theorist stands to strong AI. That is, he proceeds in his theoretical work, as in his practical affairs, as if classical realism were true. He does this because like anyone caught up in the realist stance, he cannot help himself. Caught as he is, he is awash in convictions, of which the most natural expression is a substantial commitment to realism. If we think of commitments in Quine's way, as the values the theorist assumes for the variables bound by his quantifiers, then his ontology is what those values commit him to. Even so, it is a fallacy of considerable folly – known to Aristotle as *secundum quid* – to infer the reality of what your quantificational behavior commits you to. It is a fallacy akin to the *Heuristic Fallacy* discussed earlier. It may well be true that the theorist cannot do his business, cannot do the business, even, of thinking up his abstract theories without conforming his behavior to realist assumptions as classical as you please. But, as the *Heuristic Fallacy* warns us, what is necessary to believe in order to construct one's theory is one thing; and what the theory must honor among its own derivations is often another thing. So, at a minimum, the *Can't Help It Realist* is at risk for the *Heuristic Fallacy*.

PLURALISM AGAIN

Pluralism in the abstract sciences is problematic when a) its theories are presumed to conform with the method of analytic intuitions, and b) there is inadequate reason to suppose that the theory's target concept admits of the *reconciliation* strategy – in particular, the strategy of ambiguation. In the chapter to follow, we shall take up the important point that the procedures of philosophical analysis do not exhaust the intuitions methodology. A related task will be that of showing that in all its variations, the intuitions methodology fails the requirements of conflict resolution. If we succeed in this purpose, this will be further grist for the mill of stipulationism in the abstract sciences. This, in turn, will encourage an attractive approach toward pluralism.

Pluralism comes in two forms: benign and rivalrous. The benign form, as we have seen, is a natural concomitant of playful theories; and playful theories in their turn are theories tailor-made for hyphenation. Hyphenation itself is made of *reconciliation* and it presumes on the central principle of stipulationism in a weighty way. Hyphenation does not make stipulationism true, but it frees it from what otherwise would be the rampant and unwanted inconsistencies that inhere in stipulationism's considerable latitude concerning what it is able to make true.

Rivalrous pluralism is something to be less sanguine about. It bespeaks a form of disagreement for which hyphenation is deemed a less than satisfactory accommodation. The rivalry of rivalrous pluralism is sometimes the consequence of the failure to engage – or even to recognize – the resolution devices we have been discussing in this book. In some cases, perhaps most, rivalry persists under the three-line whip of the realist stance and the abiding encouragement it lends to the method of analytic intuitions in the abstract sciences. Conflict resolution procedures, whether in the manner of *Surrender* or by way of *Reconciliation*, cannot presume to produce settlements in every case, much less be guaranteed a secure purchase. But even as these procedures falter and fail, the largely tacit social mechanisms that make for received views may be at work, to much the same effect. The dynamics of received opinions does not guarantee a unique position that all enquiries pledge themselves to. But it does have the effect of taking the sting out of once-rivalrous pluralisms. It removes the sting by conceptual variation, which is what raw hyphenation becomes under the ministrations of received opinion. And so, once again we have relevant implication, labeled deductive implication, Lambek implication, and so on. Seen thus, the received opinion dynamic is a *Reconciliation* modality. It gets former disputants to see their pluralism as benign.

APPENDIX TO CHAPTER 6

The Mischief of Ricardo Bosque by Djaitch da Bloo

He had lived in the house in the rue Bounin for more than eleven years. In the early days *madame* had been civil and "correct." But she turned vinegary by degrees, like a slowly collapsing wine, and before anyone realized it, Ricardo least of all, she had dominion over all five tenants. It was met with a passively odd servitude. They endured her calumnies and her scorn, which they were unable to resist, and they were never quite able to act on the intention to leave.

Like the others, Ricardo hated the concierge and feared her. She possessed all the warmth of a du Maupassant peasant – mean, shifty, and stupid – but with preternatural anticipation of a tenant's smallest truancy, his slightest defection from good domestic order, of which she always made the most. It was unsupportable. Things had to be done.

When at liberty from desultory and menial employment, which was often, Ricardo indulged an interest born of a quirky talent. He could build things. A device to turn on the radio at the snap of a finger and thumb; a calculator of transfinite reach driven by a single double-A battery; a scrambler to disarrange undetected the concierge's television signal. Lately he had worked on his *chef d'oeuvre*, a semantic cancelation device, which he called Negator.

Negator took as input ordered pairs of propositions and truth values and yielded ordered pairs of the same propositions and the opposite truth values. That anyhow was its design. Negator was intended to serve Ricardo's meliorism, his interest in making things that are true false, and false true, depending on prior disagreeability. Negator was a device for improving things with complete circumspection. For it to do this, Ricardo had to exploit the nonlocality of physical reality and, surprisingly enough, it seemed that he had managed to do this quite competently.

On the day of Negator's trial, Ricardo locked his door and draw his blinds. Dewy-browed with anticipation, he inputted ⟨"The concierge is healthy," T⟩ and awaited the hoped for ⟨"The concierge is healthy," F⟩. In due course, his satisfaction in her failed health abated, he would offer Negator the pair ⟨"The concierge lives," T⟩, such was the murderousness of his disaffection.

Something went wrong. The inputted ⟨"The concierge is healthy," T⟩ did not produce the anticipated ⟨"The concierge is healthy," F⟩. It produced instead ⟨"The concierge is healthy," TF⟩. Startled by Negator's indiscrimination, Ricardo tried other inputs. He tried ⟨"Ricardo has a minor criminal record," T⟩, ⟨"Ricardo has acne," T⟩, and even ⟨"Paris is the capital of France,"T⟩, and in every case Negator yielded to semantic overabundance – TF withal. Ricardo was perplexed and very, very annoyed.

In the next fortnight he lost himself in the task of making Negator behave. In this he failed. Sleeping badly, eating little, never seeing the light of day (for such was the intensity of his preoccupation), Ricardo did a desperate and ill-considered thing. He put Negator into its transfinite tracking mode (this was its cleverest feature by far – and a novel piece of engineering); he then switched the universal-simultaneous intake valve and dialed the appropriate closure codes. And *viola*! All true propositions without exception were true and false alike.

Such was not what Ricardo had wanted, although it was a powerful and intriguing result worthy, no doubt, of the attention of the French Academy. He would see to it in due course. For now he had to solve little domestic problems, for example, how to shave. Standing before the mirror and not, lathering his brush that stayed dry, inserting a blade in a razor that remained bladeless, gazing into the sharply reflecting glass that was no mirror, problems were encountered that, although they defeated his tonsorial intentions, left him clean shaved and darkly shadowed like the late Mr. Nixon. His life, now entirely without complication, became paralyzingly complex.

Not a strong youth in the best of times, Ricardo began to falter. His decline was attended by radiant good health and clear-headedness. As his fever mounted, the thermometer remained steady at 98.6 degrees. He abandoned all domestic chores, the little tasks of self-preservation, which he performed with timely and cheerful competence.

One night Ricardo died in a troubled sleep of perfect composure and restfulness. He died in the knowledge that he had spoiled God's entire creation, penalty enough for his enslavement to Madame Charcoutière (for that was her name). Madame Charcoutière meanwhile had become a solipsist, and a kicker at the Follies Bergères. Although derelict, Ricardo attended her performances faithfully. "Quite an accomplishment," he mused, "for a landlady who is a pomegranate."

©1980 The Berczy Group. By permission.

7

Currying Liars

It would be naive to underestimate the attraction exerted by the role of intuitions in the lives of theories. Whatever else they are, intuitions are beliefs strong enough to carry epistemological presumptions. Of course, this is precisely what strong belief is structural *for*, since a belief has its strongest hold on a subject when it is a belief in the form "I know that Φ." There is no doubting that intuitions are strong beliefs pretending to epistemological status. Intuitions have the force of what Peirce called the "Insistence of an Idea." What is not so clear, as Chapter 5 tried to show, is whether the epistemological presumption of intuitions is warranted.

There is more than one way of trying to make methodological use of what Peirce calls the Insistence of an Idea. The most important of these ways will be examined in the chapter to follow, and found wanting. It is the business of the present chapter to reflect on one of the ways of making methodological use of "our" intuitions. It is what I have been calling the method of analytical intuitions, used with differential affect by Frege, Russell, and Tarski. I am going to concentrate on points on which all three were in accord, without which they could not have thought *Frege's Sorrow* an appropriate reaction to Russell's Paradox or *Koryé's Sorrow* an appropriate reaction to the Liar Paradox. The point on which Frege, Russell, and Tarski agree is that when knowledge of something X is confined to the analysis of the concept of X and derivations therefrom, then an inconsistency demonstrates that the concept of X is fatally flawed and that no successor concept X′ can play the role in our knowledge of X′-things that the concept of X purported to play in our knowledge of X-things. In this, *Routley's Serenity* plays with orthogonal effect. Underlying both it and *Frege's Sorrow* is the notion of the Elsterian unitary concept, the analysis of which is the theorist's proper task. *Frege's Sorrow* and *Routley's Serenity* mark radically different boundary conditions on conceptual analysis. On the dialethic approach, a correct analysis of our intuitions can overrule the supposition that both what there is and what can be thought are subject to a consistency requirement. The classical position emphasizes the converse: when analysis discloses an inconsistency in the very idea of something there can be nothing whatever that conforms to it.

The method of analytic intuitions is a method for acquiring knowledge about non-empirical things. It is a method reserved for precisely those matters knowledge of which is possible in no other way, hence knowledge of which cannot be empirical.

I have already tried to say why the method of intuitions is not to be trusted, and I will return to the same theme more broadly in the chapter to follow. For the present, I want to explore the following issues. (1) Suppose that we were to give up on the method of analytic intuitions in set theory and semantics. Would this yield a significantly different triagic judgment of the Paradoxes? I will suggest an affirmative answer, an answer according to which *Frege's Sorrow* and *Koryé's Sorrow* are a serious mishandling of the Paradoxes. If this is right, it is also possible that (2) triagic misjudgment of the received opinion adumbrates a misjudgment further down the scale, at the level of diagnosis. The bulk of the present chapter is devoted to showing that, in the case of the Liar Paradox, this is indeed the case.

It is instructive to compare *Tarksi's Disappointment* with *Frege's Sorrow*. Tarski had a proof that any natural language has a sentence that is true if and only if it is false. Russell had a proof that there exists a set that is a member of itself if and only if it is not. Frege was much moved by Russell's discovery. He lamented the collapse of arithmetic. Tarksi, on the other hand, simply got on with the business of repairs, as Russell himself did with sets. The received view is that Tarski's proof established that the truth predicate of a natural language is inconsistent, just as Russell's proof established the inconsistency of what he and others thought of as the ordinary mathematical concept of set. We would do well to reconsider this opinion by comparing the Tarski and Russell paradoxes with a third: The village barber who shaves all and only those villagers who do not shave themselves. The barber shaves himself if and only if he does not shave himself. The Barber Paradox offers no hospitality to the received view. It is true that it shows to be inconsistent the expression (or "concept" if we must) "barbers who shave all and only those who do not shave themselves." It does *not* show the inconsistency of our ordinary concept of barber. There is no Frege in the wings wringing his hands at the death of our tonsorial practices. The moral that is drawn from the Barber is not that there are no barbers, but only that there are no *such* barbers. If we applied the moral of the Barber to the paradoxes of Tarski and Russell, we would not say that they establish the inconsistency of the concepts of truth (or statement, that is, truth valued sentence) and of sets. Neither would we say that these paradoxes demonstrate that there are no true statements and, similarly, that there are no sets. On the Barber model, the worst news derivable from the antinomy of The Liar is that, contrary to what Tarski and others antecedently believed, the Liar sentence is not a statement. And the worst that derives from the antinomy of sets is that something that Russell and others (but not Cantor) took to be a set is not. Or, if we prefer to speak of concepts, the worst of the Russell paradox is the inconsistency of the "concept" of the set of non-self-membered sets; just as the worst of the Liar paradox is the inconsistency of the "concept" of a statement that says of itself that it is false.

On the face of it, the received wisdom has made far too much of these things. On reflection, we see that the barber of the paradox does not exist, but lots of other barbers do, fortunately. Why do we not also say that, on reflection, the Russell set does not exist, but lots of other sets do, fortunately; and that, on reflection, the Liar statement does exist, but lots of statements do, fortunately? It may even be that in our original understanding of truth-valued sentences we took the Liar to be one such, and that in our original understanding of sets, we took the Russell set to be one such. Why, then, do we not say that the paradoxes served the useful function of correcting those original understandings, much in the way that the derivation of a surprising inconsistency will correct a prior understanding of just about anything?

In the case of the Liar, the present suggestion comes to no more and no less than this. Contrary to what we have originally supposed, "This statement is false" is not a statement, that is, not a truth-valued sentence. So why do we not say this, and have done with it, which as we saw in Chapter 1 is its fate in dialogue logics such as **DL3**?

Frege clearly thought that the Russell Paradox blew apart the very idea of set (or, for Frege himself, of courses of values of concepts). *Frege's Sorrow* so derives from his understanding of the concept of intuitive concepts and of axioms, concerning which he and Hilbert had persisting and celebrated disagreements. The axioms of set theory are a priori truths, made true by the very idea of set by "our" intuitive concept of set; but the Russell set is derivable from these axioms, and is inconsistent. So the concept of set is inconsistent; or, in plainer words, there is nothing to the idea of sets. When we set ourselves to speak about sets, we do not and cannot know what we are talking about. Contrary to what Frege initially supposed, there is nothing in sets to understand. So there are no axioms for sets.

The received wisdom about the significance of the Tarski paradox is difficult to fathom, short of Tarski's having the same kind of view. Given the quite general pre-Gentzen favoritism toward axiomatic methods in logic and the formal sciences, we might entertain on Tarski's behalf the supposition that the axioms for intuitive semantics are made true by the intuitive ideas of truth and of statement. Since those same axioms demonstrate the inconsistency of a string sanctioned by the axioms to be a true or false sentence, the axioms themselves are inconsistent. Contrary to what we had supposed, there is nothing whatever to the idea of a (semantically closed) natural language.

If this is right, it remains to remark on the difference between Frege's intense and despairing reaction to the Russell paradox and Tarski's own rather more sanguine and workmanlike reaction to the Liar. The difference is that Frege thought that with the loss of the very idea of set, there is no basis on which to found a successor-concept. Tarski, like Russell, thought differently. In so thinking, they revealed their tacit subscription to a methodological center-piece of minimalist idealism, never mind the presumption of its discreditation by Moore and others. Idealists *expected* to find our concepts rather routinely inconsistent, as Russell himself did in his one and only idealist book, specifically with respect to the concept of space. Since it was not their view that inconsistencies blew concepts apart, idealists could help themselves to a plausible notion of successor-concept. If our concept of K is inconsistent, the best candidate for its successor will be the K* that most resembles it minus the inconsistency. Although Frege was fleetingly tempted by prospects of a successor-concept to the blown concept of set, he soon settled down and gave up on set theory altogether. Frege was no idealist. It is not that he was dispirited or crestfallen. Rather, it was that he saw the job of reestablishing set theory as impossible. Having *no* idea of what sets would be, we lack any standard of adequacy for successor-concepts.

These speculations aside, it is interesting that when faced with his Paradox, Tarski wholly gave up on of intuitive semantics. This suggests that he thought that the Liar had blown apart the very idea of a semantics for natural languages. A dramatic way of saying this is that Tarski was drawn to the view that the Liar destroyed the very ideas of truth and statement for natural languages (Tarski, 1956/1983, p. 158). If this is right, *entailment* trails along into oblivion, given that entailment is definable by way of truth (Tarski, 1956/1983, p. 429). Not only are there no truths in Polish (or English, or any other natural language), there are no implications there either. So it is not surprising that Tarski not only set out to contrive truth predicates for formalized languages, he did the same thing for consequence or entailment. Truth and entailment alike would be constructs contrived for highly gerrymandered formalized languages, and measured for their fit with their destroyed counterparts in natural languages.

Tarksi's abandonment of natural language is not something that he dwelt on much; he just did it. Even so, the grounds of the abandonment are worth thinking about. If the Liar paradox does the utter damage that Tarski evidently supposes, then the *proof* of the paradox fails for want of an entailment relation in any natural language. This does not so much refute the proof as emphasize the disaster it is presumed to create.

The proof of the paradox now has something of the cachet of Parmenides's (and later, Wittgenstein's) ladder. What the proof proves is so utterly destructive that it destroys itself. So we use the proof to take us to what we are prepared to accept from it, like an unsecured ladder which (in cartoons) is scrambled up quickly, then kicked away before it falls of its own accord.

There is not much point going on too long with conjectures about Tarski's ladder. There may be sober explanations of why Tarski gave up on natural languages. All the same, doing so makes no sense unless *something* very bad could be proven about them. There is no doubt that Tarski thought that Paradox demonstrated an inconsistency in every natural language. If L is the Liar sentence, the inconsistency is that L is true if and only if it is not. Tarski was a classical logician. He subscribed to the principle of Excluded Middle and the standard negation laws. He also allowed that negation-inconsistency yields absolute inconsistency. Thinking that he had established the existence of an L for every natural language, Excluded Middle would guarantee that L is both true and not – a negation-inconsistency – and this in turn would demonstrate the absolute inconsistency of natural language. Every sentence would now be true. *Ex falso quodlibet.* Syntactic appearances notwithstanding, the proof that every sentence of English or any other natural language is true is all but what, in the received view, the Liar proved more directly, namely, that nothing is true in English (etc.); for this latter can now be taken as a demonstration that there is nothing in English (etc.) that is true *as opposed to false*.

It might be objected that we have overstated – as we did in the Prologue – the magnitude of Tarski's disaster in failing to keep in mind the distinction between a semantic theory and the language or languages that the theory is a theory of. By these lights, the brunt of Tarski's disaster falls not on (say) English but on the semantic theory, **ST**, of English. It is not that no sentence of English is true, or, what comes to the same thing, that none is true as opposed to false; rather these things hold in **ST**. Thus, the worst that the Liar Paradox does to English is to deny it a semantic theory. The Paradox does not demonstrate the incoherency of the idea of a true sentence of English but, rather, the incoherency of the idea of a theory of truth for English, hence for such sentences as may be true in English. In denying to English an **ST**, it is well to note that on the present view the Paradox denies to English all its own metalinguistic sentences including all sentences of the form, "'...' is a statement of English." It is a heavy price, and is one exacted by the semantic closedness of natural languages. Depending on how we conceive of details of Tarski's disaster for **ST**, every such sentence will be false (though also true), or will not be true, period. Either way, no sentence "'...' is false," is true except when false. But truth is disquotational. So no sentence in English is true, except when also false. We can pick our poison. No sentence of English is true, or every sentence of English is true and also false. There is, in **ST**, no prospect of "'...' is true" implying "...," since nothing in **ST** implies anything in **ST**. It is the same with the metalanguage of English. No sentence of English "..." will imply any other, unless it also does not. Thus, every sentence of English implies every sentence of English, provided that none implies any; or none implies any, full stop. This, too, is because English contains its own semantic theory. It is the price of semantic-closure.

This is certainly not to say that Tarski is indifferent to the trouble caused by the Liar.

A characteristic feature of colloquial [i.e., natural] language (in contrast to various scientific languages) is its universality. It would not be in harmony with the spirit of this language if in some other language a word occurred which could not be translated into it; it could be claimed that if we can speak meaningfully about anything at all, we can also speak about it in "colloquial language." If we are to maintain this universality of everyday language in connexion with semantical investigations, we must, to be consistent, admit into the language in addition to its sentences and other expressions, also the names of these sentences and expressions and sentences containing these names, as well as such semantic expressions as "true sentence," "name," "denote," etc. But it is presumably just this universality of everyday language which is the primary source of all semantical antinomies, like the antinomies of the liar or of heterological words. These antinomies seem to provide a proof that everyday language which is universal in the above sense, and for which the normal laws of [classical] logic hold, must be inconsistent. This applies especially to the formulation of the antinomy of the liar which I've given... and which contains no quotation-function with variable argument. If we analyze this antinomy in the above formulation we reach the conviction that no consistent language can exist for which the usual laws of logic hold and which at the same time satisfies the following conditions: (I) for any sentence which occurs in a language a definite name of this sentence also belongs to the language; (II) every expression formed from, (2) [x *is a true sentence iff* p] by replacing the symbol "p" by any sentence of the language and the symbol "x" by a name of this sentence is to be regarded as a true sentence of this language; (III) in the language in question an empirically established premise having the same meaning as (α) [*"c is not a true sentence" is identical with* c] can be formulated and accepted as a true sentence.

If these observations are correct, then the very possibility of a consistent use of the expression "true sentence" which is in harmony with the laws of logic and the spirit of everyday language seems to be very questionable. (Tarski, 1956/1983, pp. 164–5)

Evidently Tarski thinks, or is inclined to think, that a language is natural language only if it admits of a consistent use of "true sentence." Hence, he thinks, or is inclined to think, that there are no natural languages (compare my claim that Tarski is committed to the view that the paradox shows that there are no statements).

The Liar-trouble has led entire armies of theorists to give up on the semantic closure of English, and to find gerrymandered artifacts in which it can be reinstated without the heavy weather of paradox. Whether it is a performable task is a nice question (see below), but it is driven by the fact, not a disconfirmation of it, that Tarski's disaster is a disaster that befalls English itself.

We wondered what reason is there to think that the Russell and Liar Paradoxes differ from the Barber. It matters greatly that there be a difference. If there is not a difference, intuitive set theory and intuitive semantics are left standing, if somewhat scathed. Just as the Barber shows the inconsistency concept of a barber who shaves "all and only those who do not shave themselves," and does *not* show the inconsistency of the concept of barber, so, too, the Russell paradox would establish not the inconsistency of the concept of set, but the inconsistency of the concept of the Russell set. Similarly, the Tarski paradox would demonstrate not the inconsistency of the concept of statement, but rather of the concept of the Tarski statement. Although there are sets, as naively construed, none is the set of all non-self-membered sets. Although there are statements, as reckoned up in naive semantics, none is a statement that says of itself that it is false.

Graham Priest proposes that there is indeed relevant difference between the Barber and the other two. He says that the Russell set and the Liar statement arise because of a regress of truth conditions. The Russell set and the Liar statement are non-well-founded. The Barber, on the other hand, owes nothing to a regress of truth conditions (Priest, 1987, p. 186, n. 14). I shall not here deal with the suggestion that the Russell set is non-well-founded except to remark in passing that Aczel and others who work on hypersets have produced theories in which the non-well-foundedness of sets bears no responsibility for the Russell paradox (Barwise and Moss, 1996, pp. 60, 301–22). It is the Liar that here pinches. We will show below that the Liar statement is not, as is frequently supposed, subject to a regress of truth conditions. If this is right, Priest's diagnostic and triagic suggestions collapse.

Another suggestion bearing on our question of the difference between the Barber paradox and the other two arises from the Elster distinction in *Ulysses and the Sirens* "contradictions of the mind" and "inconsistency stemming from a single belief" and that therefore follows from a "single unitary project." This is a distinction we have met with before; we need not tarry with it here, except to make a connection with, and reinforce a contrast between what we have been calling analytic and stipulative methodologies in the abstract sciences. We saw that by contradictions of the mind, Elster means quite ordinary situations in which "contradictory attitudes coexist passively, and it is possible for the individual to abandon one of them when the inconsistency is pointed out to him." The other kind of inconsistency may be considered either as an attitude or as a belief. Considered as an attitude, inconsistency may be "constitutive of the personality as a whole." Considered as a belief, the inconsistency may lurk in a unitary concept.

Elster's distinction mirrors a more general distinction articulated as long ago as 1764 by Kant and reaffirmed thereafter (1974a and 1974b). Kant says that synthetic concepts, which is what mathematicians trade in, are *artifacts*; they are, that is to say, constructions of new concepts from antecedently existing conceptual bits. Analytic concepts, which is what philosophers trade in, are wholly formed concepts expressed in colloquial language, whose analysis is the object of philosophy. Corresponding to this distinction is a contrast between making clear ideas (which is what mathematicians do) and making ideas clear (which is what philosophers do). This difference between mathematics and philosophy as Kant understood it is adopted wholesale and without acknowledgment by Russell (1903) (inadvertence no doubt). It is a distinction that allows us to restate Elster's contrast. A contradiction of the mind is an inconsistent synthetic concept, that is, one that we have built inconsistently. A contradiction stemming from a single belief is an inconsistent analytic concept, hence an inconsistency that *inheres* in the very idea of the thing in question.

It may be that we could imagine a role for Elster in our present debate. Perhaps he would say that the difference between the Barber paradox and the other two is that the Barber involves only an inconsistency of the mind whereas the Russell and the Liar involve inconsistencies stemming from a single unitary project. In Kant's terms, the Barber reveals an inconsistency in a synthetic concept, a concept we have constructed inconsistently, whereas the Russell and Liar inconsistencies inhere in the very idea of sets as they really are and the very idea of statements as they really are.

Perhaps this is right, but I fail to see that it engages, to say nothing of answering, our present question. Let us concede that the Russell paradox and the Liar paradox show that the very idea of *something* is inconsistent. But what? I can see no case to be made for identifying these somethings as anything other than the very idea of set of non-self-membered sets and the very idea of a statement stating its own falsity. And since nothing for our question turns on the distinction between Elster's two kinds of inconsistency, I can now admit that I do not see that distinction reflected in a difference between the concept of a barber who shaves all and only those who do not shave themselves, and the concepts brought into disrepute by the other two paradoxes.

Suffice it to say for now that Frege and Russell with regard to sets, and Tarski with regard to statements, drew a different lesson from the paradoxes. They appear to have thought that the very concepts of set and of statement are the true casualties of these proofs. I have already said that if we could show that Frege, Russell, and Tarski shared, however implicitly at times, a common conception of analytic philosophy, then we could explain why they thought that what the paradoxes showed is that there is no such thing as what sets really are and no such thing as what statements really are. What is not so easy to show, as I keep saying, is why these three thought that discrediting the idea of a barber who shaves all and only those who do not shave themselves does *not* discredit the very idea of barber, whereas the discrediting of the idea of a set of all non-self-membered sets and the idea of statements toting their own falsity *does* discredit the very idea of set and the very idea of statement. Still, conjectures are possible. One to which I confess myself attracted to this. Barbers are an *empirical* matter; they stand in no epistemic dependency on philosophical or otherwise abstract theories. Sets and truth are dramatically otherwise. What is knowable about them is wholly rooted in the kind of philosophical – or otherwise abstract – theory that can be found for them. If Frege and Russell were so minded, then all the evidence suggests that in the beginning they were *analytic* theorists about sets; and the same holds for Tarski in relation to semantics.

What has come loosely to be called "analytic philosophy" had it origins this century in efforts to connect conceptual truth and intuitions in a certain way.

Best known perhaps of these efforts are the writings of Moore (1898, 1899, and 1903). In some ways, the looseness of the name "analytic philosophy" is unfortunate. In recent and contemporary usage, it denotes a wide range of things, not always compatible. In perhaps its broadest and least fruitful employment, it refers to philosophy as practiced in the manner of virtually any theorist of any influence in (mainly) English-speaking countries in the past nine decades, and so covers everything from Moore's extreme Platonism, to Frege's notional "Platonism," to more recent enthusiasms for naturalized epistemology. I shall not dwell on the question of everything analytic philosophy has been, or has been said to be, over the decades. I wish, rather, to say a little more about Moore's way of being an analytic philosopher. Here briefly is the Moorean story. Its most celebrated advocates, though not always its consistent practitioners, are, in addition to Moore himself, Frege, and, less directly, Tarski.[1]

As we say, propositions are the world (cf. Wittgenstein, 1992/1958, p. 31). In knowing propositions one knows the world as it really is independently of the epistemic success of any knower. So idealism is false. Propositions are complexes of concepts. Some concepts are compositional, others are simple and intuitive. A proposition is known and philosophically understood when its constituent concepts are decomposed, or analyzed, into simple intuitive concepts. In having an intuitive concept a person has a direct and infallible apprehension of how an aspect of the world objectively is. Philosophical knowledge of the world thus lies in the decomposition of conceptual complexes into simple intuitions directly and infallibly apprehended. Philosophical knowledge is analysis. Decomposition has its rules; they tell us how to proceed analytically, so to speak. These rules can be likened to Grammatical rules (a source of confusion). They are rules, largely tacit, for the analysis of propositions and concepts, hence, as we may say, rules of Logical Grammar. "Logical Grammar" is a metaphor bordering on solecism. The rules of Logical Grammar are not rules for the manipulation of linguistic objects but, rather, for the manipulation of nonlinguistic objects, that is, propositions and concepts. They are rules of analytic *competence*, themselves subject to a distinction between simple and intuitive, and derived or provable. As with the concepts to which they are applied, rules are either directly and infallibly apprehended, or are demonstrable from those that are. Something of the nature and career of these rules might be thought to be revealed in the logic of *Principia Mathematica* (as witness its theory of descriptions). Hence the name "Logical Grammar." If so, that is a pleasant contingency. The competent analyst no more requires a course in *Principia Mathematica* than the competent language-user requires a course in *Syntactic Structures*.

Propositions, all or many, are linguistically codable, another happy contingency. It is happy inasmuch as those that are encoded are encoded in a language, for example, English or German, that is "entirely in order," logically speaking (Wittgenstein, 1922/1958, p. 149). English or German can assist in the identification of propositions a philosopher may wish to analyze. He is assisted further by the extent to which the grammar of his language admits of a Logical Grammar. When Wittgenstein said that our language is logically in order, he did not mean that the *Grammar* of English is logically in order but, rather, that English lent itself to Logical Grammar. Moreover, since the Grammar of English is often misleading – an encouragement to faulty conceptual analysis – the provisions of Logical Grammar can be expected to override the misconception of ordinary Grammar. Although having had a course in *Principia Mathematica* may be an efficient way of improving our analytical performance, it is not to be supposed that analytical competence requires explicit tutelage.

There is much that could be added to this basic story, and much thereupon to say for and against it. It suffices for my purposes to describe what the link between intuitions and conceptual truths is thought to consist of, and in what ways the difficulties lately examined are averted. A *basic* conceptual truth is the result of a philosophical analysis, that is, an infallible decomposition into intuitive concepts. A *derived* conceptual truth is provable from basic conceptual truths by principles that themselves are simple and intuitive or derived from those that are. Waiving the looming regress, all conceptual truths are ultimately rooted and sanctioned by simple intuitions.

Set theory is a case in point. On one way of looking at it, membership is simple, and sethood is implicitly defined by the axioms on the membership relation, by axioms that are not just categorical but uniquely specifying. Historically, this was the perspective taken by Russell and Frege, and I shall adopt it here. Thus, for Frege, sets are purely logical objects of which there is direct preaxiomatic apprehension in the absence of which the axioms would lack all legitimacy. Intuitions of this sort are *analytic intuitions*. Conceptual truths are analytic truths, and analytic truths are ultimately secured by analytic intuitions.

Set theory, as I say, is a case in point for analytic intuitions. It is more than that. Since set theory attracts the Russell paradox, set theory is a *test case* for analytic intuitions. Concerning that "thundering heptameter that shattered naive set theory," Frege is the only philosopher of mathematics of his era whose response was appropriate to his analytical proclivities, that is, the perspective of analytic intuitions. He reacted by declaring *Frege's Sorrow*: There is no intuitive concept of set. That is to say, *there are no sets*. And because there are no sets, arithmetic collapses. It is important – and it bears squarely on the point of *Frege's Sorrow* – that in his reply to Russell's fateful letter, Frege did not confine his regret to the collapse of his program in the epistemology of arithmetic. He made a point of adding that arithmetic itself appeared to have caved in. It is rather routinely thought that Frege's response to the thundering heptameter was an embarrassing overreaction. Russell, who had earlier announced his surrender to the embrace of Moore's philosophical analysis, just got on with the business of making repairs, as Tarski would later do when faced with the Liar Paradox. The process of rehabilitating sets or, in Tarski's case, of rehabilitating truth, was governed as we have seen by a principle of maximal consistent similarity. The repaired concept would be constrained in ways designed to regain consistency, but once its presumed consistency was adjusted for, it would be as much like the original concept as those consistencies would allow. Thus, new sets would be as similar to old as consistently possible. How similar is this? Not very, as we have also seen.

Let us not repress the issue. The point remains that Frege refused to do what Russell and Quine, and nearly everyone else, did do. They made repairs. We might think that Frege's refusal can be explained by the fact that even those repairs that satisfy the requirement of maximal consistent similarity give us successor concepts so different from the original as to make it true that they answer to no idea of set. To these it could be replied that this is hardly more than the obvious point that the successor concept is not a very familiar one.

This is not Frege's problem. Frege could scarcely avoid seeing that Russell's response to the inconsistency of the intuitive concept of set was precisely of a kind with his response in *Foundations of Geometry* to the inconsistency (as he saw it) of the intuitive concept of space. In each case, he did the same thing. He rehabilitated the stricken concept under way of the requirement of maximal consistent similarity. *Foundations of Geometry* was Russell's first and last book written as an idealist, as we have said. Four years later, when he undertook the rehabilitation of the stricken concept of set in his first *analytic* book, there is no difference in how inconsistencies are recovered from. Frege may have been seized by the suspicion, rather subconsciously no doubt, that in his first book after his conversion to philosophical analysis Russell reverts to a standard idealist strategy for dealing with the inconsistency of intuitive concepts. If Frege were not seized of such a suspicion, he should have been. If one is genuinely an analytic philosopher in the manner of our Moorean story, inconsistent intuitive concepts cannot *be* repaired.

Why would this be? The simple and direct answer is that no method of repair can satisfy the requirement of maximal consistent similarity. The intuitive concept of set does not exist. Since the concept does not exist there is no concept that does exist for which a similarity relation is definable. Nothing is similar to nothing. Equally, by *Frege's Sorrow*, there are no sets, that is, no courses of values of concepts. For anything that does exist, there is no definable similarity relation between it and the nonexistent concept of set. Any arbitrary concept is as good a rehabilitation of the concept of sets as any other. Nothing is similar to nothing.

We shall return to the methodology of analytic intuitions in the chapter to follow. For present purposes we have stayed with it for long enough to venture an answer as to why the Russell, Tarski, and Barber Paradoxes have received such strikingly differential treatment by philosophers. For the likes of Russell, Frege, and Tarski, the only prospects for a philosophically adequate understanding of sets and of truth lie in their respective analytic theories. Analytic theories have no immunity against paradox, which is fatal to them.

It follows that there can be no philosophically adequate understanding of sets or of truths, if the methodology of analytic intuition is correct. It is evident that Frege was seized of two convictions. One is that the method of analytic intuitions is correct for sets. The other is that in default of a philosophically adequate understanding of sets, nothing else counts as knowledge of sets. Russell we may see as conceding Frege's first point. On the second, he waffled. Sets could still be the object of mathematical knowledge. Why is this a waffle? It is so because by the lights of Russell's own conversion to analysis, and given the fidelity of mathematical analysis to our minimal idealism, it can hardly be supposed that room is left for the fruits of this methodology to count as *knowledge*. The same may be said for Tarski. His reaction to the paradox that bears his name is of a piece with that of Frege and Russell toward sets. Like these, his triagic judgment was harsh, and his treatment option was as radical, as was Russell's for sets. In such a case, they would resort to what Frege could only disclaim, that is, to *prosthesis*.

Whether one's sympathies run to Frege or to Russell and Tarski, it is clear that the three would have a common thing to say about the Barber, and an answer to the question as to why the Barber did not put our concept of barber out of business. Knowledge of barbers, they would have said, is not philosophical knowledge; it is not reserved for abstract theories. The art of barbering is an empirical discipline. Besides, barbering is a *trade*.

Even so, Frege and Russell may be right in their view of the damage done by the Russell paradox, and Tarski may be right in his view of the damage done by the Tarski paradox. I want now to try to characterize that damage. There are two views of it. One is that the damage was the death of discourse. If there are no statements, how can the rehabilitation of semantics, in particular, even be considered? How would it be stated? This is the damage of constantive nihilism, and the closest that Tarski ever got to acknowledging it was in his rather breezy dismissal of natural language, which is the language in which he made his repairs. On the second estimate of the Paradox's damage, the problem is semantic promiscuity. Since the Liar proves a contradiction, and since *ex falso* holds, every sentence is true. I shall leave it to others to reconcile constantive nihilism (nothing is true) with the semantic promiscuity (everything is true), but a point also worth mentioning is that in the promiscuous damage done to truth the Liar makes true every sentence and its negation about every subject matter whatever. Similarly, the promiscuous damage done by the Russell paradox is done not only to set theory, but to any (otherwise) consistent extension of set theory. Although arising from the very idea of *set*, the promiscuous damage done by the Russell paradox is damage done to *truth*. The damage done to truth is the collapse of alethic distinctions. The same applies to the Tarski paradox, except not so vividly. Nihilistic damage is damage to the very idea of statement. Promiscuous damage is the damage that *every* paradox exacts, namely, the damage to truth constituted by the collapse of alethic distinctions. With most paradoxes, the nihilistic damage they do appears to be consistent with their promiscuous damage. That nothing whatever is a set may appear to be compatible with the collapse of alethic distinctions in set theory and every otherwise consistent extension of it. In this respect, the Liar paradox stands dramatically apart. Its local damage is that nothing whatever is either true or false. Its promiscuous damage is that everything is true. Even the apparent consistency between the two types of damage to *sets* may even now be slipping away. For how can it be that if there are no sets, every sentence about sets is true?

As stated, Tarski's disaster is a disaster for natural languages. It stands or falls on whether each Liar-sentence L is a truth-valued sentence and on whether supposing otherwise would block paradoxes of the same character and just as problematic as the Liar. I myself am attracted to simple solutions. Better to deny truth-valued sentencehood to L. Of course, this risks begging the question against Tarski, as would any subsequent insistence that paradoxes of the same character and just as bad could not be derived from the presumed nonsentencehood of L. Not wanting to beg questions, I am minded to proceed more directly with a program of attack on the proof of the paradox itself, that is, with an ambush argument against received opinion. This is the business of the sections to follow. I want this to be properly motivated business. However he produced it, whether by way of a Fregean understanding of the link between concepts and axioms, or by way of standard provisions of classical logic, Tarski's disaster is so intractably problematic that it demands the "smoke and mirror" escape by Tarski's ladder. Tarski's disaster is, on reflection, at least as ruinous as any of the most ravaging paradoxes of the ancients, including Gorgias's proofs that nothing exists, that if something did exist it could not be known to exist, and if it could be known to exist, this knowledge could not be expressed. We may have our doubts about the cogency of Gorgias's proofs, but we seem to have been too sanguine by half about the cogency of *Tarski's* proofs. I doubt that English is in as bad shape as such proofs purport to show; and this is what I shall endeavor to establish in what follows.

If the endeavor is brought to successful completion, we will have shown ourselves in a significant and justified *diagnostic* disagreement with that of the received opinion. This is a godsend. In a way, the paradox of sets does not matter. This is because sets do not matter. That this is so is amply attested to by the fact that everything that sets are wanted *for* can be supplied by stipulative accounts that meet with elite consensus. That sets do not matter, and how, is attested to by the fact that we can get away with postmodernism in set theory.

It matters utterly that the proof of the Liar Paradox be defective if the methodology of analytic intuitions is sound. If so, the concept of statement is lost and with it comes the death of constantive discourse; and all that we said late in the Prologue of the consequent postmodern devastation is so. There is another, and better, way to capture the present contrast. If the methodology of analytic intuitions is correct, the damage done to sets by the Russell Paradox does not preclude postmodern recovery. On that same assumption, there is no recovery whatever from the damage to discourse done by the Tarski proof if it is valid. The proof shuts down the postmodern enterprise every jot as decisively as it shreds pompous Enlightenment assumptions. The damage of nihilism is unrepairable.

It is consequential that the Tarski proof is defective. It allows us to escape the annihilation otherwise guaranteed by it if the methodology of analytic intuitions is sound, without having to determine whether the methodology of analytic intuitions *is* sound (I shall say in the next chapter why I think it is unsound).

TARSKI'S SOLUTION

Tarski's celebrated solution of the Liar Paradox is not an attractive solution (1956/1983). It involves the partitioning of indicative sentences into disjoint sets that inhabit the levels of the Tarski hierarchy, one set for each level.[2] To these partitions there applies the

Truth-Extension Rule: At each level n ($n > 1$), the extension of the predicate "true" is precisely the true sentences lodged at level $n - 1$.

It follows from the fact that the Tarski hierarchy reaches up into the transfinite, together with the fact that "English" is now held to the satisfaction of the Truth-Extension Rule, that there exist in this regimented language, infinitely many nonequivalent truth predicates. This is gerrymandering of precisely the sort that spares formal languages the nuisance of paradox. Whether applied to a formalized language or to a regimented natural language (to Renglish, as we might say, or **R** for short), in neither case does the Liar sentence occur at any level. Thus

(1) The Liar sentence is not a sentence of **R**.

It follows from this that

(2) The Liar sentence is not true at any level.

Of course, Tarski's solution of the Liar costs us the intuitive concept of truth. "True" is now transfinitely ambiguous. The unqualified "true" possesses only the use, and the virtue, of expressive economy, for there is nothing for "true" to mean but $true_1$, or $true_2 \ldots true_n$, or $\ldots true_\omega$, or $\ldots truew_\omega$, or $\ldots$. Even so, it is natural that we should want to know how good a solution Tarski's is, and so to know whether (2) is *true*.

The theory itself obliges us, under pain of begging the question, to take our interest in (2) to be its $truth_1$, or $truth_2$, or $\ldots$. It is easy to see that (2) will be true in no sense unless "The Liar sentence is not true at any level" is itself true at some level. Showing this to be so – if it is so – requires the heavy machinery of transfinite induction – surprising perhaps, given the apparent simplicity of the claim at hand. In fact, one wonders whether the theory itself will *let* (2) be true (in some sense). Certainly (2) will not be true at some level unless

(1*) (n) (n) is not true

is not true at any level. For this to be so, the "not true" of (1*) must be read as "not true$_{k+1}$ where the level of (n)(n)... is k. But, again, the theory itself would not sanction the *production* of (n) (n)...; there is *no* level on which it occurs, hence nothing for "not true$_{k+1}$" to be true *of*. Let us waive these difficulties for now. Let us suppose that (1) does indeed solve the Liar. Since it does and since (1) implies (2), the predicate "is not true at any level" is part of the solution of the Liar for **R**. I shall say that the predicate "is not true at any level" is a *solving term* for the Liar. It is an interesting predicate and standing occasion to consider

(3) (3) is not true at any level.

Those who fancy the present solution of the Liar will be quick to point out that

(4) (3) is not true at any level

which is surely so. Hence by (3) itself, (3) is true at some level; but this contradicts "is not true at any level." Whether applied to formalized languages or to Renglish, the Tarskian solution of the Liar induces a solving term on which a variation of the Liar is constructible.

A good many semantic theorists have regretted the application of the Tarski solution to Renglish. They have done so for reasons different from the present problem. Their complaint has been that a Tarski-solution for **R** requires us to plump for the semantic openness of English. This they say is *ad hoc* or, worse, that it cannot be done (that is, that **R** is not English). With this complaint firmly in mind, we have had a flurry of theories all dedicated to the proposition that the Liar will know no philosophically satisfying solution until it is solved for semantically closed languages such as English.[3] Our problem suggests that whether "a language" is open or closed is not the only decisive consideration in the matter of semantic paradoxicality.

Consider now:

(6) (6) is true

as a version of Truth-Teller. Since (6) is an indicative sentence of English, we can lawfully produce the conditional compound

(7) If (7) is true then Bigfoot exists

or, more abstractly,

(8) If (8) is true then Φ

for arbitrary Φ. Curry's paradox is a proof of arbitrary Φ, hence a proof of what the paraconsistent logician desperately wanted *not* to follow from true negation-inconsistencies. The moral of Curry's paradox is that Φ is provable without reliance on negation-inconsistent premises. This is interesting. It purports to show that absolute inconsistency is not *essentially* parasitic on negation-inconsistency. Here is the proof.[4]

Suppose that

[i] (8) is true.

Then by substitution on "8" in sentence (8).

[ii] "If (8) is true then Φ" is true

Hence, by the disquotational character of truth,

[iii] "If (8) is true then Φ" is true iff if (8) is true then Φ.

From [ii] and [iii], *modus ponens* gives

[iv] If (8) is true then Φ.

And so, from [i] and [iv], we have

[v] Φ

Suppose now that

[vi] (8) is not true.

Then it is possible both that

[vii] (8) is true

and

[viii] Φ is false.

But if (8) is true, its truth implies Φ. So it is not possible that [vii] and [viii] are jointly true. Hence [x] (8) is true and so too is [xi] Φ.

What the Curry Paradox shows is that absolute inconsistency has nothing intrinsically to do with negation-inconsistency (except by implying it) or with hostile semantic substitutions into environments such as

(n) (n) is ....

It does the paraconsistent logician no good, therefore, to contrive solutions for Liar sentences and to think up ways of quarantining or disarming demonstrable negation-inconsistencies. The Curry Paradox demonstrates that it does not matter whether the Liar is resolvable. For if the proof of Curry is good, every natural language (and every regimentation of any natural language adequate to the formulation of the exact sciences and enriched by a truth predicate) is demonstrably inconsistent.

If *ex falso* is true, every language is also absolutely inconsistent. But every language is absolutely inconsistent even if *ex falso* is false. The preoccupation of paraconsistent logic with the management of negation-inconsistency is, fatefully, beside the point.

There is a curious asymmetry in the triagic assessments attending the set theoretic and semantic paradoxes. In the case of the paradox of sets, the received opinion is that the paradox destroyed the intuitive idea of set. The Liar Paradox draws a different assessment. It is said to have destroyed the intuitive idea of truth. Details of the Liar Paradox's proof tell a different tale. The prime victim is the concept of statement. Since statements are bivalent sentences, truth is also a casuality, of course. If it were the sole casuality, the illusion would be created that the paradox is intractable in a certain way. Here is an illustration. If we try to disarm the Liar, say by claiming that it does not express a proposition, we have a further paradox, as follows. For consider

(1) (1) does not express a proposition.

Now again:

If (1) does not express a proposition then, given what it says is true and, not being a proposition, is not true. If (1) is both true and not true then, by *ex falso quodilbet*, it expresses a proposition. Hence (1) expresses a proposition if and only if it does not.

Let τ be a *solving-predicate* for the Liar. It operates in the manner of "does not express a proposition," or "is meaningless," "is neither true nor false," "is not a statement," and so on. It is easy to see that τ cannot succeed as a solving-predicate for the Liar unless it entails the semantic predicate "is not true."

This is an awful result. If left standing, it shows the Liar to be unresolvable. Any solving-predicate that brings local relief to a Liar construction can be slotted into a Liar-like construction of its own, and it produces a contradiction every time. This is a bad enough result to qualify for a name. I propose the *Libertine Liar Theorem* (**LLT**). This being so, then there exists a structure

(1) (1) is τ.

Hence,

(a) If (1) is τ, then given what it says, it is true and, by the meaning of τ, is not true.

(b) If (1) is both true and not true then, by *ex falso quodlibet*, it is not τ.

(c) So (1) is τ if and only if $\sim \tau$.

Now, of course, if **Koryé's Sorrow** is correct, *LLT* fails. If we think that the Liar's victim is the intuitive concept of truth, and if one also thinks that the victimization of truth is consistent with certain lexical strings *saying something*, then we have a view of the Liar in which the very idea of statement is not its victim and is left in good standing for real work (see lines (1) and 1(a) above). This misses what the Liar's proof does indeed prove. Its central target is the concept of statement, that is, bivalent sentence, and it does it the damage of constantive nihilism. This makes truth and falsehood secondary victims, as we have seen. But now there is no logico-semantic space left in which lines (1) and (a) of our proof of the libertine Liar Theorem can do their intended work.

No doubt it will be thought that we have not succeeded in showing that it is the concept of statement, not of truth, that is the Liar's primary victim that the damage it does is the damage of nihilism. Certainly the paradox's modern progenitor appeared not to see things in our way. This is right, but unavailing. Tarski just missed the boat. This is suggested by the following line of argument:

(1) The concept of statement is the primary victim of the Liar paradox or it is not.

(2) If it is not, then the Libertine Liar Theorem is true. The paradox has no solution.

Some will find *LLT* dismissible on its face, and reasoning *modus tollendo tollens* will attribute primary victimhood to statements. On the other hand, others will notice that if this is indeed the case, then the Liar makes itself unstateable, hence unprovable, as a result they will take to be no less absurd than *LLT*. I think that we should take their point. Something is wrong somewhere. If what we have been saying so far is correct, we seem to have landed ourselves in a dilemma.

*Either the Liar is unstateable, the Liar is irresolvable, or the Liar is unstateable since everything is.

Fearing the exhaustiveness of the alternatives, a goring each time, we seem landed in a paradox – a metaparadox – as nasty as the ones we have been trying to disarm. Perhaps this is so. But if it is so, we have not yet shown it, since we have overstated our dilemma. Our true dilemma is a conditional affair:

*If the proof of the Liar is correct, then either the Liar is unstateable, since everything is, or the Liar is unresolvable.

Not liking the alternatives, an attack against the shared condition offers a certain tactical allure. I shall accede to it in the sections that follow, where the *proof* of the Liar will be our primary target.[5]

OPACITY

I do not propose to persist with **LLT**. There is a good chance that the argument that abets it makes unjustified, that is, question-begging, use of the intuitive truth predicate. Even so, **LLT** leaves us with a nontrivial question. How are we to judge Tarski's theory without some recourse to intuitive semantic predicates, the very ones which, if *true* (so to speak), the Tarski theory disqualifies from use?

My purpose here is rather to take on the Liar rather more directly. I shall show that there are vitiating difficulties with all standard variations of it, except possibly one. Then I shall show that the existence of the exception does not matter. That is, it does not damage our intuition that semantically closed languages, such as more-or-less ordinary English, are not rendered inconsistent by Liar difficulties.

Consider the Liar construction,

(1) (1) is false.

Call this construction $\mathcal{L}$. In $\mathcal{L}$, the left-hand occurrence of "(1)" confers a number name on what follows, and the righthand occurrence of "(1)" refers by name to that same thing. This is what I shall call "the Standard Interpretation" of $\mathcal{L}$. It would appear that (1) is either true or false. It is a statement containing a referring expression in subject position and the semantic term "false" in predicate predication. Since we know what left "(1)" names in $\mathcal{L}$, we have access to the true identity statement,

(1) = "(1) is false."

This being so, substitutions can be made via the substitutivity of identicals. Thus, from $\mathcal{L}$ we have

(1) "(1) is false" is false,

which gives us a new identity statement,

(1) = "(1) is false" is false,

which itself is false if the former is true – on the principle that things equal to a given thing are equal to one another. This is interesting. It shows that $\mathcal{L}$ is an opaque context, since substitutivity of identicals in $\mathcal{L}$ cannot be given unfettered sway.

Let us now rederive the Liar paradox. We say that $\mathcal{L}$ is either true or false. If true, then it is false. It could be well to pause and ask: "If *what* is true?" The answer appears to be "what left '(1)' names," that is, "(1) is false." So if "(1) is false" is true, what can be concluded about (1)? On the received view, it is that (1) is false. But what does (1) here refer to? It refers to what left (1) names, viz., to the string

(1) is false.

Thus, in saying that (1) is false we are saying that

"(1) is false" is false.

So we have a truth and a falsehood on our current suppositions. The *truth* is

(1) (1) is false,

whereas the *falsehood* is:

(1) is false.

But this is no paradox. What we required was that, from the supposition of the truth of

(1) (1) is false,

the falsehood of

(1) (1) is false

would follow. But what actually follows is the falsehood of

(1) is false.

There is need to re-ask the question. To what in this string does "(1)" refer? It refers to what, on the Standard Interpretation, left "(1)" names in $\mathcal{L}$, viz.,

(1) is false.

Thus what is false on the supposition that $\mathcal{L}$ is true is

"(1) is false" is false.

This leaves us unprovided with what is needed, namely, the falsity of

(1) (1) is false.

On reflection, it is odd to be thinking of this structure $\mathcal{L}$ in such terms, that is, in terms of its truth or falsity. Having done so may have had something to do with our present difficulties, our inability to show that "(1) (1) is false" is true if and only if it is false. Of course, as we see we have said, it emerges that

(1) (1) is false

is not a statement, or what speech act theorists call a *constative*. Rather, it is a baptism:

"(1)" [hereby] names "(1) is false,"

which produces in turn the identity-statement

(1) = "(1) is false."

The assertion that "(1) (1) is false" is not a constative, hence not an appropriate vehicle for the predicates "true" and "false," is important enough to be clear about. What is true is that, just as it comes, that is, without the aid of supplementary context,

(1) (1) is false

is not a constative, but a performative, as all baptisms are. What is not true is that there is no context in which

(1) (1) is false

involves the making or presentation of a statement, a constative. For consider:

"I hold the following to be true:

(1) (1) is false."

What does seem apparent is that whenever there are contexts in which

(1) (1) is false

involves the making or presentation of a statement, it is always the case that in those contexts it is also a baptism. I am surer of the latter point than I am of the former, as will presently become clear.

We may take it then that our misplaced interest in the truth or falsity of (1) (1) is false should be directed to "(1) is false," which, given the contextual specifications of "(1) (1) is false," is the *statement* that (1) is false. Here the received wisdom is right, "(1) if false" is either true or false. On the supposition of either the opposite follows. Thus if "(1) is false" is true, then (1) is false; and if "(1) is false" is false, then (1) is false; and if (1) is false, "(1) is false" is false; so,

(1) is not false, hence true. Paradox.

Extremely important as it is, that "(1) is false" is an opaque context, it is quite another thing that it does not or cannot give rise to paradox, which is the heterodoxical suggestion that I wish to explore. It may be that the two facts are connected, but it is not transparent that they are. Indeed, if we stay with the received wisdom, we will be satisfied with saying that even though "(1) is false" is an opaque context, that does not stop it from being a paradox-generator.

In what, then, does the opacity of "(1) is false" consist? It is worth emphasizing that if "(1) is false" is also paradoxical, the paradox cannot be derived without the information imparted by $\mathcal{L}$, on $\mathcal{L}$'s Standard Interpretation. It is necessary to know that left "(1)" in $\mathcal{L}$ bestows itself as the number name of what follows, and it is necessary to know that right "(1)" refers by that name to that same thing. We must now revisit the derivation of the paradox.

(a) Suppose, we said, that (1) is true.

(b) Then, since

(1) = "(1) is false"

"(1) is false" is true, by substituting identicals.
(c) Disquoting,

(1) is false

(d) Suppose now that (1) is false.
(e) Then, by quotation, the converse of disquotation, and one half of Convention T,

"(1) is false" is true

(f) But

(1) = "(1) is false."

(g) Hence, substituting identicals,

(1) is true.

As we see, the derivability of the paradox turns indispensably on substitutions licensed by the identity statement

(1) = "(1) is false."

It may strike us that something has gone amiss with these substitutions, involving as they do the opaque context "(1) is false." This would be a mistake, however. The opacity of "(1) is false" precludes truth status-preserving substitutions *within it*, that is, on its own occurrence of "(1)." But in our derivation of the paradox no such substitution is made. The derivation's substitutions are of "(1)" and "(1) is false" into the presumability transparent context "... is true." So the opacity of "(1) is false" appears to give no trouble for the proof of the paradox.

Still, the identity statement

(1) = "(1) is false"

is key to the derivation. We would do well to linger a little. $\mathcal{L}$, we said, is a baptism. It has the force of

"(1)" [hereby] names "(1) is false."

Consider another such baptism, improbable though it may seem:

"Charlie" [hereby] names "Charlie is bald."

Just as the first baptism produces the identity

(1) = "(1) is false,"

the present one produces

Charlie = "Charlie is bald."

We now ask whether left "Charlie" has an occurrence in "Charlie is bald." Suppose that we said that it does. Then we might think that

Charlie = "'Charlie is bald' is bald."

which is not true since it violates provisions of the baptism of "Charlie is bald." This is a mistake, as we have seen. The opacity of quotation-contexts precludes the fatal substitution.

We might think, however, that the preclusion is fatal in another way. Quotation being what it is, *right* "Charlie" cannot behave as *left* "Charlie" does. The two "Charlie's" are not coreferential in

Charlie = "Charlie is bald."

This might lead us to think that, since the "Charlie" of "Charlie is bald" is not an occurrence of that string's baptismal name, the logical form of

Charlie = "Charlie is bald"

is not

N = "N is bald"

but, rather,

N = "X is bald."

If this were so, it would be the same way with

(1) = "(1) is false."

Its logical form would not be

N = "N is false"

but, rather,

N = "X is false."

If this were so, the proof of the paradox would collapse.

It is not so. It is true that in

(1) = "(1) is false"

the second occurrence of "(1)" is nonreferring. If this meant, in turn, that the logical form of (1) is

N = "X is false"

then from the fact that *right* "(1)" is nonreferring in

(1) = "(1) is false"

it would follow that it is not an occurrence of the term of which *left* "Charlie" is an occurrence. But if that were so, the disquotation of "true" would not be truth-preserving. Consider an unproblematic case, for example,

"The cat is on the mat" is true.

Disquoting,

The cat is on the mat.

If each time "the cat" were a different term, then the disquotational inference would exhibit the invalid form,

"S is P" is true.

Hence,

X is P.

What the disquotational character of truth provides is an operation that transforms a nonreferring occurrence of a term into a referring occurrence of it (ignoring the problem of empty terms). What it does not provide is an operation that converts an occurrence of one term into an occurrence of another.

So far we have seen nothing that counts against the correctness of the proof of paradox. It is true that "(1) is false" is an opaque context, but nothing in the proof requires substitution into it. The proof does require the substitution *of* it for something, as well as the substitution of something *for* it. Although this involves the manipulation of a context in which "(1)" has a nonreferring occurrence, reference is restored when needed by disquotation.

DISQUOTATION

Even so, it is best to stick with the proof. At line (a) we entertained the supposition

(1) is true

and at line (g), we concluded from prior lines, another token of

(1) (1) is true.

The proof presumes without ado that truth is disquotational. The presumption provides that from lines (a) and (g) something follows in a truth-preserving way. It is not derived in the proof, indeed is not needed for the proof. But if truth is disquotational, it follows anyway. It is:

(1) .

By our previous discussion, the "(1)" of this line is an occurrence of the "(1)" of

(1) is true.

It is guaranteed by the disquotational character of truth.

I mean to take seriously the disquotationality of truth. Doing so gets us to see that there is something oblique about the construction

(1) (1) is true.

It makes us see that it leaves something of its structure imperfectly disclosed. For if truth *is* disquotational, then the structure loosely conveyed as

(1) is true

is either the structure more stringently conveyed as

"(1)" is true

or, perhaps, as

That (1) is true.

No doubt it will be objected that the supposition that

(1) is true.

has an undisclosed quotational character flies in the face of established canonical practice in which this is not manifestly so. Let it be the case that there are "ideal grammars" in which the syntactic character of

(1) is true

is wholly disclosed in its surface structure. Then the string of *English*

(1) is true

will be reconstructed in this canonical notation in the manner indicated. What is not foreclosed, however, is that it is a *mis*reconstruction of the English expression. In any event, either it is or it is not. If it is, the possibility exists that the "(1) is true" of English embeds a quotational context, and it is that possibility I now wish to examine.

Let us revisit an unproblematic kind of case. We might decide to baptize the sentence "The cat is on the mat" with the number-name "9." Once baptized, there are now two way of referring to this sentence. We can refer to it by direct quotation, as in

"The cat is on the mat" is true,

or by its baptismal name, as in

9 is true.

But truth is disquotational. "9 is true" is therefore an oblique construction. What is needed is either

"9" is true

or

That 9 is true.

In the first instance, "'9' is true," the quoted expression names its interior – the number 9. And since disquotation also preserves termhood, the inference to

9

would be an inference to the the number of the same name.

In the second instance, in the passage from

That 9 is true

to

9

"9" is an occurrence each time of some same term. In "That 9 is true," "9" is a baptismal name, hence a name. In the detached line

9

"9" is either that name or not. If it is a name, then "true" in

That 9 is true

is not disquotational, since the detached

9

is not a sentence, but a name, and truth-preservation is lost. If, on the other hand, "9" in

That 9 is true

is a sentence, disquotation likewise fails to be truth-preserving.

Why should this matter? It shows that not every use of "true" to characterize a sentence as true is disquotational. For example:

What Charlie said is true.

If in the proof's tokens of

(1) is true

"true" fails to be disquotational, perhaps it is because "(1)" the refers in the manner of "Whatever Charlie said." Perhaps not, too. The point at present is simply that finding places in which truth is not disquotational does nothing, just so, to discourage our confidence in the proof.

I am not so sure. In the proof of the paradox we made a substitution *into* the context

(1) is true.

All seemed right, since "…is true" is a transparent context. The proof substituted for a term in

(1) is true

viz., "(1)," a coreferential term, licensed by the identity

(1) = "(1) is false,"

to obtain

"(1) is false" is true.

Compare this with cat-case. Let us say that

9 is true.

Then since

9 = "The cat is on the mat,"

we have

"The cat is on the mat" is true.

Simple as it seems, the argument trades on the obliquity of

9 is true,

which is either directly quotational or baptismally so. Consider each in turn. First,

"9" is true.

The would-be substitution is blocked, since the quoted expression is an opaque context. Second,

That 9 is true

is unavailable for substitution since that would give

That "The cat is on the mat" is true,

which is not a sentence. Oddly, then, while

"…is true"

appears to be a transparent context,

"9 is true"

is not. So the appearance is deceiving. What is more, the peculiarity of this situation is not tied to pathological structures such as $\mathcal{L}$. It is a feature of reference by baptismal name quite generally. Even when it is known what the baptismal name names, substitution into contexts such as

"...is true"

is precluded because it is not a transparent structure. Truth is disquotational; it is also *quotational*. The context that

"...is true"

is "trying to be" is one that downplays that fact. The context that

"...is true,"

truly is

"'...' is true,"

which clearly (no pun) is not a transparent structure.

It is the same way with the proof of the paradox. The proof requires the legitimacy of

(1) is true,

which appears to be a lawful closure of the lawful context

"...is true."

But truth is quotational. What is needed is either

"..." is true,

or

That...is true,

and neither of these when filled by "(1)" yields the substitution that the proof requires.

Parallel considerations disarm the Curry Paradox:

(10) If (10) is true, then Φ. (Baptism).

Suppose, we said, that

(10) is true.

Given the identity,

(10) = "If (10) is true then Φ".

substitution yields

"If (10) is true then Φ " is true.

Disquoting

If (10) is true then Φ.

By *modus ponens*,

Φ.

On the other hand, suppose that (10) is false. Then it is possible that

(10) is true

and

Φ is false.

If (10) is true, its truth implies Φ. Hence, it is not possible for (10) to be true and Φ false. In other words,

(10) is true.

But, as we have seen, if (10) is true, so is arbitrary Φ. Paradox?

No. Here, too, the would-be proof carries tokens of (10) is true.

Here, too, it conceals its structure. It embeds the structure of direct quotation,

"(10)" is true

or indirectly,

That (10) is true.

Either way the proof fails.

Some paragraphs ago, we mentioned what may seem an obvious objection to the present line of argument. Why, it was suggested, is it necessary to suppose that

(1) is true

be taken quotationally, either directly or indirectly (by baptismal name)? True, if this is the only way in which "(1) is true" is lawfully construable, the proof of the paradoxes is wrecked. But, as everyone knows, not every predication of truth to a sentence is one in which "true" is or is required to be disquotational. Why, then, can the attribution of truth in

(1) is true

not be construed on that model? Let us see.

Suppose that "The cat is on the mat" is true and that
What Charlie said = "The cat is on the mat."

Substituting,

What Charlie said is true.

But "What Charlie said is true" is not available for disquotation. We do not get and do not want,

What Charlie said.

Let us now determine whether the problem-case can be reconciled to this model. Assume that

What Nixon said = "What Nixon said is false,"

and that what Nixon said is true. Then, substituting

"What Nixon said is false" is true.

Disquoting,

"What Nixon said is false."

Requoting,

"What Nixon said is false," is true.

Substituting,

What Nixon said is true.

So, what Nixon said is true if and only if what Nixon said is false. Paradox? On the present suggestion, it is proposed that we construe

(1) is true

not as

"(1)" is true

or as

That (1) is true,

but, rather, along the lines of "what Nixon said," that is, as

What (1) refers to is true.

So taken, the paradox derivable from

(1) (1) is false

goes through in the manner of the paradox of what Nixon said. Thus,

What "(1)" refers to = "What '(1)' refers to is false."

Suppose that

"What '(1)' refers to is false" is true.

Disquoting,

What "(1)" refers to is false.

Requoting,

"What '(1)' refers to is false" is true.

Substituting,

What "(1)" refers to is true.

Hence, what "(1)" refers to is true, if and only if what "(1)" refers to is false.

The present proof turns indispensably on the fact that what '(1)' refers to is referred to by "(1)," that is, that "what '(1)' refers to" and "(1)" are coreferential. Substitution into the presumably transparent context

What "(1)" refers to = "What '(1)' refers to is false"

gives

(1) = "What '(1)' refers to is false."

Suppose, again, that

(1) is true.

Substituting,

"What (1) refers to is false" is true.

Disquoting,

What (1) refers to is false

By the coreferentiality of "(1)" and "What '(1)' refers to"

(1) = what (1) refers to

we have it that

(1) is false.

Substituting

What "(1)" refers to is false.

Substituting again,

"What '(1)' refers to is false" is false.

This disquotes to

What "(1)" refers to is not false.

Bivalence and coreferentially yield

(1) is true.

For ease of reference, let us call this proof (i.e., the proof "in the manner of the Nixon example") the **N**-*proof*, and our original proof from the structure $\mathcal{L}$, the $\mathcal{L}$-*proof*. As we have seen, the **N**-proof is good only on the assumption that what "(1)" refers to is referred to by "(1)." But the $\mathcal{L}$-proof follows principles no different from those required by the **N**-proof *together with* a sentence (the coreferentiality claim) without the presumption of whose truth the **N**-proof collapses. Accordingly, the **N**-proof is good only if the $\mathcal{L}$-proof is. But the $\mathcal{L}$-proof is defective. It shows tokens of

(1) is true

which connive to conceal its quotational character. Of course, we tried to deny its quotational character by taking "(1)" in the manner of the Nixon example, that is, in the manner of

"What '(1)' refers to."

Doing so was unavailing. What licensed the **N**-proof also licensed the $\mathcal{L}$-proof, which, in turn, produced

(1) is true

which, so I say, is quotational. We could set out to show again that it is not quotational by taking

(1) is true

in the manner of the Nixon example, that is, as

What "(1)" refers to is true.

As we now see, this will only serve to redeliver

(1) is true,

which will have or lack a quotational character *independently* of our construing it as

What "(1)" refers to is true.

Is there, then, anything about

What "(1)" refers to

that would suggest a quotational character for

(1) is true?

There is. It is the manifestly quotational character of "what '(1)' refers to." The sentence "(1) is true" comes from a proof that itself is licensed by principles that drive the **N**-proof, which is a proof on constructions such as "what '(1)' refers to." The tightness of the connection between the **N**-proof and the $\mathcal{L}$-proof, together from the fact that in the $\mathcal{L}$-proof

(1) is true

comes from principles that operate solely on quotational structures such as "what '(1)' refers to," constitutes a further reason for supposing that in the $\mathcal{L}$-proof

(1) is true

imbibes the overtly quotational character of "(1)" in the **N**-proof.

A THIRD PROOF

Perhaps this is so. If it is so, the failure of the $\mathcal{L}$-proof brings the **N**-proof down. However, a critic could say, and would be right to say, that in our discussion to date there was a third proof, in fact, the second proof in order of appearance. This was the proof of the Nixon paradox itself. Suppose we call this the *RMN*-proof. The critic will say that even if the *N*-proof is defective, nothing that gave it trouble affects the *RMN*-proof and that the *RMN*-proof appears to have been impeccable. If this is so, we have found a proof that delivers the Liar paradox, and our evasions to date have been unavailing. So we will need to reexamine the *RMN*-proof.

The *RMN*-proof turns on the identity-statement.

What Nixon said about Watergate = "What Nixon said about Watergate is false."

Identity being what it is, anything that Nixon said about Watergate is "What Nixon said about Watergate is false." As a matter of fact, this is false, even if (also contrary to fact) Nixon also did say, "What Nixon said about Watergate is false." We could try to suppose that there is a world in which everything that Nixon said about Watergate was "What Nixon said about Watergate is false," and only it. With that assumed, we might think that paradox goes through for what might have been the case and only contingently is not the case. Whereupon,

What Nixon said about Watergate = "What Nixon said about Watergate is false."

If what Nixon said about Watergate is true, then by substitution,

"What Nixon said about Watergate is false" is true.

Disquoting,

What Nixon said about Watergate is false.

Requoting,

"What Nixon said about Watergate is false" is true.

Substituting,

What Nixon said about Watergate is true.

The Nixon paradox is telling. In two places in its proof, once as an assumption and thence as a conclusion from prior lines,

What Nixon said about Watergate is true

has an occurrence in which truth is not disquotational and, yet, in which there is no reason to suppose that "What Nixon said about Watergate" has a quotational character, whether direct, or baptismal, or any other.

Some people may not like a proof that trades on the counterfactual truth of an objectively false identity-statement. Their disquiet can be accommodated. We could set out to find a bright six-year-old who has never heard of Watergate. Suppose this youngster is named "Dixon." We could pay Dixon a dollar, or perhaps five, to say only this about Watergate:

"What Dixon said about Watergate is false."

We could then adjust the proof accordingly.

Even the Dixon example is troublesome. Dixon could break his word and say some further things about Watergate. Or, having forgotten his puerile pledge, he might grow up and become one of the great experts on Watergate. The identity needed for the Dixon proof is fugitive in a way that the others are not.

(1) = "(1) is false"

is an identity guaranteed by the baptismal rites of the structure $\mathcal{L}$,

(1) (1) is false.

As long as its provisions are in place,

(1) = "(1) is false"

will not falter.

We could try saying the same thing about the Dixon case. As long as it remains the case that

What Dixon said = "What Dixon said is false,"

The Dixonized *RMN*-proof goes through. So it does, but only if the Dixon of "what Dixon said" and the Dixon of "What Dixon said is false" are the same Dixon, and only if, further, the "said" of "What Dixon said," and the "said" of "What Dixon said is false" are cotemporal "saids." What interests us is not what Dixon said, but what Dixon said on some occasion. And what interests us is not what some or other Dixon said on the occasion, but the person who said what this very Dixon said on that occasion said that what he himself said on that occasion is false. "On that occasion" is not, of course, some occasion prior to his utterance of what is referred if by "what Dixon said on that occasion," but, rather, the very occasion on which Dixon himself utters what he utters. The required identity, therefore, is

What Dixon said on that occasion = "What I, Dixon, am now
saying about Watergate is false," that is, "This very sentence is false."

Let us retry the proof.

What Dixon said on that occasion = "This very sentence about
Watergate is false."

Suppose that what Dixon said on that occasion is true. Then, substituting

"This very sentence is false" is true.

Disquoting,

This very sentence is false.

But just as it stands, "This very sentence is false" does not contradict "What Dixon said on that occasion is true." For that, we need the identity

This very sentence = What Dixon said on that occasion.

Then, substituting, we would have

What Dixon said on that occasion is false.

Commentators who press the Namely-Rider objection tend to press it promiscuously. For example, pressing it against

(1) = "(1) is false"

is a mistake. For in the $\mathcal{L}$-proof there is no need and no occasion to substitute into the quotation context. If there were, it might rightly be said that "(1)" has no fixed reference in "(1) is false" and that "(1) is false" is dismissible on that account. Here, however, the Namely-Riders have an *oddly* perfect case. The Dixonized *RMN*-proof will not go through without the identity

This very sentence = What Dixon said on that occasion.

Here there is no prospect of protection by way of opaque contexts. The expression "This very sentence" is obliged to refer to itself if the identity is to hold. It refers to itself only if it is a sentence. It is not a sentence in fact, but a singular term (and an empty one). The proof fails. There is no paradox. In saying that this is an *oddly* perfect case for the Namely-Riders, we see now that what Namely-Riders allege is so, namely, reference failure. But it fails not because "This very sentence" produces "namelys" without end, as in:

This very sentence, namely, this very sentence, namely, this very sentence, namely . . .

but, rather, because its reference failure is of a type that precludes even the *first* application of "namely."

INDEXICALITY

The present argument is not without its own appearance of trouble. Earlier we made much of the fact that in,

(1) = "(1)" is false,"

the quotational context on the right denied to the two occurrences of "(1)" a coreferential role. But in the argument we have just made, it was insisted that the Dixon of "what Dixon said" and the Dixon of "'What Dixon said is false'" be the same Dixon. This is right. It is a requirement of any reading on which

What Dixon said = "What Dixon said is false"

is genuinely paradoxical, that the Dixons involved here be the same Dixon. If we thought that the requirement could be honored only by the fact that "Dixon" is coreferential in both occurrences, we would have overlooked the coreference-canceling feature of quotation contexts. In the *L*-proof, it will be recalled, the failure by "(1)" to be coreferential in the identity-statement

(1) = "(1) is false"

does not adversely bear on the requirement that here, too, the (1)s involved must be the same (1). For subsequent disquotation restores to right (1) its coreferential role with left (1). In the present case, there is no such subsequent disquotation on "What Dixon said is true."

Still, there are differences. In the Dixon-argument, we required that the Dixons of

What Dixon said = "What Dixon said is false"

be the same Dixon, not just that the *what-Dixon-said*s be the same *what-Dixon-said*s (although that is required, too). Since disquotation is unavailable for these ends, other devices had to be tried. In the end, we opted for the elimination of *right* "Dixon" together with the device of sentential indexical self-reference. Thus

> What Dixon said about Watergate on that occasion = "What I
> Dixon, am *now* saying about Watergate is false," that is, "This very
> sentence about Watergate is false."

This suffices to achieve the required identities short of having to plead the sameness of the Dixons, never mind having to do so on the basis of the coreferentiality of "Dixon" in

> What Dixon said = "What Dixon said is false."

So the trouble the present argument gives is only apparent.

The failure of the Dixonized *RMN*-proof bears on another common version of the Liar Paradox, the so-called "Following-Preceding-Paradox." Let "FS" abbreviate "the following sentence," and "PS" "the preceding sentence." Consider the pair

> [a] FS is true.
> [b] PS is false.

Suppose that FS is true. Since

> FS = "PS is false,"

substitution yields

> "PS is false" is true.

Disquoting,

> PS is false.

But,

> PS = "FS is true."

Hence

> "FS is true" is false.

By the negation rule and bivalence,

> FS is false.

If indeed FS is false, substitution yields

> "PS is false" is false.

By negation and bivalence,

PS is true.

Substituting,

"FS is true" is true.

Disquoting,

FS is true.

FS is true if and only if it is false. Paradox.

Of course, we have allowed ourselves a little looseness in our abbreviations. "The following sentence" does not refer to any sentence that follows; neither does it refer to any sentence that follows some sentence or other. The proof requires that "FS" mean "the sentence immediately following this very sentence," and that "PS" mean "the sentence immediately preceding this very sentence." The propinquity of the pair [a], [b] makes such precision unnecessary for informal purposes. But since the proof threatens with Tarski's disaster, attention to precise meanings is advisable. Whereupon we see that

FS is true

is

The sentence immediately following this very sentence is true,

and that

PS is false

is

The sentence immediately preceding this very sentence is false.

Precision of interpretation provides us with two occasions to ask after the identity of *this very sentence*. Is it or is it not the case that in the first instance,

This very sentence = "The sentence immediately following this very sentence is true."

and, in the second instance

This very sentence = "The sentence immediately preceding this very sentence is false."

Either these identities hold or they do not. If they do,

"The sentence immediately following this very sentence is true" is identical to "The sentence immediately preceding this very sentence is false,"

an obvious falsehood. So the identities do not hold. What would explain this? It is the same thing that afflicted the Dixonized *RMN*-proof. In

This very sentence = "The sentence immediately following this very sentence is true,"

left "This very sentence" refers to itself if it refers at all. But it is not a sentence, rather an empty singular term. So is does not refer, and the identity collapses. Similarly for the other case. Collapsing with them is the Following-Preceding-Paradox.

Identical considerations bring to grief the Box Paradox. Suppose there is a box in which is inscribed "The sentence in the box is false." Thus

> The sentence in the box is false.

As before, a certain precision is called for. The box of "The sentence in the box is false" is *that* box, the box we have drawn here. And the box of

The sentence in the box is false.

is *this* box, the one in which that sentence appears. The paradox is derivable only if the sentence in *that* box is the sentence inscribed in *this* box. But we will not get the intended identities, unless *that* box is the very box in which that sentence is inscribed and where identity is fixed by the sentence it contains. The sentence

The sentence in the box is false.

does not fix this identity. It must somehow refer to the very box, and none other, in which the sentence occurs. How is this done?

Suppose we say,

The sentence in that box = "The sentence in this very box is false."

If a paradox is to be got from this, we must also have it that

That box = this very box.

The identity is true, rightly enough, but it cannot be delivered by the coreferentiality of "that box" and "this very box" in the identity-statement

The sentence in that box = "The sentence in this very box is false."

As we saw in the Following-Preceding Paradox, paradoxes do not spring from contingent and fugitive identities. The judgment of paradox is a harsh judgment. If sound, it begets Tarski's Disaster. This lays tough interpretation conditions on purportedly paradoxical constructions. The Principle of Charity is nothing unless it commands nonparadoxical interpretations except where inescapable. Paradoxes exist when there is no means whatever of contriving them away. It is easy enough to dispose of

The sentence in that box = "The sentence in this very box is false."

It suffices to plead the opacity of the quotation context and so to call into question that it conveys the identity of that box and this very box. As before, a sentence is paradoxical when there is something wrong with it independently of the contingent devices employed to refer to it. People who think of the Box Paradox and the Following-Preceding Paradox as parlor games – as edifying recreations for the notionally clever – are on to this point. Contingencies of reference do not produce paradoxes. In this they are right, but they make too much of it in concluding that there are no plausible candidates for paradoxes in the Following-Preceding and Box paradigms.

This argues that we should give up on box-identity as inessential to whatever plausibility there is to the claim that the Box Paradox is genuinely problematic. What is needed is a sentence within that carries the appearance of trouble short of inessential ways of referring to it. Such a sentence is

"This very sentence is in this very box and is false."

The box-clause being a purely contingent rider, can be supposed without loss. Thus,

The sentence in that box = "This very sentence is false."

Whereupon, the identity-statement itself becomes otiose. It matters not for whatever trouble that "This very sentence is false" may give that it is the sentence in that box. What is not inessential, if the sentence is to be a plausible candidate for paradox, is what this very sentence *is*, so we must have the identity if paradox is to be given a run for its money,

This very sentence = "This very sentence is false."

But, as we have seen, this is a purported identity only. It collapses with the reference-failure of its left-hand term. "This very sentence" refers, if it all, to itself. It is not a sentence, however; so it does not refer to itself. Hence, it does not refer at all. The "Box" Paradox is disarmed.[6]

Hintikka has developed a first-order logic that incorporates independent quantifiers and the game theoretic semantics for which he is well known (1996). In this first-order IF logic ("IF" for "independence friendly"), the negation of Φ is the absence of a winning strategy for verifying Φ. Similarly, the truth of Φ consists in there being a successful strategy of verification. Truth, T, is a property of closed sentences in an IF logic. When we activate the Diagonal Lemma, we have it that there exists a number i that is the Gödel number of

$$g(\neg T[\mathbf{i}])$$

which in turn is the Gödel number of the string resulting from $\ulcorner \neg T[x] \urcorner$, by replacement of "$x$" by "$\mathbf{n}$." Obviously, if

$$\neg T[\mathbf{h}]$$

is false, it is also true; and if true, is likewise false. However, $\ulcorner \neg T[\mathbf{h}] \urcorner$ need not be either true or false. Given the game theoretic interpretation of "not," it does not follow from the fact that there is no winning strategy for Φ that there is a successful verification strategy for Φ. The paradox is blocked on these grounds alone. But Hintikka has another card to play. The string $\ulcorner \neg T[x] \urcorner$ is ill-formed in an IF logic. T is not a property of open sentences.[7] Hintikka's second complaint resembles the position developed in the present chapter. Each position can be read as saying that the Liar sentence does not say anything, or has no identifiable content. Whereas my solution did not take into account generation of the Liar by diagonal means, Hintikka's argument takes diagonalization head-on. Although the structural reasons for the "does not-say-anything" treatment of the Liar differ for the ranges of cases considered, especially the cases I have considered and the diagonal case considered by Hintikka, conceptually the solution is the same.

The Liar is underivable in first order IF logics. Hintikka holds that, for various reasons, such a logic can plausibly be seen as the base logic of natural languages. This is encouraging. On the other hand, for technical reasons pure IF logics cannot quite deliver the goods in the foundations of mathematics. The logic needs to be enriched or, as Hintikka says, extended. Alas, a version of the Liar is provable in extended IF logics. So we need to pause over the issue of diagonalization.

DIAGONALIZATION

Even if the arguments made so far are correct, I have not proved that no paradox of the Liar or Curry sort is generable in a natural language as English. Perhaps there are constructions in English that give us these unstoppably. Let **n** be the numeral for the number n. The Diagonal Lemma asserts that for any formula $\ulcorner \Phi(x) \urcorner$ of elementary arithmetic with x as its sole variable, there exists a number n named for **n** such that the following condition holds:

$$g(\Phi[\mathbf{n}]) = n.$$

As we saw, the left-hand expression is the Gödel number of the expression arising from $\ulcorner \Phi(x) \urcorner$ by replacement of x by n. In any language for which the Diagonal Lemma holds, the Liar Paradox is easily derivable if the language contains its own truth predicate, T. Applied to "$\neg T[\mathbf{m}]$" the Diagonal Lemma produces an m such that the Gödel number of "$\neg T[\mathbf{m}]$" is m. This sentence says that the sentence with this Gödel number is false. But the sentences are one and the same. Contradiction. I have not proved that the Diagonal Lemma holds only for formalized languages. But our discussion in this chapter suggests that this is so nevertheless. Hintikka is right in saying that "the status of the diagonal lemma strictly speaking needs a review" (Hintikka, 1996, p. 142). But he allows that it does hold for a first-order game theoretic logic that he develops in *Principles*. He also holds that this logic makes a better claim on being the logic of natural languages than standard first-order logic. So, presumably, he is ready to accept that the Diagonal Lemma holds of natural languages.

Does it? Hintikka says with some emphasis that the Lemma simply registers a combinatorial fact about basic arithmetic. This suggests that he may think that in any language containing a truth predicate and rich enough to carry the resources for its arithmetical coding sentences, such as "$\neg(T[m])$," to "say something." This might be thought to confuse two questions. (i) Is "$\neg(T[m])$" a sentence? Does it have *any* semantic status, much less the semantic superstatus that constitutes paradox? It is clear that the answer to (i) is straightforwardly the kind of combinatorial fact of which Hintikka speaks; but it is not clear that the answer to (ii) is determined in that same way.

Perhaps it will be thought that by the somewhat heterodox diagnosis of this chapter, the ground has been prepared for denying the diagonalization function for languages such as English; or more exactly to deny its existence in contexts affected by Convention T. Perhaps, on the other hand, this is unstoppable. If so, this would be a black mark against such constructions and against English if it contrived to tolerate them. It could be that expunging them would harm the cause of things we like, such as recursion theory. In that case, we would be well advised to concede, what is all but true in any case, that the recursion theory and the other diagonalization theorems wanted for mathematics come off not in natural languages, but in formalized languages. To the extent that this is so, English cannot be said to be our best language for the abstract sciences. Who would have thought otherwise? Nobody ever supposed that the set theoretic paradoxes put English out of business. English can get by without talk of sets, just as it can get by without the upper reaches of transfinite arithmetic. But it is hard to see how English can get by without *semantic closure*. Tarski supposed that the Liar established that the cost of semantic closure is the inconsistency of English, and thereupon by classical logic – the truth of all its statements. The same result comes from the Curry Paradox directly. Paradoxers are strikingly sanguine about such difficulties. They fail to attend to what they supposed they have proved; that the very existence of semantically closed languages is an illusion.

I concede that if there is any style of construction in English that gives the paradoxes (for example, the diagonalization structure of paragraphs ago), it must be expunged from English on pain of contradiction. Let us assume the expungement. What is left of English? What is left is enough to resist entirely the attacks from paradoxes in the forms considered in this chapter. Left intact, in particular, are the structures

    " ... " is true
    it is true that ...

and

    that ... is true,

as well as structures such as

What so-and-so said is true.

What is left, therefore, is semantic closure, the very thing that Tarski thought English could not have, except in the absurd world in which all is true. What I have attempted to show is that Tarski was mistaken.

Diagonal arguments are important.[8] In what has recently been said, perhaps I have made a waffling concession. I said that there may be constructions in English satisfying both the Diagonalization Lemma and Condition T. If so, a paradox is derivable in English. This led me to suggest that English could retain its status as a semantically closed language even if we made a reform in which English no longer admits of diagonalization in those contexts to which Convention T is applicable. There would be a cost, of course, matching the importance of diagonalization.

Gödel's theorem is representable as a consequence of a diagonal argument, as are such things as the indenumerability of the real numbers, the fact that some algorithmic functions are not primitive recursive, Tarski's indefinability theorem, and the power set theorem. However, it is also true that diagonal arguments generate Richard's paradox, the heterological paradox, many versions of the Liar, and the so-called cycling paradoxes in the theory of sets and in formal semantics. Russell once asked why it is that some diagonal arguments prove important theorems, whereas other leads to paradox.

So far as is known, the diagonal argument was invented by Cantor for use in his celebrated proof that the real numbers are indenumerable. Suppose that $\{x_1, x_2, x_3, x_4, \ldots\}$ is an arbitrary denumerable set of reals. Real numbers have binary representations; that is, any real number is representable as a unique sequence of 0s and 1s. Suppose then that the following array gives the binary representation of $\{x_1, x_2, x_3, x_4, \ldots\}$.

	1	2	3	4	...
x_1	0	1	1	0	...
x_2	1	1	0	0	...
x_3	0	0	0	1	...
x_4	0	1	0	1	...
...	...	...	...	...	...

The denumerable sequences in each row are the binary representation of the real number occurring at the left side of the row. The integers running along the top identify positions in rows. Consider now the diagonal sequence, running from top left towards bottom right. This is the sequence 0101 .... Suppose now we transform this diagonal sequence by replacing each occurrence of "0" with an occurrence of "1," and each occurrence of "1" with an occurrence of "0." The result of performing this transformation on our diagonal sequence is an antidiagonal sequence. Cantor proved that the antidiagonal sequence cannot occur as a row of our array (Cantor, 1890/1891). It differs from the first row in at least the first place, from the second row in at least the second place, and, for all n, from the nth row in at least the nth place. But since the antidiagonal sequence is a binary representation, there is a real number r that it and it alone represents. And since our array is an array of *any* denumerable set of reals, and r is not a member of it, r is a real number of no denumerable set of reals. Hence, the reals do not constitute a denumerable set.

As we can see by examining the outline of Cantor's Theorem, an *array* is a certain arrangement between side-items and values of these. Let R be a ternary relation and let S (side) and T (top) be sets. Then (following Simmons's limpid exposition) (Simmons, 1993, p. 27),

DefArr : R is an *array* on S and T iff $\forall x \forall y (x \in S \wedge y \in T \rightarrow \exists! z\ Rxyz)$.

R provides that for any pair of elements of S and T there is a unique value.

We can think of these values as occupying boxes. Boxes, in turn, are the constituents of diagonals. Thus,

DefDiag: D is a *diagonal* on S and T iff D is a 1-1 function from S into T.

By **DefDiag** diagonals pass through every row but need not pass through every column. Accordingly, if the following are arrays:

x			
	x		
		x	

x			
	x		
		x	
			x

each is diagonalized, but only (2) exhibits what is called a *leading diagonal*, that is, one from the upper left to the lower right of the array that passes through all rows and all columns. (An array supports a leading diagonal only if both S and T are ordered and the order is reflected in correlation between them.)

Arrays fix *values* and diagonals fix sequences of *boxes* for values. Even so, we can speak of the value of *a diagonal in an array*.

DefValDiagArr: V is the value of diagonal D in array R iff
$\forall x \forall y \forall z (Vxyz \leftrightarrow Dxy \wedge Rxyz)$.

Intuitively, the value of a diagonal is an array in a sequence of boxes which are unit sets of members of the counterdomain of R, the array in question.

Now we must remember that in order to make our diagonal proofs work, it is necessary to replace diagonals with their "flipsides," so to speak. So it is necessary to introduce the concept of a *countervalue*.

DefCountVal: Let R be an array, and D a diagonal, on S and T.
Then K is a *countervalue* of D in R iff
(a) $\forall x \forall y (\exists z Kxyz \leftrightarrow Dxy)$.
(b) $\forall x \forall y \forall z \forall z' (Kxyz \wedge Kxyz' \rightarrow x = z')$
(c) $\forall x \forall y \forall z (Kxyz \rightarrow z \in \text{range})$
(d) $\forall x \forall y \forall z (Kxyz \rightarrow \neg Rxyz)$.

Condition (a) says that there is something z to which x,y bear the K relation if and only if the diagonal function D takes y as value for x as argument. Condition (b) provides that K has only unique values. And although, as condition (c) specifies, the unique value of K must be in the range of R, condition (d) makes it clear that it cannot be in R's counterdomain.

We now require the idea of a value or countervalue *occurring as a row*.

DefOcRow: Let R be an array on S and T and let U be a value or a countervalue of a diagonal D of R. Then U *occurs as a row of R* iff $\exists s \in S \forall x \forall y \forall x (Kxys \rightarrow Rsyz)$.

We can now state the *Basic Diagonal Theorem*.

BasDiagThm: Let R be an array on S and T, and let D be a diagonal on S and T. Let K be a countervalue of D. Then *K does not occur in R*.[9]

Simmons distinguishes direct and indirect diagonal arguments. A diagonal argument abstractly considered is a *triple* of (1) a set of set-theoretic assumptions, (2) some deductive manipulations of those assumptions, and (3) a resulting conclusion. A direct diagonal argument specifies set theoretic structures whose existence it asserts. Indirect diagonal arguments also specify such structures; but rather than asserting their existence, they assume the existence of at least one of them for the purpose of a *reductio*. Simmons calls a diagonal argument *good* just in case it is either direct or indirect, and *bad* just in case it has the structure of an indirect argument *minus* the contrary assumption of set theoretic legitimacy (Simmons, 1993, Ch. 2). The definition of good and bad diagonal arguments turns on an idea of Richard. We might call it "Richard's Thesis." Fundamental to Richard's Paradox is the existence of the set S of real numbers definable by an expression of French. On reflection, Richard came to think of S as not "totally defined"[10] (Richard, 1967, p. 143).

A diagonal argument assumes the existence of a number of components: a side, a top, an array, a diagonal, a value, and a counter value. In bad diagonal arguments, one or more of these sets is not well-determined.

Accordingly, we have what might be called "Simmons' Corollary."

This characterization of a bad diagonal argument requires the notion of a well-determined set. This is supplied if we suppose we are working with some standard set theory, say **ZF**. (Simmons, 1993, p. 29).

Consider the following two diagonal arguments: **A** and **A***. By the lights of Richard's Thesis, each is a triple of set theoretic assumptions, some deductive manipulations of those assumptions, and a conclusion. In the case of **A**, the conclusion is inconsistent. In the case of **A*** the conclusion is not (known to be), inconsistent, though it might be terribly surprising. Since **A*** is clearly not bad in the way **A** is, there must be a fault in **A** that **A*** does not share. Inasmuch as (i) **A** and **A*** exhibit the same internal deductive profile and yet (ii) have different set theoretic assumptions, it is sensible to suppose that the fault in **A** is to be found in its set theoretic assumptions. Hence, it is reasonable to suppose that **A**'s set theoretic assumptions are illegitimate and that **A***'s set-theoretic assumptions are legitimate. After all, **A*** does not lead to a self-contradiction.

Simmons's Corollary is an interesting suggestion, well worth reflecting on. I shall not do that here, except to note in passing, in regards to the suggestion that **A*** is a good diagonal argument if it does not lead to contradiction, that students of logic will have long since learned a healthy hesitation in the face of *argumenta ad ignorantiam* in the modern, rather than Lockean, conception of them. In the present case, we are offered an argument grossly in the form

[i] Argument Σ does not produce a contradiction.
[ii] Therefore, argument Σ is sound.

Whether this argument pattern is a fallacious *ad ignorantiam* is a nice question. Whatever its answer, it prompts an irresistible *ad hominem*. The argument with which we have been trying to dispatch the Liar and the Curry has the following gross form:

[α] There is a solving term for Liar and Curry, which is *consistently* applicable.
[β] Therefore, it is a correct solution.

I have several times volunteered, in effect, that argument [α]–[β] is not valid. Why then should we accept the validity of [i]–[ii]?

This should give us pause, and is occasion to wonder why the sheer oddness of the conclusion of **A***-type arguments, together with the fact that occasionally (namely, when they are **A**-type arguments, *as well*) they lead to inconsistency, does not justify a skepticism about what **A** and **A*** undeniably have in common – the structure of the internal diagonal logic itself.

There is no stopping such skepticism (and no good reason to try) short of ZF's pronouncing *independently* on the set-worthiness of the structure of diagonal arguments deemed good and on the set-unworthiness of the structure of diagonal arguments that are manifestly bad. But, as is notorious, ZF cannot easily meet this test. It is unable to pronounce on the set-theoretic legitimacy of totalities short of having a reason to disbelieve them to be paradoxical. But it is precisely this that we are trying to determine.

Hintikka takes a tougher stand, as we saw. He doubts that set theoretic axioms have intended interpretations. But suppose they do. Then for "each model of axiomatic set theory, one can find a sentence S in the language of a set theory that is true interpretationally but false in that model. In brief, axiomatic set theory cannot be true on its own intended interpretation. When we try to construct the liar paradox for axiomatic set theory, the Liar turns out to be axiomatic set theory itself."[11]

The main burden of this chapter has been to construct an *ambush* of the received view of the Liar Paradox. The *ambush* had a motivation. What could be said, we wanted to know, for the view that if the analysis of an intuitive concept proved a contradiction, the concept is irretrievably lost? Our answer was that theorists for whom this is a plausible view are those who have already pledged to the method of analytic intuitions. If, then, a conceptual analysis of the unitary idea of X proves a contradiction, then a logical falsehood inheres in the very idea of X hood. Short of dialethism, there is no option but to disclaim X hood itself. Chapter 5 was an effort to call to account the method of analytical intuitions. In making the effort, we found the accounting to be unacceptably thin. We went on to wonder how the untenability of the method of analytic intuitions would bear upon the distinction between the Barber Paradox and those we have here been examining. Given that we no longer react to the Paradoxes in the manner of *Frege's* and *Koryé's Sorrow*, what now can be said of the received view on which what the Barber shows is the nonexistence of that particular barber, whereas what the other Paradoxes show is the nonexistence of the intuitive concepts of set and of bivalent sentence? The short answer is that nothing can be said for sticking with the received view. Without this underlying epistemology of analytic intuitions, the received view of the Paradoxes is a lazy habit, and a bad one, too. In this, we derive encouragement for the position we took in the Great Delta Debate of Chapter 5. We said that on the question that divides the dialethist and his opponent, the onus falls on the dialethist and it is an onus he lacks the dialectical means to discharge. It fell to the dialethist to show that an inconsistent concept of something that one is philosophically interested in is better than no concept of it at all. Thinking so presupposes on the classicist's behalf, and on the dialethist's own as well, the tenability of the epistemology of conceptual analysis, at least in the form presupposed by *Frege's Sorrow*. But if that presupposition is now canceled, there is no reason to say that the Paradoxes of Russell and the Liar cost us the very idea of set and the very idea of statement. Once the presupposition is withdrawn, we have it at once that in the very idea of X there is nothing decisive for the epistemics of theories of Xhood. Even if it were true that the only way of retrieving the very idea of set and of statement is to allow that some sets and some truths are actually inconsistent, doing so is no longer a virtue if that presupposition has lapsed. If analytic theories are dashed hopes in general, then analytical theories that promise actual inconsistencies are more dashed still. If theories of sets and of truth now have to be formal theories in the manner of Chapter 5, the debate between the dialethist and the nondialethist now turns on whether they share sufficient interests and enough of a common understanding of the economics of those interests to settle the question one way or the other.

If what we have been saying about the received review is correct, the standard *triagic* judgment of the Liar is a mistake. It is also possible that it is not the only mistake that has attained the status of received opinion concerning the Liar. In making our ambush argument, we took this possibility seriously. We took on the received view at the level of *diagnosis*. We were guided in this by suggestions that up to now could not be made to succeed. Lots of people (although still a minority among semantic theorists) have been drawn to the idea that the Liar construction disqualifies itself in some way – that it is meaningless, or non-truth-valued, or does not express a proposition, or some such thing. The trouble with this good idea is that it attracts its own version of the Liar. The Libertine Liar Theorem both explains this fact and gives us hope. Either there can be *no* paradox-free solving term for

(1) (1) is false

or there is something about this structure that blocks the proof of the paradox. Although

(2) (1) is not a statement

is a true statement,

(3) (3) is not a statement

is not a true statement. It is not a statement.

Another insight offered by semanticists over the years was that the Liar Paradox commits the Namely-Rider Fallacy. As we have seen, this is both right and wrong. It is wrong in its charge of an infinite regress; but it is right inasmuch as the substitutions that the complainant saw as generating are infinite regress do not, in fact, survive substitution in the first "namely"-context.

Ambushes are at their best when they bring down quite a lot of entrenchment as cleanly and leanly as possible. In the abstract sciences, the clean and the lean are what approximate to the crucial experiment. If a huge misconception can be laid at the feet of a small misunderstanding, the ambush can claim impressive, perhaps decisive, economic advantage. Here, as we said, is the little confusion on which the misdiagnosis of the received view reacts: A condition can satisfy itself without saying that it does; a condition can satisfy the condition that it does not say anything without saying that it does not say anything.

Loss of the method of analytic intuitions is the central blow for philosophers, dialethic and otherwise, who have it in mind to ply their trade *realistically*. Whether one is drawn to *Frege's Sorrow* or *Routley's Serenity* there is something common to them both that is placed at serious risk if the method of analytical intuitions is called credibly into question. I keep emphasizing the utter attraction the theorist has toward his intuitions. If the method of analytical intuitions à la Moore were the sole example of the intuitions methodology, we would have done serious damage to it. But, of course, there is much more to the intuitions methodology than a Moorean analytical platonism. In the next chapter, I mean to renew and broaden the attack.

8

Normativity

One should not pay lip-service to fallibilism: "To a philosopher there can be nothing which is absolutely self-evident" and then go on to state: "But in practice there are, of course, many things which can be called self-evident...each method of research presupposes certain results as self-evident."
Imre Lakatos, *Mathematics, Science and Epistemology*, 1980.[1]

We said in the previous chapter that on the face of it, the various forms of *Reconciliation* are compatible with presumptions of the analytical intuitions methodology in the abstract sciences. Strictly speaking, one can defang an ugly pluralism about concept K by finding (or pleading) the relevant conceptual multiplicity, $K_1, \ldots, K_n$, each K_i of which might still qualify for how K_i-hood really is, as revealed by a conceptual analysis of K_i. Against this, however, is that, short of provably preexisting ambiguity that warring theorists just happen to miss, various of the instances of Reconciliation, especially those that figure in the received opinion dynamic, do not sit well with the realist presumptions of conceptual analysis taken as unearthers of antecedent conceptual fact. Rather than press this point, however, I shall attempt in this chapter to discredit all forms of the intuitions methodology. Doing so will disclose dialectical procedures of some importance, which we shall meet with in due course. One of these maneuvers I call the *Can-Do Principle*. Another is a degenerate variation of it, the *Make-Do Principle*. I want to be able to show that much of what continues to be admired (and trusted) in this methodology of intuitions is sanctioned by this degenerate dialectical principle.

The tenacity of the method of intuitions is almost impossible to overestimate, as I keep saying. There is a joke making the rounds about how best to infuriate a postmodernist. Tell him that he is just making it up! The intuitions to which he, and we, too, are most closely drawn are those he holds with conviction enough that his state of mind takes on an epistemic significance. Speaking more behaviorally, our intuitions are those beliefs we know not how to do without. They dispose us to think that they are what any reasonable theory of

277

such things must absolutely preserve. We should not for a moment doubt the causal significance of this picture of the theorist at work. When he is an abstract theorist to whom observations are of no direct, if any, benefit, the tenacity of his beliefs, of his uttermost subjective certainties, is what his emerging theory must look to for guidance. Intuitions are the abstract theorist's observations.

The issue of conflict resolution in the abstract sciences requires us to have a serviceable notion of the abstract. In these pages, abstraction is marked by the absence of empirical checks, by the unavailability of sensory confirmation or discouragement, as the case may be. If one's tastes run to holism, the standard of empirical uncheckability has to be conceived of with a careful indirection. My own view is that enough remains of the basic idea to give it a load-bearing role in the reflections that have occupied us in preceding chapters. In those places, the abstract sciences have been exemplified by systems of logic, set theory, and semantics; and correspondingly, the idea of conflict resolution in the abstract sciences has been exemplified by what we have been trying to say about the conflictual dynamics of logic, set theory, and semantics. We saw in Chapter 1 that if we take the lack of empirical content to mark a theory as abstract, then a verdict of abstractness is required for "hard" sciences such as those we have been examining here. It is clear that by that test, some of the "soft" sciences also qualify as abstract. Such theories can be characterized as those to which the "is/ought" distinction applies under classical presumptions of the failure of interderivability. Theories so conceived are *normative*. They include, but are not restricted to, the entire range of theories of rational per-formability, including theories developed by economists, moral philosophers, decision theorists, and discourse analysts, among others. Also on this list are theories of inference; and so the "soft" has already entered our discussion in what we have had to say earlier, especially in Chapter 5. Here, too, we see the pull of intuitions. No set of observations will lawfully pull the theorist over the divide from "is" to "ought." Something else is needed for the job. Intuitions not only track down the abstract; they are the natural markers of normativity.

We have been making much of the twin ideas of intuitions and counterintu-itions in our discussion of conflicts that arise in logic, set theory, and semantics. We have found that appealing to them is all but useless as a conflict resolution strategy, with question-begging an endemic problem. We also have seen some-thing of the damage that accrues to a reliance on the methodology of analytic intuitions – a practice we associated with Moore, but to which Frege, Russell, and Tarski also were drawn. It leads us to make too much of the paradoxes; it leads us to suppose that the paradoxes destroyed the very idea of what they purported to be about, with direct consequences for the realism/antirealism controversy in the foundations of mathematics and formal semantics.

The methodology of intuitions and counterintuitions is also vigorously in play with normative theories of human performance. Here, too, conflict is en-demic. Paul Churchland is famous for his impatience with the dissensus to be found in "folk psychology," in his own pejorative description. Churchland counseled its wholesale abandonment (Churchland, 1981); one can only wonder why he stopped there, for much the same discord afflicts *les sciences humaines* quite generally. It is perhaps premature to say that the link in the methodology of such theories between intuitions and normativity is the diagnostic key to their troubled state, but it is a link worth examining if only for its pervasiveness. So we turn now to the "soft." This is not to say that normativity is reserved for the soft. For a view close to my metatheoretical heart, see Hintikka: "Tarski's procedure is generally taken to be *normative*. However, it is seldom, if ever, explained what the reason for this normative character of Tarski's definition is, or even what Tarski's own rationale was in proceeding in the way he did" (Hintikka, 1996, p. 106, emphasis added).

IDEAL MODELS

What are the procedures and methods by which an agent performs rationally? Under what conditions is it rational to reenter a burning building, or to buy a used Pinto (or a used Hansen and Pinto), or to remind your father that he too smokes, or to detach a proposition on the warrant of *modus ponens* (Hansen and Pinto, 1995)? What would it be like to have a theory that answered such questions persuasively and authoritatively? A common suggestion is that the theorist of rational performance proceeds in two main stages (Cohen, 1986). At stage one, he marshals his "intuitions" about what it would be rational to do in certain cases, or rough ranges of cases. Intuitions are "untutored"; they are pretheoretical beliefs that the theorist holds with conviction and that, he assumes, are widely shared by others. They are what the theorist supposes everyone with the relevant curiosity and attentiveness already knows about the issues at hand. They are the *data* for his theory.

At the second stage, the theorist ventures beyond. He tries to generalize on his data, to link them up in nomically attractive connections. However this is done, and whatever else the theorist is doing at stage two, it is essential to the enterprise that cases not reported in the data be dealt with, so that the theorist's generalization can be taken as covering unobserved instances and predicting future instances. Stage two also provides the wherewithal to reconsider, or to re-construe, the data in a principled way. If, as they emerge, the theory's provisions appear to require the revision or abandonment of a pre-theoretical belief, the theorist is at liberty to do so. This is a third function performed by the theory's laws (or, more circumspectly, "laws"). They enable us to correct pretheoretical beliefs, if not retire them wholesale.[2]

It is rather striking that the theoretical "laws" of accounts of rational performance fail the prediction test rather substantially. People do not actually do what the "laws" can be taken as saying they will do. This presents the enquirer with a genuinely interesting problem in the methodology of theory construction. There are situations in which the theorist feels himself justified in refusing the status of counterexample to actual instances that contradict his "laws." The problem is one of showing that he is justified in this refusal. To that end, the theorist will commonly invoke the notion of *approximations*. In mechanics, to take a common example, the laws of frictionless surfaces fail even on the pregame ice of Maple Leaf Gardens. The physicist says that such laws are true at a mathematically describable, but physically unrealized, limit of the slipperiness of real life, and to which the slippery world approximates only more or less well. In probabilistic inference theory – an example we have seen before – laws relating to the conjoining of probabilities are routinely dishonored in practice. Rather than abandon the calculus of probability, the theorist invokes the device of an ideal agent, and describes his behavior in an ideal model in which, he deems himself free to say, the conjunction rule is universally obeyed. The theorist then ventures that an actual reasoner is rational to the extent that his behavior approximates to that of the ideal agent.[3] Here is Keynes on this point:

Progress in economics consists almost entirely in a progressive improvement in the choice of models. . . . But it is of the essence of a model that one does *not* fill in real values for the variable functions. To do so would make it useless as a model. . . . The object of statistical study is not so much to fill in missing variables with a view to prediction, as to test the relevance and validity of the model. (Keynes, 1973, p. 296)

There are, of course, more ideal models than one can shake a stick at. How does the theorist choose among rival models? How does he test for "relevance and validity"? Consideration of two cases might throw some light on the matter. In the first case, as with probability theory or first order logic, the theorist already knows (or thinks he does) that his axioms are true, that his derivation strategies truth-preserving, and so on. This creates an abiding temptation to specify as ideal any model preserving those axioms and strategies. They are the "laws" of his theory, obeyed without exception by the ideal agent – by the ideal probabilizer, the ideal logician, and so on. The second case is trickier. The theorist's judgment as to which theoretical "laws" are true is inseparable from his choice of the model in which they *are* true. It is interesting that in neither case should the theorist be worried about the underdetermination of theories by observation, or by the fact that no (first order) theory determines its objects up to isomorphism. The issues raised by our cases are not primarily ontological. In part, they are semantic. They are the problem of finding models in which favored "laws" come out *true*. But the more central case-two problem is that of validating the model independently of having to demonstrate the truth of the "laws" that describe it (and which actual agents routinely shirk). Suppose that the theorist is trying to construct a theory of argument. Waiving the utter implausibility of it, what if he were drawn to the following "law":

(L): If X and Y are opponents in an argument, and if X asserts that P, Y will beg the question with respect to P.

L is sometimes honored in actual practice, but for the most part, it is not. L fails empirically. What is the theorist to say for himself? What if he says that, while false in practice, it is true in his ideal model? Critics will not like his choice of ideal model, of course; and their dislike of it will turn entirely on their conviction that L is a bad law – a faulty norm – conformity to which, or approximate conformity to which, would ensure an agent's lack of rationality (Cohen, 1986, p. 201).[4] So, again, how is the case-two theorist to pick his "laws" from sets of empirically false generalizations, short of already having at hand a validated ideal model in which to rescue him from that very contingency? How, in turn, is he to determine a model's validity, short of stocking it with empirically false generalizations that, even so, have the property of "deserving" to be "laws" against which rational agency is measured? This is a difficulty serious enough to have a name. Let us call it the *bootstrapping problem*.

Stipulation – another recurring theme – offers promise of a way out of the bootstrapping problem. The theorist now stipulates that his ideal model is *constituted* by the perfect conformity of its objects with those "laws" to which the theorist is drawn. They are now analytically true in the model (waiving for now philosophical reservation about analyticity). But, with careless disregard for *secundum quid*, the fallacy of omitting a qualification, the theorist forgets the qualification "in the model." He puts it that his favored "laws" are analytically true, and he concludes that an actual agent is rational to the extent that his behavior complies with them, for how can an agent be rational if he dishonors analytic truths?[5]

Such seems to be the position in which ideal model theorists find themselves in certain branches of discourse analysis and the theory of argument. Consider:

In order to clarify what is involved in viewing argumentative discourse as aimed at resolving a difference of opinion, the theoretical notion of a critical discussion is given shape in an ideal model specifying the various stages, in the resolution process and the verbal moves that are instrumental in each of these stages. The principles *authorizing* these moves are accounted for in a set of rules for the performance of speech acts. Then together, these rules constitute a *theoretical definition* of a critical discussion (van Eemeren et al., 1996, p. 280, emphasis added). Moreover, [a]nalyzing argumentative discourse amounts to interpreting it from a theoretical perspective. This means that the interpretation is guided by a *theoretically motivated model* that provides a point of reference for the analysis, emphasis added. As a consequence of the adoption of such a theoretical perspective, specific aspects of the discourse are *highlighted* in the analysis (van Eemeren et al., 1996, p. 191).... [And so], [e]ven a discourse which is clearly argumentative will in many respects not correspond to the ideal model of a critical discussion (van Eemeren et al., 1996, p. 299, n. 49).... In this model, the rules and regularities of *actual* discourse are brought together with *normative* principles of goal-directed discourse. The model of critical discussion is an abstraction, a theoretically motivated system for ideal resolution-oriented discourse. (van Eemeren et al., 1996, p. 311, emphasis added)

Case one stands out in apparent relief from these difficulties. We do indeed know (or so it is said) that the laws of logic and probability theory are analytic or are analytic if anything is; and we know these things without the nuisance of having to specify ideal models in which they are true by stipulation. Here all talk of ideal agents in ideal models is idle. An actual agent is rational to the extent that he obeys, for example, the analytic truths of logic and probability theory. Full stop.

It will not work. The proposal is afflicted by well-rehearsed problems discussed earlier in Chapter 4, and they need not detain us long (Harman, 1986). The present story forces on us rational performance *regulae* that, due mainly to the license they give to complexity, provide a conception of rationality that defeats any hope of a credible approximation relation connecting actual performance to it. The *regulae* count as rational the ideal agent's perseverence with problems that are intractable, problems whose solutions involve a computational explosion. The presumed conception of rationality is one full fidelity to which would be irrational for an actual agent even to aspire. Most case one theorists see the difficulty well enough to reconsider the device of ideal models. An ideal logician is no longer one whose behavior wholly conforms to the laws of logic. His ideality consists in perfect fidelity to softer norms, which are restrictions of the laws of logic. In many writings, this is something to welcome, and make something of, viz., the difference between implication and inference. The theorist will grant that the laws of logic correctly describe the (or some designated) implication relation, but he will now say – in the spirit of Chapter 5 – that few if any "laws" of deductive *inference* will be mere *simulacra* of the laws of implication. So he will propose restrictions and add qualifications. So placed, the case-one theorist encounters the same difficulty as his case-two colleague. As with his colleague, his judgment about which are universally correct "laws" of inference is indistinguishable from his judgment about which norms define and validate his ideal model. Finding himself at risk for the bootstrapping problem, the case-one theorist is beset by similar temptations regarding a way out. If he yields, he will stipulate an ideal model as one in which his favorite "laws" are made analytically true. Again, "analytically true" here means "true in the model," that is, "stipulated in the ideal model." If the theorist concludes that his favorite "laws" are analytically true (full stop), he, too, has committed the howler of *secundum quid*. He need not, of course, be sprung in this trap. He may simply say that an actual agent is rational to the extent that his deductive inferences conform to the analytic laws of the model. He could say this, right enough, but without independent reason to like the model he will not have provided us with the slightest reason to believe him.

By and large, social scientists are not much drawn to epistemological questions, even to those that bear on their own theoretical practice. When I say that an economist's idealizations or simplifications have the effect of making his laws analytic in the model, I do not mean that this is a consequence that the theorist always or even typically recognizes expressly.[6] Indeed, he will often, in effect, deny the analyticity of his laws by insisting on their descriptive character. This is explicable. Such laws are true in the model, but they are not true.

It would be wrong to leave the impression that simulators of ideal models are tendentious cynics; that they are simply making it up as they go along. Rarely is it so. The theorist of ideal models is no trifler; his exactions are too heartfelt for that. He is not even taking cover in Quine's amusing dictum that theories are free for the thinking up, or in Eddington's that they are put-up jobs. The "laws," norms, and performance standards with which our theorist stocks his model are precisely those that he thinks are sound. He puts them in because he believes that they are objectively *correct*. If he is philosophically minded, he might explain himself this way: These "laws," norms, and standards arise from, and are validated by, an analysis of the concept of rational performance. The analytic truths of his model are now *conceptual truths* about rationality.

Perhaps this is so. Perhaps the canons of the model are indeed the correct ones; maybe they do express something of the conceptual content of the very idea of rational performance. What remains is to show that they do. Short of this, the theorist has surrendered his whole stake in the theory to the category of "*data* for theory," that is, statements confidently believed without benefit of tuition. This matters. If the theorist's "laws" are no more than what he takes to be so, we have no theory, and are met instead with the phoney imperiousness of what the theorist believes. Having no theory, he is left with his intuitions.

In speaking of the phoney imperiousness of what the theorist believes, there is an understandable inclination to resist the imperiousness. But it is not intended that we make light of the beliefs. Often enough our theorist is sharper than the rest of us and sees what we have missed. Sometimes a hithertofore unnoticed distinction will clear away some damaging confusion. There are those who think that, so far as the conditions on human flourishing are concerned, the best that can done is done dialectically, preferably at the elbow of a master. So I may think that I should always try to close my beliefs under consequence. You might point out how profligate and pointless that would be. You will have pressed me with consequences of my own view and, not liking them, I may start seeing things your way. If this were the canonical way of proceeding across the whole spectrum of normativity, we could say with certain of the ancients, that theories are not possible for such matters.[7] It could then be proposed that we end the pretense of supposing otherwise – that we abandon all talk of norms and ideal models as empty.[8]

Philosophers are unlikely to welcome this suggestion. For what is a philosophical theory about rational performance if not a theory that tells us in a principled way what people should and should not do? On the present suggestion, there are no such theories; hence there are no philosophical accounts that tell us in a principled way what people should and should not do. Philosophy, that is, normative philosophy, is out of business. There are, of course, philosophers galore for whom the suggestion of the impossibility of philosophy is absurd. Perhaps this is right. Perhaps it is an idea to be scorned. Even so, its disreputability is not self-guaranteeing. So the prophilosophy theorist will be obliged to construct a case. One thing to try is a revival of the intuitions methodology *minus* the encumbrance of ideal models. Let us see.

MATHEMATIZING

The fact that the bootstrapping problem is a problem for ideal model theorists may have little to do with how serious a problem it is, and nearly everything to do with their unawareness of it. In the present section, I consider two exceptions to this rather entrenched indifference. The first exception – or, rather, class of exception – are ideal model theories such as neoclassical economics or decision theory, theories that are heavily mathematical in character. The second exception is that of a philosopher. A metatheoretical account advanced with a *panache* that equals, across the grain of classical discouragement, the derivation of *ought* from *is*.

As for the first, we might consider Subjective Expected Utility theories of rational choice in economics and decision theory. In these theories, rational deciders maximize their expected payoffs. A decider's utility function is something derived from patterns of choices under risk, or "gambles," and subjective probability is, in turn, defined in terms of these utilities. The *locus classicus* of Subjective Expected Utility theories is the axiomatization of Savage (1954). Of the Savage axioms, perhaps the most contentious are those proclaiming transitivity and independence as norms of decisional rationality or economic agency.[9] The Subjective Expected Utility theorist shares in the difficulties already discussed. In particular, he tends to be drawn, however implicitly, to the following argument:

(1) Certain Axioms on ideal agency constitute a definition of rationality.
(2) An actual agent is irrational if, or to the extent that, his behavior disconforms to the axioms (by the definition of rationality).
(3) Consequently, axioms on ideal agency are norms for rational agency.

As before, the mistake occurs in (1). It may be that the axioms on Subjective Expected Utility define rationality in the model. It does not follow that this is rationality for actual agents, that is, rationality *outside* the model against which actual agency should be appraised. Of course, it is true that something gets to *be* an axiom in the model because it encodes the theorist's prior belief that the content of the axiom is indeed normative for actual deciders. But making it an axiom in the model is not a way of making the prior belief true. It is just a way of reexpressing that belief sententiously. So here, too, the bootstrapping problem recurs.

Does the decision theorist have a principled way to avert the bootstrapping problem? Can he find or secure his idealizations independently of his pretheoretical belief that they constitute norms for actual economic agency? There is an old dig to the effect that economists cannot get the success of physics out of their minds, and that they will not rest until they are able to do for socioeconomic nature what the physicist has done for nature herself. In reconciling his theories to the paradigm of physics, the economist assigns a load-bearing role to mathematics. Like the physicist, he is struck by the fact that "...beyond a minimal level, we do not know how to do natural science without mathematics" (Hart, 1996, p. 53).

Neoclassical economics is an instructive case in point. As is widely known, the neoclassical theory replaced the law of diminishing marginal utilities with the law of diminishing marginal rates of substitution. With the additional "simplification" that goods are infinitely divisible, the theory had direct access to the firepower of calculus and could be formulated mathematically.

Consider, as a second example from economics, the so-called semantic view of theory construction. On this view, theories are set theoretic predicates (Beth, 1949; Suppes, 1957). It sees the physicist's theory of motion and gravitation not as statements about the world, but rather as an analysis of the predicate, "is a gravitational system." Such predicates, may of course, apply to the world. Let Φ be a consequence of the gravitational system predicate. Then Φ is assertible about the world under the empirical assumption that the universe is a gravitational system. It may seem to some that the theorist's attention to the analysis of predicates rather than to the description of nature is ultimately a diversion, since nature ends up getting described after all.

Why, then, bother with theories constrained by the semantic view? A good part of the answer is that it makes it "easier to reconstruct the claims of science in a rigorous formal or mathematical way if one employs the semantic view" (Hausman, 1984, p. 13).

Indeed,

This kind of endeavor is particularly prominent in economics, where theorists devote a great deal of effort to exploring the implications of perfect rationality, perfect information, and perfect competition. These explorations [are]...what economists (but *not* econometricians) call "models." One can thus make good use of the semantic view to help understand theoretical models in economics. (Hausman, 1984, p. 13)

It is apparent that the mathematical economist sees things in one of two ways:

Ontologically. Nature, or the natural world, is to a considerable extent the physical realization of mathematical systems. That is, a mature scientific theory could be "formalized only as an extension of some part of mathematics" (Hart, 1996, p. 53). Socioeconomic nature is part of the natural world. So it too, to a considerable extent, is a sociophysical realization of mathematical systems. In both cases, the more the theorist is able to mathematize his accounts, the closer they will bring him to the true nature of things. Therefore, if shaping the theory's idealizations in a certain way – even with apparently *ad hoc* latitude – enlarges the theory's mathematical power, then, other things being equal, the idealizations are to that extent justified.

Pragmatically. Let T be a mathematizable theory. Suppose that a particular idealization enables T to engage some mathematics in ways that make possible simpler and deeper laws, and more transparent, encompassing, and unified nomic connections. Other things being equal, the idealization is justified because, or to the extent that, it facilitates the evolving of a better theory, a theory that handles its subject matter economically, deeply, and with a high level of comprehensiveness.

There is much that can be said against these ways of thinking. Critics will bring charges ranging from neopythagoreanism to the fallacy of division, and from there to complaints of reductionist simple-mindedness. I shall press none of these objections here. I am interested only in the suggestion that when idealizations are made in order to facilitate mathematical power, then, other things being equal, the idealizations are justified. They take on, so to speak, the objective validity of the mathematics that they were contrived for.

Whether he thinks of his deployment of mathematics ontologically or pragmatically, the theorist chooses and adjusts his idealization to facilitate the mathematics. But this is not done in isolation from the requirements of empirical adequacy. On the score of empirical adequacy even quantum physics does marvelously well (as it would have to, given its vexing conceptual *arcana*), but economics does rather abysmally.[10] Of course, there is an important difference between physics and economics. Physics is a descriptive enterprise, and economics of the sort under review is normative. This all but guarantees a certain differential importance attaching to the empirical failure of theoretical pronouncement. A theory under empirical attack is not, just so, put out of business. Various compensatory adjustments are available to the theory's loyalists. But there is one plea that the working physicist cannot even consider entering, which for the economist is a metatheoretical commonplace. What the physicist cannot say in the face of recalcitrant experience is that nature is not operating as she *should*. This is precisely what the economist will sometimes say, and may be right in saying. Bill and Sue and Fred are not acting as they should. When this is the case, it will also sometimes be true that their behavior contradicts a theorem in the theorist's ideal model. The contradiction will not matter unless those theorems also confer norms upon actual agency. We might even suppose that the theory brims with rich mathematical content, and that, for this to have been so, the behavior of the ideal model had to be shaped to that end. The two facts, however, are not linked as the theorist may have wished. If the theory's laws are indeed normative for actual economic agency, then the norms of actual agency will enjoy mathematically rich assays. In this there may be a metaphysical connection worth remarking on. It might be true that the norms for actual agency have lots of mathematical content. But it cannot be true that having lots of mathematical content suffices for normativity. Neither is it true that idealizations are justified *because* they facilitate engagement of the mathematics. So the economist and decision theorist, no less than the pragma-dialectician, is remet with the bootstrapping problem.

It may be that no economist has made errors as brazen and obvious as these. Even so, avoiding them is essential for any theorist disposed to rest the justification of his idealizations on their readiness for commerce with the theory's mathematics. Perhaps, too, I have been making rather too much of the theorist's affection for mathematics. The story of that affection instantiates a more general metatheoretical pattern. Let T be a theory of human agency, an account, say, in cognitive psychology. If T is to be a contribution to bona fide science, it must produce laws. Laws are what the theorist is able to make of raw correlations. It is an abiding question, and a problematic one in its details, as to how the theorist is able to nomicize correlations. An altogether common answer is that psychological laws are the outputs of correlations filtered through the theory's idealizations. The idealizations – some of which may invoke mathematical variables – are justified to the extent that the laws are descriptively true. The empirical adequacy of a psychological theory is thus justification of the theory's idealization precisely because it is indispensable to the formulation of the theory's laws. What we were saying about mathematical economics is reconcilable to this picture. If mathematics is necessary for the nomicization of economic correlations under various constraints in the ideal model, then two things are true. The idealizations need to be shaped to accommodate the mathematics; and, thus mathematicized, they are necessary for expression of the theory's laws. If there is a fault in this by which the economist would be rightly troubled, it is not the fault of adjusting idealizations merely to accommodate the mathematics (indeed, he is shaping his idealizations for the broader and wholly legitimate purpose of deriving laws – of nomicizing correlations). Rather, the fault, if there is one, is the failure of empirical adequacy.

Even so, there is empirical adequacy and there is empirical adequacy. Every empirical theory can expect its laws to fail in the real world, whether on the ice of Maple Leaf Gardens, or in decisions or inferences taken by Bill, Sue and Fred. It is necessary for the theorist to have an explanation of why these failures do not matter when indeed they do not. The physicist may plead the approximate conformity of the skating surface at Maple Leaf Gardens, pointing out that the slippery surfaces of real life are to some extent contaminated by factors that interfere with slipperiness. The economist or the pragma-dialectical theorist must also say why raw misperformance does not matter for empirical adequacy; but as we now see it is never enough for the theorist to try to privilege his rawly disconformed laws simply by announcing their ideal provenance. This is what he cannot or should not do. What, then, *can* he do? He wants now to be able to make good on two claims. First, that raw misperformance is not a defeater of empirical adequacy; and second, that if raw misperformance is indeed irrational, it owes it irrationality to its noncompliance with norms grounded in the theory. But grounded *how*?

COLLAPSING A DISTINCTION

There is a clever answer to this by Jonathan Cohen (1979; 1981). Consider two psychological theories, a descriptive theory D, and a normative theory N. As we have seen (and details aside), D will contain idealizations that enable the conversion of correlations into laws. The laws of psychology are descriptive in the way in which any law of nature is, whether Ohm's Law, or the law that phylogeny recapitulates ontogeny.

Rationality, of course, has a normative sense. An agent's rationality is a matter of his complying with certain norms. Suppose that n is such a norm, proclaimed in a theory N. Is it plausible to suppose that n might contradict a descriptive law d? Some people will demur from the suggestion. How could an agent's rationality consist in his being required to do the psychologically impossible? So let us say, for now, that an agent is rational only if he complies with the laws of psychology. Or better,

X is rational $\rightarrow$ X discomplies with no d

Is this a reversible implication? Do we also have it that

X discomplies with no d $\rightarrow$ X is rational?

If so, we would have grounds to assert that

For all $d \in D$ there is a $n \in N$ such that $d = n$; and for all $n \in N$ there is a $d \in D$ such that $n = d$.

Whereupon,

For all D there is an N such that $D = N$; and for all N there is a D such that $N = D$.

In other words, an empirically adequate psychological theory is also a normatively sound psychological theory. The theory's descriptive laws are also its norms, and its norms are justified or validated, no less and no more, by whatever validates its descriptive laws, for they are the same thing. Idealizations are descriptively essential for lawhood in D and are justified by the empirical adequacy of the laws they abet. Norms have the same justification, and can have no other. A norm is a descriptive law of an empirically adequate psychological theory. *Ought* does indeed follow from *is*.

It is an attractive idea, almost too good to be true. It flies in the face of what we tend to think of as common experience. In other words, it is *counterintuitive*. This need not be a bad thing, for it may merely be a surprise. We have insisted all along that any theory worthy of the name must have the capacity to correct pretheoretical beliefs. Indeed if Cohen's story is true, it secures us three desirable results. It gives an account of psychological theories in which (1) theories discharge a requirement of all theories, namely, to be able to correct pretheoretical assumptions; in which (2) ideal models play an indispensable role; and in which (3) the bootstrapping problem seems to have been solved. Whether indeed it *is* solved depends on what we make of the misperformance of actual agents.

Misperformance is a commonplace of life. Cohen is notorious for holding that, just as such behavior is not a defeater of empirical adequacy, neither is it behavior that qualifies its performance as irrational. Although such behavior appears to be *wrong*, it does not constitute a violation of the norms of rational performance. Compare this situation with the "misperformance" of the ice at Maple Leaf Gardens. The laws of frictionless surfaces are untroubled by these defections, precisely because the icy surfaces of real life have properties that interfere with their being frictionless. Given what interferes, NHL rinks can only approximate frictionlessness; and to the extent that even under interference they are frictionless surfaces, they will obey the law of physics. Although NHL rinks misperform under the laws of physics, it is not that they do so insofar as they are frictionless surfaces, but in so far as they are not.

Cohen likens all psychological defections to defections occasioned, indeed necessitated, by interference. To be sure, human beings commit fallacies in hefty multiplicity. These, too – in a technical expression current in the linguistics and psychological literature – are "performance errors." They are committed because of interference – fatigue, intoxication, inattention, and so on. Let us imagine our having at hand a representative list of these "performance deficit variables," $pd_1, \ldots, pd_n$. If we now examine the laws of an empirically adequate cognitive psychology, a theory D of inference say, we will notice laws of up to considerable complexity, laws, that is, involving lots of variables. What we will not notice are laws in which some of the variables are pd_i, that is, performance deficit variables. There will be in D no law of the form

If X, Y, Z and $pd_i, pd_j, \ldots$, then W.[11]

Why would this be so? It will be instructive to compare the factor of fatigue (say) with the factor of complexity. Even an ideal agent cannot be expected to compute intractable problems. If we liken an ideal agent to one who performs as perfectly as humanly possible, there will still be problems too complex for such a being to handle. Accordingly, the laws of D will not say that he can handle them; nor will they say that he should. It does not lie in the nature of human beings to solve intractable problems. Human beings are not intractable problem solvers as such. Still less are they tired as such, or drunk as such, or in a rage as such, and so on. D aspires to tell the story of human cognition, valid for human beings as such. Hence, D's laws do not hold places for performance deficit variables.

Performance deficit variables are causal interference with what it is for a human being to be a reasoner. One does not enter the causal interference factors afflicting the slipperiness of the ice at Maple Leaf Gardens into the frictionlessness laws of mechanics. These deficits apply not in virtue of the ice's slipperiness, but in virtue of its not being slippery enough. Psychological performance deficits are the same. They apply to actual reasoners not in virtue of their being reasoners, but rather in virtue of what prevents them from being reasoners.

Psychologists commonly distinguish between two kinds of performance error. These are errors having to do with the agent's *situation* (e.g., his being drunk or tired), and errors having to do with the *psychological constitution* of human agents (e.g., their being unable to solve intractable problems in their lifetimes). It is not obvious that an agent's failure to compute intractable problems should be deemed an error of any kind. But if the theorist makes it true in D that ideal agents always, for example, close their beliefs under consequence, then, on Cohen's account, the idealized descriptive law is always a norm, disconformity to which by actual reasoners is indeed an error; or it is indeed an error if the agent's behavior is correctly construed as noncompliance. I return to this point below.

How credible is it that human rational misperformance is always a matter of interference? Consider a psychological theory of probabilistic reasoning. Let K be a piece of reasoning behavior that is an apparent counterexample; and let us also grant that, on the face of it, K involves no performance deficit errors. Still, if K *is* a counterexample, an error of some kind will have occurred, an error of probabilistic reasoning. Something is an error of probabilistic reasoning only if it violates some law, for example, a law of the probability calculus. If we cannot find a way of pinning the performance error on K, we must discredit K itself. We do this by claiming that the laws of probability, while perfectly valid in the applied mathematics of chance, are *psychologically invalid*. So K is not a counterexample after all.

Consider the following inference:

(1) It's not that Fred hasn't yet been unprepared for the pleasure of his not being uninvited to parties.

(2) So Fred is prepared for the pleasure of his being invited to parties.

It is a mistake, of course. The Fred-inference violates the Double Negation Law of classical logic. On the face of it, the Fred-error does not depend on the presence of a performance deficit variable, and it does not present the inferer with a problem that comes anywhere close to intractability. That is, its complexity is not such that the Fred-inference lies beyond the reach of human reasoners as such. The Fred-inference has all the signs of a counterexample. Can we disarm it in the way the we disarmed K? If so, we will say that the laws of logic, which are perfectly valid as conditions on the implication relation, are psychologically invalid. After all, was this not Harman's point about the failure of *modus ponens* to be a rule of inference? Was this not also Harman's point about our K-case, that is the failure of the laws of the probability calculus to give rules of inference?

Confusions lie in wait. When it is said that a law L of logic is not a law *d* (or norm *n*) of psychology, what is meant is that if L is said to sanction psychological behavior B (belief-revision, for example), it need not follow that B is permissible. What is not meant is that behavior is permissible that *contradicts* L. For then logically invalid inferences would be normatively sound. This is just the trouble with the Fred-inference. It violates Double Negation, it is not a performance error, and it does not lie beyond the computational reach of a human agent as such. So it is a counterexample.

There is, it may be thought, a way out. It is a way offered by philosophy's old friend: "When in trouble, *ambiguate!*" Why, then, do we not say that the Fred-inference trades instances in a concept of negation different from the one in which the validity of Double Negation is rooted? While we're at it, why not also say that not only is the maker of the Fred-inference trading in a different concept of negation, but that it is a concept of negation appropriate to his purposes in drawing inferences such as this one, and that the "classical" concept of negation is not appropriate to those purposes?

I take it as given that these proposals warrant dismissal out of hand. They are laughable. If this is to be our view of how *not* to handle the Fred-inference, it is worth pointing out that this is precisely the way Cohen handles errors in probabilistic reasoning, that is, errors that derive neither from performance deficits nor from difficulties with computational complexity. Of course, lots of inferences endorsed by the calculus of probability do exceed the complexity-reach of human beings. But not all do, as for example, violations of the conjunction axiom. Cohen opts for the ambiguation strategy. When subjects of experiments where the conjunction axiom appears to be violated perform their assigned tasks, not only are they failing to use the concept of probability fashioned by the probability calculus, but the probability concept that they employ is both appropriate for the tasks assigned and such that the performance of these tasks was error-free.

Why is it the case that, as applied to the Fred-inference, the ambiguation strategy would be laughed out of court, whereas in its application to the conjunctivitis problem, it has attained the status of a clever though controversial possibility? The answer is that there is independent evidence that human beings make use of at least two different concepts of probability, the aleatory concept and the Baconian concept, as Cohen calls them. Cohen's views are well known, and it is unnecessary to dwell on their details here. Suffice it to say that aleatory probability pertains to the laws of chance in gaming situations, whereas Baconian probability pertains to induction. That we have two concepts here rather than one is amply attested to by the scandal of inductive logic, that is, by the murderous difficulty in getting the (aleatory) laws of the probability calculus to serve as principles or strategies of induction (Gillies, 1993). For the purposes at hand, I am entirely happy to concede to Cohen the two-probabilities thesis. This matters, for it lends some plausibility to the ambiguation strategy when applied to cases of apparent probabilistic error. What is not made plausible, indeed what remains dismissible out of hand, is the ambiguation strategy as a wholly general technique for disarming counterexamples, as we saw in our review of aggregative and preservationist paraconsistent logics in Chapter 6. Concerning the Fred-inference, there is no independent reason for supposing the existence and employment of a different concept of negation

from that which makes Double Negation a valid law.[12] Nor is there any reason to suppose that for every "error" of reasoning that is not a performance
deficit error and is not a matter of our complexity-management incapacities,
there will nevertheless be an independently establishable and, so to speak,
preexisting ambiguity or conceptual multiplicity guaranteed to quash the verdict of *error*. Cohen's handling of what I have been calling the bootstrapping
problem stands apart for its metatheoretical sophistication and its cleverness.
For all of its attractions, it is a failure. It fails not just in matters of detail, but
in a major way; it fails in principle. In so doing it leaves the bootstrapping
problem menacingly undealt with. It also offers the theorist what appears to
be equally unattractive options. Either he can just restate his own normative
convictions, or he can take false comfort in the model-relative analyticity of
his laws. Cohen himself would opt for a reinstatement of intuitions. He sees
something in intuitions that we ourselves have so far missed. So there will be
need to return to Cohen when we direct our attention to the subject of reflective equilibrium. As for now, something called *linguistic* intuitions demand a
hearing.

LINGUISTIC INTUITIONS

There has been a large investment by philosophers and other theorists in the
probity of what "we say," that is, in our linguistic intuitions. What is meant here
by "probative?" It is that our linguistic intuitions are fact-constituting, or, in
greater strictness, that certain of our linguistic intuitions are for certain classes
of properties fact-constituting with respect to them. In several ways, this is still
a loose way of speaking. A tightening up is called for. It takes some doing to
get the story of linguistic intuitions straight. For present purposes, it suffices to
concede that the straight story is tellable, and to presuppose it here. To this we
need only add four brief points. (1) Any one person's intuition, rooted as it is
in his fluency, has a high presumptive distribution in the community of fellow
fluents. (2) Linguistic intuitions can be supposed embedded in fluent practice,
and need not always be accessible to articulation. (3) Linguistic intuitions are
not decision-procedures. Certain intuitions are general in their formation in
ways that can only be called defeasible. (4) Fluent speakers will not always
have the same intuitions.

As is said, some facts are fixed by linguistic intuitions, as for example, the
fact that "Seven by of" is not grammatical in English, or the fact that "The
Sheraton desk filleted by the Prime Number Theorem is eligible for crop ro-
tation" is not meaningful in English. How does it come to be that such facts
exist? A common answer, which I will not challenge here, is that language-use
contains mechanisms for solving coordination problems. For a language to be a
language, there must be – up to some minimum – constancy in use. Solutions of
coordination problems are sometimes called "conventions" (Lewis, 1969), and
I shall use the word with that meaning here. Conventions have an interesting
peculiarity: they are not recorders of facts, but creators of them. It is a fact that

driving on the right is the right way to drive in Canada, the Netherlands, and Germany; and it is a fact that driving on the right is the wrong way to drive in South Africa, Hong Kong, and Japan. It might be supposed that these facts are facts deriving from the Highways Traffic statutes of the countries concerned, but this would be wrong. The *lawfulness* of such arrangements proceeds from the statutes, but not the properties of being-the-correct-side-of-the-road-to-drive-on-in-country *x*. When the byways of Japan started to be vehicularly active in the long ago, the Imperial Court did not establish a Committee of Experts and charge it with the task of determining which was the objectively correct side of the road to drive on in Japan.

Linguistic conventions operate similarly, although with greater latitude. There are no Committees of Experts (except in France) to determine with objectivity whether

"The cat is on the mat"

answers to some use-independent fact about sentence-grammaticality. Even in France, the heroic efforts of l'Academie Francaise are little more than wishful thinking. The Academy may *say* that "le drugstore" is not French, but it is French all the same; and saying that it is not is rhetorical overkill for something else, namely, a plea that speakers of French change their ways so as to *make* it a fact that "le drugstore" is not French (any longer). A more common sort of intervention is that routinely found in dictionaries. Dictionaries perform both a descriptive and a prescriptive function, reporting what facts the conventions constitute and recommending for and against the retention of them.

At a certain point ambiguity creeps into the word "grammar." If the grammar of a language (at a time) is what its conventions fix as linguistic facts, a Grammar of that language (for that time) is a record of those conventions and of the facts that they constitute. Both grammar and Grammar traffic in facts, but with a difference. In the first instance, a grammar constitutes, but does not record its facts, whereas the reverse is true in the second. Uppercase Grammars often take a normative or prescriptive turn; they urge, as we see, the status of conventions on facts now lapsed or yet to be. The normative Grammarian falls prey, like our case-two theorist, to serious misconceptions. Too often, for example, he is taken in by the following argument (putting "L" for the theorist's target language):

(1) It is a grammatical fact of L that P.
(2) The Grammar of L reports the fact that P.
(3) Hence, it is a Grammatical fact that P (a theorem of the Grammar, so to speak).
(4) There is some evidence that L-users are in noncompliance with P.
(5) These individuals have made and are making a mistake; they are transgressing a theorem.

There is not likely to be an easy accord as to the number of ways that this argument goes wrong – or as to where its deeper errors lie. But will anyone doubt, upon simple inspection, that it is wrong? Will anyone doubt that for certain Grammarians, it is precisely the whole focus and impetus of their urge to normative pronouncement?

The existence of fact-constituting conventions give rise to all sorts of questions. We are right to wonder about the extent to which, and by what mechanisms, such conventions are learned rather than innate. We also should wish to know something of the neurological design required for human language use, and so on. I shall not be concerned with such questions. Rather, I restate a central point:

Linguistic intuitions are fact-constituting for certain ranges of properties.

If grammaticability and meaningfulness are members of such ranges, could the same be said for the property

...means — ?

For lexical meaning, an affirmative answer seems attractive. That bachelors are men who have not married comes to be a fact because the linguistic intuitions of English-speakers concur on the application of the predicate "...means —" to the lexical pair ⟨"bachelor," "man who has not married"⟩.

If truth is disquotational, then by its converse, quotationality, lots of *truths* will be constituted by our linguistic intuitions. It is a point worth noting that these are fugitive truths; they have no purchase other than in the conventions that beget them. Conventions come and go, and although there are *conditions* under which this happens, when it does happen it does so in utter indifference to what the facts are; for the only facts there are in such cases are the facts that come and go with them. The fugitivity of linguistic truths is not, of course, this distinctive feature. Most truths are fugitive – little but laconic comments on the passing scene, as someone has said (perhaps it was Donald Davidson) – going in and out of retirement as the scene renews or displaces itself. The more striking feature of linguistic truths is that their fugitivity lies wholly in the ebb and flow of linguistic practice. This is worth emphasizing:

In the passage of time and circumstance in which linguistic intuitions retire fact F in favor of its successor fact F* there can be no fact of the matter as to whether this change was in *any* way a mistake, that is, a misrepresentation or misidentification of what the facts really are.

For quite a long time in the present century, English-speaking philosophers were minded to take linguistic intuitions seriously. At its most emphatic, they constituted a kind of school or movement with common names, "Oxford Philosophy" or "Linguistic Analysis" – subjects of Russell's wry crack about the "Philosophy-Without-Tears" School (see, for example Ryle, 1945, 1953, 1954/1972, 1961; Urmson, 1956; Flew, 1960; Russell, 1959). Even to this day, moral philosophy is nothing without its intuitions, and it, along with the harsh mistress of old – logic – has tendered itself to the mercies of the method of reflective equilibrium. I shall return to the matter of reflective equilibrium. Before that there are further things that require saying about linguistic analysis.

The thesis of earlier in this chapter was that in purportedly normative theories of human agency all the load is borne by the theorist's intuitions. This being so, talk of ideal models is theoretically idle, nothing short of an empty loop from and back to the theorist's pretheoretical enthusiasms. It also was proposed that this was a fact that retired from service the very idea of a normative *theory*. Perhaps this was too fast. It may be that there are further conceptions of theory to which the linguistic philosopher could lay justified claim. I, for one, do not doubt it, and shall assume so here, subject to this proviso:

In any conception of theory in which intuitions are given an indispensable role, a theory – if it is to be anything other than a mere rehearsal of its intuitions – must in principle provide the theorist with the wherewithal to amend, refine, and retire prior intuitions. If the purported theory lacks the capacity to do even that, the concept of theory is itself idle.

Consider, then, a philosopher's attempt to make theoretical hay of his (and, he presumes, our) moral intuitions about some matter or other. For ease of exposition, suppose the theorist's intuition is that what is good is always and only that which the gods command. If he gives expression to this intuition and his language is English he will assent to

S: "What is good is always and only that which the gods command."

S is a sentence of English, a language in which we presume the theorist to be fluent. There will be conventions in English not only about S but also about its constituent terms. In particular, there will be conventions in English about the expressions "good" and "that which the gods command." If it were the case that as English is used there is concurrence on the application of the predicate "... means – " to the pair ⟨"good," "that which the gods command"⟩, then, given the conventions of English, it is a fact that what is good is always and only that which the gods command. This is telling. The theorist began with the *moral* intuitions, the belief that what is good is always and only that which the gods command. He saw his job as either confirming or disconfirming that belief (undoubtedly hoping for confirmation). He then consulted his *linguistic* intuitions and (in our admittedly counterfactual story) discerned that they constituted as fact that self-same thing. His moral intuitions were confirmed by his linguistic intuitions. Moreover, since the meaning-connection between "good" and "that which the gods command" is established in just the way as that between "bachelor" and "man who has not married" is – that is, by appeal solely to how we use our words – the theorist's original moral intuition is not only true, it is necessarily true or, as may be preferred, *conceptually true*. The job of the linguistic analyst is thus to test his nonlinguistic beliefs against what his linguistic beliefs tell about them. His nonlinguistic (philosophical) beliefs will be justified precisely if his linguistic intuitions confer on them the status of conceptual truths.

There is something to be said for the present conception of philosophical theory, not all bad. It honors a central requirement of theories, for they allow for the correction of pretheoretical beliefs. If we held our present example to the standard of realism, we would have it that, as is indeed the case, the conventions of English do *not* constitute it as a fact that what is good is always and only that which the gods command. Not only is the theorist's pretheoretical belief not confirmed by our linguistic intuitions, nothing else of the kind *would* confirm it; so it is disconfirmed. Our theory corrects a pretheoretical belief.

As presently conceived, philosophical theories also give the appearance of elegance and simplicity. There is nothing more to them than the sorts of fact-recognition procedures that theories of Grammar marshal in their interaction with grammar, supplemented by – what any theory worthy of the name begins with – inventories of "what everyone already knows." Theories of this sort thus evade two of the worst things that could be said against ideal model theories. Talk of ideal models is otiose, but there is no such talk here. The ideal model theorist's "laws" are just disguised forms of what the theorist already believes; but here they are genuinely independent claims that hold prior beliefs to objectively independent protocols of confirmation and rejection.

In fact, linguistic theories seem able to do precisely what ideal model theories could not, namely, provide independent justification of the soundness of an ideal model short of having to "privilege" by stipulation the laws that would end up being true in it. It appears that such theories provide a solution to the bootstrapping problem. The independent justification comes by way of facts constituted by linguistic practice. We can make models of such facts if we like, but there is no need. The facts verify beliefs directly and in ways that raise them to the standing of theoretical laws, for the linguistic facts confer on those beliefs the status of conceptual truths.

Attractive as its virtues are, the method of linguistic analysis is a failure in principle. Many complaints against it can be laid. Linguistic protocols do not in fact retire normative disagreements in communities of fluents; moral terms do not always have synonyms; there is disagreement about what our linguistic intuitions actually are; and so on (Moore, 1903). I do not make light of these difficulties, but they do not seem to me to be the heart of what is wrong with linguistic analysis. This calls for some simplification. In particular, not every variation of the method of linguistic analysis wholly conforms to the description of it here. In some versions, our linguistic intuitions can confirm a conceptual truth short of constituting any fact of synonymy. For example, many people will be ready to suppose that

"No daughter is her own father"

is a conceptual truth. So it might be; but if it is, what the present theory requires, namely, a conceptual truth confirmed by some fact constituted by our linguistic conventions, that fact, while it need not be a fact about synonymy, must be a fact with regard to some or other linguistic property. We have borrowed the idea of linguistic intuitions for our further metatheoretical – indeed, philosophical – purposes. The theories we wish to construct are about nonlinguistic matters; they are about morality, or rationality, or whatever else. This should not blind us to the provenance of what we have borrowed. Linguistic intuitions are constituters of linguistic facts only. So, if the conceptual truth of "No daughter is her own father" is sanctioned by any fact yielded by our linguistic intuitions, it must be a fact with regard to a linguistic property, even if that property is not one of synonymy. It is true that sometimes a philosopher, claiming allegiance to the method of linguistic analysis, will simply say that *given the way we use our language*, it is a conceptual truth that no one is her own father. If his allegiance is genuine, his claim must be rooted in a fact about linguistic properties that "the way we use our language" constitutes as fact. Anything short of this is smoke-blowing. For expository economy, I shall confine my remarks to synonymy.

Let it be agreed that linguistic conventions constitute facts of the form

(1) "X" is synonymous with "Y."

Do we then have it as a fact that

(2) Xs are Ys?

Let us say, for now, that we do. What we do not have is the *necessity* of (2). Without necessity, there is nothing to say for (2) as a conceptual truth, and the linguistic analyst has failed to secure what is essential to his way of doing things. The fugitivity of (1) passes to (2) itself if (2) is a fact deriving from the linguistic fact reported by (1). What is more, since there is no fact of the matter to which (1) is accountable apart from the fact constituted by the relevant linguistic practice, there can be no fact that the nonnecessity of (2) transgresses. That is to say, there is no fact that makes (2) necessary, not even the fact that makes it true. Indeed, the fact that makes it true is a fact of a type that guarantees its nonnecessity, and that ensures it eventual retirement.

It is significant, therefore, that linguistic analysts are not much drawn to contemporary skepticism against *propositions* (Quine, 1981). Propositions appeal to analytic philosophers; they offer promise of redemption (Ryle, 1930; Cartwright, 1962, 1968). We may grant that the passage of time and circumstance may, and likely will, retire the synonymy of "X" and "Y" and, with it, the truth of "Xs are Ys." Even so, given our present conventions, it is a fact that Xs and Ys. Consider now a person who began life in France. (He might be a cousin of Hillary Putnam.) Before long he masters the conventions pertaining to "La neige est blanche." Still young, his family emigrates to Alberta and, as events turn, the lad loses his French in an unstoppable flood of Edmontonian English. In time, he takes charge of "Snow is white," and "La neige est blanche" sinks without a trace. For all this change, what he used to say by way of "La neige est blanche" he now says with "Snow is white." He is saying the same thing throughout. This constancy of what is said under linguistic variation philosophers call "propositions"; they say that the two sentences *express the same proposition*. Although related (or relatable) to linguistic objects, propositions are not linguistic objects. This also matters. It was the fugitivity of facts about synonymy that barred the truths they constituted from the desired standing of necessary or conceptual truths. If propositions are not linguistic objects, there is some chance that they will evade the contingencies that inhere in linguistic objects. So the theorist finds himself drawn to the following picture.

Truth, he says, is not primarily, in fact is only notionally or honorifically, a property of sentences. In greater strictness, it is a property of propositions. Because of the synonymy reported by (1), we allow ourselves the lazy convenience of supposing (2) true. What is really true is the proposition, P, which (2) chances now to express. Suppose that over time linguistic practice alters in such ways that (1) lapses, and with it (2). What does *not* lapse is P. We have in this a fateful concurrence of facts. The first fact is that linguistic conventions made it the case that (2) expresses P. The second fact is that the state of affairs reported in (1) makes (2) true, or less honorifically as we can now say, makes it true that P. The third fact is that no subsequent changes in linguistic conventions can create facts that contradict P. Such changes might even verify (honorifically) some sentence that is (2)'s syntactic negation, as when future conventions displace the fact of (1) with the new fact constituted by the anatomy of "X" and "Y." This would give us the new sentence, honorifically true, "Xs are not Ys"; but that would not be a sentence expressing a proposition which is P's negation. We have it then that any P born of linguistic conventions is immune from retirement by any variation in them. Even if this does not endow such Ps with necessity of the S4 or S5 sort, it still provides enough for conceptual truth. Thus, the great charm, if not the utter mystery, of our linguistic intuitions is that they beget truths that subsequent intuitions can do nothing to undo.

It is a seductive picture, but fatally stricken all the same. It is at once too accommodating and not accommodating enough. If ever it were true that linguistic conventions constituted the synonymy of, say, "good" and "commanded by the gods," the proposition, P, that what is good is always and only that which the gods command was made true then and is true evermore. Of course, as we have it presently, there is no such synonymy. In present practice, "What is good is always and only that which the gods command," expresses no conceptual truth, indeed no truth at all. We might think it worth our while to dwell on details of the baffling semantic lineage borne by their sentence to ours, one that in virtue of which the proposition ours expresses for us lacks precisely what the proposition expressed by theirs possesses. Why do we not say instead that the ancients had a faulty understanding of such things, that they were wrong in holding the synonymy of "X" and "Y"? The answer is clear enough. If we said that, the methodology of linguistic analysis would implode. We would have denied the distinctive character of linguistic conventions, namely, that when they constitute facts there are no prior or independent facts to which their constitution is answerable. What then is the advantage? It is to reinstate the distinctive feature of the method of linguistic analysis, and to pay the price of doing so. Because there is nothing in our linguistic practice that even remotely suggests the linguistic truth of "What is good is always and only that which is commanded by the gods," the proposition that this sentence now expresses gives no guidance as to the proposition that it expressed in days gone by. If this is thought too harsh, translation manuals may be summoned. I have no wish to mire our present discussion in contemporary controversy about translation manuals, except to explain why they do no good here (Quine, 1960, Ch. 2).

Translation manuals belong to Grammar, not to grammar. More carefully, they straddle the divide between grammar and Grammar (this indeed is the locus of the vaunted thesis of the indeterminacy of translation). No linguistic practitioner has access to the grammar of yore, since it will have lapsed with the death of its fluents. His sole guide is such Grammars of it as may chance to be available. Translation manuals trade in synonymies: S in that language is synonymous with S* in this; and so on. Where S is from a language distant and dead, it is easy to see that translation synonymies are not constituted by any grammar that includes the linguist's own. It cannot be by the conventions of his linguistic practice that the synonymy of S and S* is constituted as a fact. Even when S and S* are drawn from languages with respect to which the Grammarian is fluently bilingual, there is no convention in either that constitutes the synonymy of S and S*. Rather, the linguist *judges* the concurrence of conventions pertaining to S with conventions pertaining to S*. In this, he will often be right, but if and when he is, it is not because his judgment is synonymy-constituting.

Whatever its details and mysteries, the synonymy that warrants the translation of S by S* is not a synonymy constituted by any synonymy-constituting set of linguistic practices, such as may be. Reverting to our example of a distant and dead language, the translator endeavors to say that what "they" meant by S is what "we" mean by S*. If what we mean by S* is a proposition that is not true, what they mean by S is likewise a proposition that is not true, that is, that self-same proposition.

Something has gone awry. We introduced the idea of translation manuals to help with the problem of identifying conceptual truths no longer expressible, owing to intervening changes in linguistic conventions. Since we now find that translation manuals take what our ancestors meant by S as what we mean by S*, and S* is false, we have lost what we were after, namely the conceptual truth that S once expressed. What we required was a mapping from old S to some current sentence S* that preserves the conceptual truth of S, that is, of the proposition expressed by it. Given that such mappings are not underwritten by translation, we will need to match S with an S' that is not a translation of it. It is an understatement that there are no such mappings worthy of the name;

S′ must be a sentence that does not mean what S means. But if the mapping from S to S′ is from S to a sentence that does not mean what S means, not only is S not translated by the mapping, it is untranslatable. The conceptual truth that S is purported to express will be some proposition other than what S means.

If this is bad news for conceptual truths of ancient origin, it is no better for those that arise from our own linguistic practice. Let P be a favorite, and current, conceptual truth, for example, that Xs are Ys. We may hope that P is what is expressed by the sentence "Xs are Ys," for how else are we to say what our conceptual truths are? Everything we told of the fraught history of sentence S, and of the passage from ancient times to now, applies equally to the conceptual truths of present-day conviction and affection. The conceptual truth that we take "Xs are Ys" to express is not what "Xs are Ys" currently means. One culture's present is another's antiquity. In this fateful identity ruin awaits. Concerning our own conceptual truths, we can have it that they are necessary, or we can have it that they are accessible. We cannot, however, have it both ways. On the score of accessibility, there can be no guarantee that the sentences now expressing (as we suppose) our conceptual truths will have their truth preserved in translation. A sentence meaning the same in a successor culture may be false, or to be stricter, may be held false with no less justification than that which attaches to our taking it for true. On the score of necessity, our Ps will have it only if it can be guaranteed that no current expression of them will have eventual translations in which they are false. This *can* be guaranteed only if any sentence expressing a P is in principle an untranslatable sentence. Translation being what it is, we are left to conclude, *modus tollendo tollens*, that P is inexpressible. We may note in passing that there is now no bootstrapping problem. Conceptual truths are their own guarantors and they arise without the need of valid ideal models. Evasion of the bootstrapping problem is secured at a cost. The cost is that conceptual truths now go into a "black hole." It is not a happy trade.

Perhaps we have made rather free with the idea of "translation being what it is." Some have held that what it is is nothing. The introduction of the device of translation manuals worsened the problem it was meant to solve. Could we not take this as occasion to shoot the messenger, to give up on talk of translation? If, as indeterminists aver, there is no fact of the matter about translation, there can be no fact as to whether that S translates as this S* or any other. Why not say the thing plainly: sentences are untranslatable. The thesis of the indeterminacy of translation has long rankled critics who see no good in it. As it arose historically, the indeterminacy thesis was launched to bring propositions to heel. Propositions were sometimes disliked on nominalistic grounds, and sometimes on grounds more broadly ontological. It was said that they lacked satisfactory identity-conditions, and that they were insusceptible to individuation, and so on. There has been long standing disagreement as to whether the indeterminacy thesis warrants a serious hearing at home, that is, whether it can be marshaled against contemporary intralinguistic sentence-synonymies. Our linguistic analyst can go this far if he is so minded, but no further. Short of intralinguistic *term*-synonymies he has no case to make for his methods. If this leaves him with the task of denying sentence-synonymy while proclaiming, indeed making everything of, term-synonymy, so be it. Desperate remedies are for desperate times. What the linguistic-analyst desperately requires is the existence of necessary truths expressible in sentences of the language in which he is fluent. It may be that, strictly speaking, *propositions* are dispensable for his purposes. But what he cannot do without are *sentences* of his language made true by linguistic convention. Let "Xs are Ys" be such a sentence, secured by the term-synonymy of "X" and "Y," secured in turn by linguistic convention. Whether to the proposition P expressed by "Xs are Ys" or to "Xs are Ys" itself, access is no problem so long as the requisite conventions remain in force. Beyond that, access lapses and our conceptual truths will eventually be lost to our successors. Of course, it is entirely tempting here to add a qualification, namely, "short of their having in their different conventions the different means of expressing them." The attractiveness of the qualification is evident. It invites us to downgrade indeterminacies *of fact* to indeterminacies *of access*. It offers us the view that any such truth that chances to persist across cultures, or to recur after temporary retirement, will genuinely exist and genuinely be true, notwithstanding that nothing of our own linguistic practices can tell us whether, for any given successor culture, that truth does indeed recur there. What counts is that it is accessible to *us*, never mind what we cannot know of our distant forbears and remote descendents. We at least, in our time and place, will know the conceptual truth of "Xs are Ys" (in fact if we held our noses and declared the reflexivity of translation, there would be exactly one sentence that translates "Xs are Ys," namely, itself).

The qualification that buys us this supposed comfort does so in counterfeit coin. On the suggestion presently in view, there can be no question of successor conventions giving expression to the very thing expressed for us by "Xs are Ys." Let S* be the sentence in the successor culture in which, it is presumed, what "Xs are Ys" expresses for us is given the same expression. Then, although we need not know it, S would be a translation of "Xs are Ys," contrary to what we have assumed. It may be that this is enough to slip the tug of the black hole problem, but, if so, it, too, has its costs. It makes for a deep and radical culturocentrism. Each community of fluents is bound to conventions that generate conceptual truths for it alone. If we opt for talk of propositions, these are propositions that enter into Plato's heaven as permanent residents; but neither their residency nor their permanence is anything to anyone other than those fluents. Since these colinguists have access short of permanent residence in Plato's heaven, talk of propositions brings them no advantage not already attaching to those of their sentences made true by the conventions of their language. Either way, what is conceptually true for any such community is so for it only. There is little comfort in a shoe that pinches so. The traditional analyst will not like the harshness of the constraints that now fall on the doctrine of conceptual truths. Thus fettered, it is a doctrine tailor-made for snide dismissal of precisely the sort and intensity that Gellner gave it, and for which in turn he was excoriated by offended dons the world over, from Oxford, England, to Oxford, Mississippi, not to overlook little Oxford on the Otonobee (Gellner, 1968).

If we step back a little from the discussion of the past few pages, a pattern is discernible. We began with a problem. It looked as if the linguistic analyst's quest for conceptual truth could only meet with failure, owing to the utter contingency of the linguistic facts presumed to underwrite them. Various repairs were then considered. Propositions were introduced, then translation manuals, and then lexical meanings. Quine has made a career of discouraging these ways of talking. For the most part, he dislikes propositions, meanings, and the like robustly enough to challenge their very intelligibility. At times, he is more circumspect. He says only that these "intensional" notions play no load-bearing role in science or philosophy. Perhaps even this, softer though it is, is a bit overstated. But as I have tried to show, with arguments that I have not myself seen before, Quine's more moderate skepticism is wholly on the mark in the metatheory of linguistic analysis.

The linguistic analyst finds himself back where he began, faced with the problem of getting conceptual truths out of linguistic contingencies. This is a juncture at which another skein of Quinean skepticism could claim a hearing. From 1936 on, Quine has abjured talk of truth by convention. His focus has been on the truths of logic (Quine, 1936). But if he could be brought to acknowledge conceptual truths, his disdain of truth by convention would apply here as well. It is not my purpose to examine the details of Quine's arguments, interesting as the task would be. I will say only that there are places galore in Quine's writings where he seems to be pointing to, without quite getting at, what is fundamentally wrong with the idea of truth by convention. The core mistake turns on misplacing the dependency relation. The analyst argues that the statement

(1) "X" and "Y" are synonymous

implies that

(2) Xs are Ys.

This attracts two related mistakes. One is the mistake of thinking that (2)'s truth owes itself to (1)'s truth. The second is that of thinking that since (1) has the property of being constituted by linguistic conventions, and since (1) and (2) are vitally linked, the link is a relation that transmits that same property. So now (2) is also constituted as true by linguistic convention. Let us try to see what has gone wrong here.

Imagine a society, Σ, in which all fluents believe that Xs are Ys. Since concurrence is not vouchsafed by direct observation, the irresistibility of the

Y-hood of Xs shows up in certain patterns of behavior. No one in Σ ever met, or heard of, an X that was not Y. The suggestion that someone did or might have such an encounter is met with widespread derision. The elders of Σ reflect on this. They ascertain (as they suppose) that nothing would make their fellow fluents, and they too, give up on the belief that Xs are Ys. In the upper reaches of the academy of science, logicians lounge about. After much cogitation, they determine that even making free with possible worlds would not crush this belief, whether their own, or their colinguist's of whatever rank or station. They e-mail this intelligence across town to the Grammarians. Being a slow day, the news is welcomed. A new entry is made in the Grammar of Σ: "X" is synonymous with "Y." Down below, in the streets, where language sustains itself and is remade, there is simply the persistence and pertinacity of the belief that Xs are Ys. When a scholar descends to enquire, "Is it a law of nature, or is it perhaps a conceptual truth?" he will be asked to move on, no doubt. For he asks a question that linguistic practice cannot answer.

That "X" is synonymous with "Y" has nothing directly to recommend it but the persistence and pertinacity of the common belief that Xs are Ys. In some cases, namely, when nothing substantive hangs upon it, there is room for synonymy by overt stipulation ("let us call these things 'X-rays'"). But it cannot be supposed that baptisms are canonical for the synonymies in which the Grammarian prefers to trade. The pertinacity, persistence, and commonality of the belief that Xs are Ys is underdetermining even so. That "X" is synonymous with "Y" requires a leap of scientific imagination. It is a matter of judgment, a matter for Grammar; it is not a matter for grammar. Nothing in the grammar of Σ constitutes it as a fact that "X" is synonymous with "Y." Perhaps in time his fellow fluents will come to avow the Grammatical principle that "X" is synonymous with "Y." They may even see it as common knowledge, but it will be common knowledge of the sort that attends other scientific claims, that $e = mc^2$, for example, or that certain human traits are naturally selected for. No doubt "X" is indeed synonymous with "Y," but if it is, it is a matter of scientific conjecture, of the Grammarian's reckoning up the significance of the persistence, pertinacity, and commonness of the belief that Xs are Ys. Crucial to his enterprise is his supposition that the persistence and pertinacity of the belief has a high presumptive distribution in Σ. Supposing so is telling. He attributes to the Σ population at large the (no doubt largely tacit) belief in the unfalsifiability of the belief that Xs are Ys. Anyone judging Σ's commitment to the unfalsifiability of its belief that Xs are Ys may think that he has solid occasion to call in the Grammarians, but it unnecessary. If he and his fellow fluents are already seized of the conviction that the belief that Xs are Ys is unfalsifiable, this is already a conceptual truth for them, if anything is.

At the beginning of our examination of the method of linguistic analysis, we said that the central objective was to find a way of confirming or disconfirming philosophical (and other nonlinguistic) beliefs by appeal to facts independently ascertainable as such. Furthermore, since such facts, including synonymy facts, are constituted by linguistic conventions, concerning which there are no other facts of the matter, the facts that confirm or disconfirm our philosophical beliefs have an independence and an objectivity passed on to the beliefs they confirm or to the negations of the beliefs they disconfirm. Thus, the method of philosophical analysis was an answer to an ancient philosophical question: Are there means to go beyond mere belief to genuine knowledge? Or is there a way of subduing the *subjectivity* of even our most widely shared and confident beliefs? The verdict of linguistic analysis is "Yes" each time.

It is apparent that the Grammarian is similarly situated. His synonymy claims are nothing to go on unless rooted in persistent and pertinacious belief for which he is right to claim a high presumptive distribution in Σ. His judgment of synonymy awaits upon a prior judgment that the pertinent belief is, in effect, believed unfalsifiable in Σ, that it is taken for a conceptual truth in Σ. Suppose that the Grammarian's judgment is correct in a given case, specifically that Xs are Ys is in effect judged a conceptual truth by the fluents of Σ. The Grammarian's determination of the synonymy of "X" and "Y" is but a description of what he has already ascertained, unless it chances to be the case that the fact of synonymy suffices to convert the belief that Xs are Ys is unfalsifiable into an objective fact. He may well think this, but he would be wrong. He might think this because he might also think that synonymy-facts are constituted by linguistic conventions. If this is indeed his view, the Grammarian shows himself careless in two ways. For one thing, he allows himself not to attend to what is right under his nose, that synonymy verdicts are always after the fact, and that synonymies are technical artifacts of the Grammarian-Scientist. For another, he betrays a confusion about conventions.

So we must ask, "How did conventions enter this picture?" Conventions are solutions to coordination problems. If everyone in Σ believes that Xs are Ys, it will not do to have fluents giving expression to it in inequivalent idioms. When the Grammarian records in his lexicon the synonymy of "X" and "Y," he is reporting a convention that bears on the belief that Xs are Ys. The convention no more verifies that belief than the conventional symbol for "poison" verifies the toxicity of the stuff in the bottle it labels. The convention that the Grammarian captures with his synonymy-entry is the convention in Σ: "X" and "Y," rather than "X" and "Z," are adequate for the expression of that belief, which itself is believed unfalsifiable. It is a fact constituted by the linguistic practices of Σ that the uses of those expressions are objectively adequate for the expression and communication of that belief. Beyond that, the objectivity cannot flow. If it did, conceptual truths would be held hostage to the radical contingency of linguistic conventions, as we have lately argued. With that said, the linguistic-analyst is dispossessed of any coherent rationale to bring linguistic conventions into his methodology. Intended to be objectifiers of belief, their sole abiding role is as falsifiers of conceptual truths. More plainly still, the theorist wanted linguistic intuitions because he thought that they would objectify his philosophical beliefs by showing them to be conceptual truths. What they do is far short of that. They conspire to show that, on this account, nothing is a conceptual truth.

What befell the method of linguistic analysis should not happen to a dog, as the saying has it. Introduced as a way of solving the bootstrapping problem, it tripped the snare of the black hole problem, only to recover (or show the appearance of it) and succumb to the radical culturocentrism problem. Rarely has an idea so loved had such trouble. On reconsideration we see that linguistic intuitions *rob* us of conceptual truths, which is what linguistic analysis cannot do without. Linguistic intuitions are nothing to linguistic analysis, that is, linguistic analysis is nothing to them. Linguistic intuitions cause the method of linguistic analysis to implode. It is the final insult.

ANALYTIC INTUITIONS

As is apparent, the linguistic-analyst is struck by what he sees as a link between the *linguistic* intuitions of the sort in which Grammarians are interested, and his own *conceptual* intuitions, as we might now call them. In a later section, I shall say something about a way of construing this link that we have not yet discussed. On that construal, the link is *analogy*: Linguistic intuitions stand to linguistic theory, or Grammar, as conceptual intuitions stand to conceptual theories (e.g., moral theory and so on). For the present, it is worth emphasizing that, as we have been describing him so far, the conceptual analyst places a different construction on the link between linguistic and conceptual intuitions. He sees a link as validation, in which the theorist's subjective philosophical, or other nonlinguistic beliefs, are transformed by corresponding linguistic facts into objective conceptual truths. Inasmuch as the validating linguistic facts are ultimately sanctioned by linguistic conventions or linguistic intuitions, he sees the link as one in which linguistic intuitions overcome the subjectivity of nonlinguistic belief, and hence validate it.

It was the burden of the previous section to show that any such alliance is paradoxical. If a theorist's conceptual truths are all and only those statements grounded always and only in such linguistic facts, then either his conceptual truths go into a black hole, or they are defaced by a radical relativity, or they do not exist at all. If this is right, then any theorist wishing to reclaim the idea of conceptual truth must dissolve the partnership with linguistic intuitions. What is more, if he wishes to insist on a vital link between conceptual truths and intuitions he must sever the link between these intuitions and the intuitions we have been calling linguistic.

Conceptual truths are analytic truths, and analytic truths are delivered by analytic intuitions. Analytic intuitions are linguistically encodable and are expressed by the sentences of the coding language. But, strictly speaking, analyticity and conceptual truth are not linguistic properties. Consequently, neither are they properties constituted by intuitions that are fact-constituting for linguistic objects. Analytic and conceptual truths are not constituted by the grammar of any language. The linguistic analyst sees linguistic intuitions as indispensable to his purposes; but linguistic intuitions are nothing to the essential workings of philosophical analysis. So the philosophical analyst averts the fateful problem for the linguistic analyst of the radical contingency of linguistic facts. Thus, it seems that the philosophical analyst might be able to exploit *analytic* intuitions without exposing himself in these difficulties.

I think that we may say with confidence that analytic and linguistic intuitions are wholly disjoint, notwithstanding that analytic intuitions are linguistically codable. That no more makes them linguistic intuitions than the encodability of the laws of mechanics makes them laws of linguistics. Even so, the theorist's analytic intuitions are certainly the wherewithal of some of what he believes. Frege's intuitions about sets are preaxiomatic, hence pretheoretic. Thanks to them, Frege had the pretheoretic belief that sets having the same members are the same set. This is a belief preserved in the ensuing theory, but it was not established by it. So it can be supposed that Frege's belief was untutored. If so, there is an evident overlap of someone's analytic intuitions and his or her untutored pretheoretical beliefs. It may be that to every analytic intuition there corresponds an untutored pretheoretical belief. Of course, if true, their concurrence is not perfect. Not every untutored belief is analytically true.

Referential promiscuity makes for confusion. Theorists have fallen into the habit of hallowing too many disparate things by the name of intuition. Doing so encourages bad inferences. Here is one:

(1) Intuitions are constituters of facts, hence are reliable guarantors of truth (linguistic sense of "intuition").

(2) We have intuitions about, for example, philosophical matters (untutored belief sense of "intuition").

(3) Hence we have reliable guarantees of these, for example, philosophical, propositions of which we have intuitions (from (1) and (2), equivocating on "intuition").

(4) But truths guaranteed by our intuitions have the appearance of radical contingency and hyperrelativity (linguistic sense of "intuition").

(5) Intuitions guarantee analytic truths (analytic sense of "intuition").

(6) So, contrary to appearances – and a good thing – truths guaranteed by our intuitions are neither radically contingent nor hyper-relative (linguistic sense of "intuition"; from (5), equivocating on "intuition").

(7) Since what is disclosed in our intuitions is neither contingent nor relative, but analytic, our intuitions are sound constraints on theory (untutored belief sense of "intuition"; from (6) equivocating on "intuition").

I leave it to the reader to decide how much of a caricature this inference is. My own view is that, alarmingly, it is hardly a caricature at all.

Analytic intuitions were the business of Chapter 5. They did not fare well there. Better that we let them be, and move on.

REFLECTIVE EQUILIBRIUM

In our examination to date, we have had occasion to consider four different conceptions of intuitions: pretheoretical beliefs, linguistic intuitions, conceptual intuitions, and analytic intuitions. Intuitions entered our story as part of an attempt to elucidate theories. The theories in question include not only normative theories of human agency but also philosophical theories having application to nonnormative concepts (e.g., sets). A common theme was that a successful theory would generate laws or conceptual truths that hold objectively, and, indeed, necessarily. In this way, theories would solve an ancient problem, that of showing how it is possible to proceed from the subjectivity of belief to the objectivity of knowledge. In our discussion of this, we found that in every case the introduction of intuitions failed to produce satisfactory results. Things went wrong in each case, and in ways that reflected peculiarities of the concept of intuition then under review. But the failures, different as they are in detail, also exhibited a common trait. There is a pattern to their collapse. Each of our theories was an attempt to overcome the subjectivity of what the theorist believes, an attempt to secure principles that in their objectivity and their necessity would validate such pretheoretical beliefs as happen to *be* objectively true, and that would extend these beliefs to new ones, also objectively true. The common pattern of collapse was that in none of the cases examined does this attempt succeed. The theorist begins, as he must with his own subjective beliefs, but at the end of the day they are all that he is left with. Each time, the collapse of the theory in question can be likened to an empty loop from subjective belief back to subjective belief.

We might call theories of this sort *transcendent*. They seek to rise above subjective belief in a manner that permits an objectivity loop back to pretheoretical belief. If the criticisms we made are sound, they suggest that transcendent theories are a mistake in principle. If this is so, it faces us with a Peggy Lee problem: *Is that all there is?*[13] Are we stuck with our subjective beliefs, bereft of any theoretical mechanism for validating them? It would be imprudent to surrender to the Peggy Lee problem without a fight. Indeed there is a possibility we have yet to examine. Although we may grant the failure to transcendent theories, might there not be theories of a different character that succeed in their mission of validation? If there are such theories, they will be nontranscendent, or *immanent*. So our question is, what are the prospects of immanent theories of human agency?

At *Nicomachean Ethics* 1145^b 2–7, Aristotle says that an account of something is correct when it solves all the questions left open by the received wisdom concerning it, and yet preserves as many of those opinions as possible. Thus, before we can produce an adequate account of some matter Q we must first see the puzzles (*aporiai*) that Q generates; and later, when we do produce the account, it is subject to an adequacy condition. It must resolve the *aporiai* and it must do so conservatively; that is, it must leave undisturbed as many of the received opinions about Q not directly in doubt as possible. It is ancient adumbration of Quine's maxim of minimal mutilation, M^3. It is the first allusion in Western Philosophy to the method of reflective equilibrium, a method that underwrites theories we are calling "immanent." Centuries later, here is Goodman to the same effect:

Principles of deductive inference are justified by their conformity with accepted deductive practice. Their validity depends upon accordance with the particular deductive inferences we actually make and sanction. If a rule yields unacceptable inferences, we drop it as invalid. Justification of general rules thus derives from judgments rejecting or accepting particular deductive inferences. This looks flagrantly circular.... But this circle is a virtuous one. The point is that rules and particular inferences alike are justified by being brought into agreement with each other. *A rule is amended if it yields an inference we are unwilling to accept; an inference is rejected if it violates a rule we are unwilling to amend.* The process of justification is the delicate one of making mutual adjustments between rules and accepted inferences; and in the agreement achieved lies the only justification needed for either. All this applies equally well to induction. (Goodman, 1955/1965, pp. 63–4, emphasis in original)

Coinage of the term "reflective equilibrium" rests not with Goodman but Rawls (1971, p. 20). We arrive, Rawls says, at "principles which match our considered judgments duly pruned and adjusted" (1971, p. 20). The matching is reflective since "we know to what principles our judgments conform and the premises of their derivation" (1971, p. 20); and it induces an equilibrium "because our principles and judgments coincide" (1971, p. 20). It is all a matter of "everything fitting together into one coherent view" (1971, p. 21).

As was said, Cohen too is a proponent of the method of reflective equilibrium (1979, p. 407, 1981). Two things come together in Cohen's writings. First is the doctrine, earlier met with, that the descriptive laws of a psychological theory coincide with the norms of reasoning. The second thing is the method of reflective equilibrium. They are brought together by the following argument:

(1) Normative principles of reasoning arise from the operation of reflective equilibrium the inputs to which are our untutored beliefs about what good reasoning consists of.
(2) A descriptive theory of reasoning is a theory of reasoning competence arising from the operation of reflective equilibrium, the inputs to which are our untutored beliefs about what reasoning consists of.
(3) Consequently, on the principle that an operation with identical inputs produces identical outputs, reasoning competence must satisfy the normative principles of reasoning.

Looking at the reflective equilibrium thesis in the form presented by Goodman, we may find ourselves drawn in by his rather casual-seeming assertion that "in the agreement achieved [by reflective equilibrium] lies the only justification needed for either [i.e., our principles and our accepted inferences]." A critic might find that Goodman has confused his modalities, that, confusing necessity with possibility, he should have said, "this is the only justification possible." This, of course, is one of the classical positions. It is open surrender to the Peggy Lee problem. It allows that the best we can do by way of justification is not good enough to subdue the subjectivity of belief and, hence, is not good enough to count as justification. Intersubjectivity is just a lot of individual concurrence; agreement, even widespread agreement, does not convert *endoxon* to *episteme*. Goodman thinks that the best we can do is good enough. He may be right, but many people will think that he is not. Those who think that human beings, as such, are somewhat irrational, or are prone to nonperformance errors, will likely think that he is not right. Goodman and his opponents risk begging one another's questions.

This is what makes Cohen's way of being a reflective equilibriumist so attractive. He is a reflective equilibriumist who brings to the table a thesis asserting the identity of norms with descriptive laws. Were he right about this, he could argue that, contrary to what people may think, the best that we can do by way of justification *is* good enough. It is good enough because the right norms are secured by our best descriptive science. No one would say that we are unjustified in accepting the laws of physics. That they come from our best science is all the justification they have, or need. Parity argues that we not hold the principles of human conduct to a higher standard. They, too, are secured by our best science, and because of the identity of laws and norms, what we are justified in accepting as rational requires no more than what justifies those descriptive laws. Since norms imbibe the objectivity of scientific laws, and since norms arise indispensably from our untutored beliefs, there is, in this, no prospect of the Peggy Lee problem. The subjectivity of our pretheoretical beliefs is subdued by our best science; and this is not only objective enough even for the hardened critic, it is just what objectivity is.

In an earlier section, I tried to say why we should not accept the identity of descriptive laws and norms. If this is right, Cohen's position reduces to Goodman's. Goodman and Cohen alike are met with the Peggy Lee problem. The objection was that the "Fred-inference," as we called it, violates the deductive law of Double Negation. On Cohen's position, this must be either a performance error or, no error at all, contrary to appearance. Finding it difficult to ascribe the error to a performance deficit factor, such as fatigue or intoxication, or to a psychological state "error," such as one arising from intractable complexity, we examined the only option Cohen's theory leaves room for. We considered that the Fred-inference might not be an error at all. If so, it incorporates a conception of negation for which Double Negation rightly fails and, given the purposes which the inference appears to be serving (i.e., negation-cancellation), we supposed that this "nonclassical" concept of negation, rather than its classical vis-à-vis, was the concept of negation appropriate to that inference. But, as we saw, none of this is even remotely plausible. It is a story that even intuitionist logicians would not think twice about before rejecting. It is tempting to generalize this point. There is no known theory of nonclassical negation in which the present suggestion would find a home. So it looks as if the Fred-inference is a counterexample to Cohen's position.

If Cohen's position fails, we might look for evidence of difficulty apart from the Fred-inference. With this possibility in mind, I propose to revisit the argument of two paragraphs above. This will give us occasion to reconsider a further plank in the Cohen platform. It is a plank that rests on a strong analogy between linguistic and reasoning competence. In each case, as Cohen sees it, our untutored beliefs are inputs to the mechanisms of reflective equilibrium, and the principle of linguistic and reasoning competence are, in just the same way, its outputs. Since linguistic intuitions are restricted to untutored beliefs that are not subject to any reconsideration or "second guessing," it is important to Cohen's project that the same restriction must apply to our intuitions about reasoning if the strong analogy is to hold. In like manner, Cohen's reflective equilibrium mechanism is of a "narrow," rather than "wide" sort.[14] For my purposes, it is not necessary to be drawn into a discussion of the tangled issue of wide equilibrium. Instead, I shall concentrate on the strong analogy between linguistic competence and reasoning competence. It might occur to us to question the analogy. A critic might find himself drawn to the view that, whereas the principles of linguistic competence are, just so, linguistic norms, there is room to doubt that the principles of reasoning competence are, just so, norms of rationality. If the critic yielded to this temptation, then, without supplementary argument, he would have begged the question against Cohen. Better, I think, to insist on the strong analogy between linguistic and reasoning competence and to question whether it is true in *either* case that competence is inherently normative.

The argument we used to summarize Cohen's position was:

(1) Normative principles of reasoning arise from the operation of reflective equilibrium, the inputs to which are our untutored beliefs (i.e., our immediate and untutored inclinations, without evidence or inference about what good reasoning consists of).

(2) A descriptive theory of reasoning is a theory of reasoning competence arising from the operation of reflective equilibrium, the inputs to which are our untutored beliefs about what reasoning consists of.

(3) Consequently, on the principle that an operation with identical inputs produces identical outputs, reasoning competence must satisfy the normative principles of reasoning.

It is easy to see that the argument is invalid. Readers familiar with Cohen's writings will attribute its invalidity to misrepresentation. This is quite true, but my misrepresentation was purely tactical. I wanted to focus attention on what I take to be the Achilles Heel of Cohen's project. It is instructive to see where on the present version of it the argument's invalidity lies. It lies in the dropping of a qualification, a further example of *secundum quid*. In premise (1) our intuitions are described as our untutored beliefs about what *good* reasoning consists in, whereas in premise (2), our intuitions are about what *reasoning* itself consists in. Since these are intuitions about different things, it cannot be true, as (3) avers, that they are identical inputs to the operation of reflective equilibrium. Nor can it be concluded, on the principle (which is true) of same inputs-same outputs that a descriptive theory of reasoning is, just so, a normative theory of reasoning.

Let us now correct the misinterpretation and in so doing correct the argument's error. Cohen's argument is now (and always was, properly construed) valid. It fails with the falsity of premise (2), as I shall try to show. Here is the pattern of the case I shall offer against Cohen's position.

A. There is *no* normative content in the idea of linguistic competence.
B. So the principles of linguistic competence are *not* (just so, or at all) linguistic norms.
C. The strong analogy between the language case and the reasoning case holds, as Cohen insists.
D. Therefore, the principles of reasoning competence are not (just so, or at all) the norms of reasoning.

Along the way, I also shall adduce considerations intended rather more directly to undermine the idea that the principles of reasoning competence are, just so, the norms of reasoning.

There is no normative content to the idea of linguistic competence, except (and this is a stretch) for the instrumental. The principles of linguistic competence tell us what we need to do if we are to speak, for example, English. In this sense, the concept of tying your shoes are also normative. How did it prove so easy for theorists such as Chomsky to sell us on the normative character of the principles of linguistic competence? No doubt the fact that speaking English is indeed subject to normative appraisal is part of the answer. So, too, is the ready application of normative idioms without normative effect, as in "He is good at speaking English" and "He is good at tying shoes," where this is *just* to say, "He speaks English competently" and "He knows how to tie his shoes." A further possibility has to do with how competence is defined. Competence is performance, or capacity for performance, entirely free of performance error. If we embrace the concept of performance error both enthusiastically and carelessly, we may find ourselves thinking of competence as inherently error-free. If it is error-free, it is as it "should" be. So competence is inherently normative. There is something to be said for and against these suggestions, but I shall not pursue them here. A more fruitful approach would be to take a hard look at the idea of linguistic intuitions. In Cohen's writings, and in virtually all that I have examined, the concept of intuitions blurs a distinction between

*untutored testimony

and

*untutored practice.

It is a fateful distinction, inasmuch as it echoes the methodological divide between Introspectionism and Behaviorism, and it should have been a warning to us. Consider the string

(i) Single moms, they all lazy; between dah lot ah dem you would't git a days work done.

Aside from its odiousness, how does (i) fare in light of our intuitions? Judged from the point of view of practice, it is a grammatical sentence, where *that* means "is a sentence of English." From the testamentary point of view, a different verdict might be given. Certainly large numbers of speakers of English would find fault with (i)'s Grammar. But it is not clear that they would query its status is a sentence of English. What, then, of the old favorite

(ii) Colorless green ideas sleep furiously?

Not found in practice, it is nevertheless the theorist's variation on what does occur in practice, and is thus a fit subject for testamentary pronouncement. If I were asked, I would allow that it is a sentence, and a grammatical sentence at that. I would draw the line at *intelligibility*. Others might do the same thing and yet make more of the unintelligibility than I do. They might be led to think that intelligibility overrides grammaticality and sentencehood alike. They would be wrong.

Suppose that I am right. Then "Colorless green ideas sleep furiously" is a grammatical sentence of English. Even so, we will never, or not likely ever, see it win a place in English practice. This has everything to do with what sentences of English are for. They are for saying things, for communication. Competent users of English have no occasion to use such sentences as "Colorless green ideas sleep furiously." They can do without them entirely. From the point of view of *practice*, we might just as well say that they are not sentences at all. From the *testamentary* point of view, this becomes a much harder thing to say with either conviction or plausibility. So the idea of linguistic intuitions embodies different standards, to which samples of speech conform, or disconform, differentially. On the score of practice we might just as well say that (ii) is not a sentence. On the score of testimony, the sentencehood of (ii) is a hard judgment to avoid.

The moral of the case of (ii) is that, for speakers of English, there is no natural project for grammaticality. English speakers have no stake in producing grammatical sentences, or indeed sentences at all, except for the communications they abet. Trivially, sentencehood is a condition on sentential communication, and beyond certain variances grammaticality is as well. When the theorist asks a speaker of English how he judges a lexical string, he may judge it badly on the score of communication, as with (ii), and yet, if he is required to render his disapproval in categories furnished by the theorist, he may have to to judge it "ungrammatical" or even "nonsentential."

The same difficulties and obscurities apply to reasoning. Imagine a psychologist intent on constructing a theory of inference. He asks his subjects to consider the following case:

A person S concedes that p and that if p then q. He records these concessions as premises.

(1) p, and if p then q

He then draws from a hat containing three tags on which respectively are inscribed the names "q," "r," "s." Eyes shut, he picks a tag, and, eyes now open, reads it. He reads "q." He then says

(2) Therefore q.

Assuming that there is nothing more to tell of this case, with nothing salient suppressed, it is obvious that, if asked, the theorist's subjects would have no difficulty in saying that S was not reasoning, or he was certainly not drawing any inference from premise (1). If the theorist pressed with a further question of whether S had made a good inference, it is hard to imagine that his subjects would leave their eyes unrolled. What a silly question they will think (and will be right in thinking). If they *had* to answer it, they would say that S's inference was a bad one; but they would be wrong.

Our case is instructive. It establishes that the conditions constituting a bit of practice as an inference are not the conditions that constitute it as a good inference. Here, too, there is no natural project for inference. People draw inferences to get to the further truth of things, to help themselves figure out what to do next, and so on. Our interest in inference is undetachable from our interest in good or successful inference. But that should not mislead us into supposing that what makes for inference is also what makes it good. It is fair to say that our very capacity for inference owes itself to our success in putting our inferences to uses that interest us. If so, it will be true that an agent is capable of inference only if he is sometimes capable of successful inference. This would go some way toward explaining why we do not see such pseudo-inferences in the practice of actual reasoners. It would follow trivially that if the pseudo-inference does not occur in the practice of reasoners, it will not occur in the practice of reasoners who reason as they should. But, as is evident from what testifying subjects are prepared to say, there is ample room for a piece of behavior that fails conditions that constitute it as inference, good or bad.

And here, too, the testifying subject is constrained by the theorist's question. "Is this a good inference," he asks, "that is, good as opposed to bad?" If these are the choices, the subject will say "Bad" and will be mistaken.

Human beings have no occasion to privilege sentencehood or inference. There is nothing interesting in them as such. Sentencehood is needed for saying things, and inference – as opposed to drawing tags from a hat – is needed for figuring things out. So there is a concurrence in practice of sentencehood and sentential communication, and of inference and figuring things out. The concurrence is explicable, but it is also *contingent*. If this is right, a difference between sentential utterance and intelligible utterance (or beyond certain variances, between grammatical utterance and intelligible utterance) should be possible, and between inference and good inference as well. Conditions on sentencehood and grammaticality are constitutive conditions, not normative. If there are normative conditions on sentential, or grammatical, utterance, they will be something else. (Something like, "Don't say 'damn,' dear.") Equally, conditions on inference are constitutive, not normative conditions. If there are also, as surely there must be, conditions on good inference, they will be norms of reasoning, and they will be something else, never mind that we might have little knowledge of them.

We see in this the traditional rivalry between Introspectionism and Behaviorism. Worse, the Introspectionism of our cases is triggered by bad questions; by questions that are suggested by behavioral patterns. Our behavior, we might say, seeks to maximize good inferences. We ask the untutored not what we could ask, "Is this an inference?" but what we should not ask if we are descriptive scientists is "Is this a good as opposed to bad inference?" Abetting our error is the ambiguity of "good." For on the model of "good at," a good inference is *just* an inference.

Reflective equilibrium theories trade in a concept that embodies different criteria yielding differential, and at times incompatible, arguments. Human beings have a stake in sentencehood and in inference that derives wholly from what these are wanted for. What they are wanted for lies open to factors of success and failure. Thus, it is reasonable to conceive of there being norms for what sentences and inferences are wanted for. Norms will regulate the success-failure matrix in ways that favor success. Theorists of various stripes take these norms to be somehow implicit in our intuitions. I am not questioning that assumption here. What I am questioning is the view taken by Cohen, and perhaps implicitly also by lots of others, that the intuitions in which the norms of reasoning inhere are the same intuitions in which the principles of reasoning competence inhere. Perhaps the very word "competence" should have tipped us off. "Competence" is a technical term borrowed from Generative Grammar. It denotes the ability to speak languages such as English. In the theory of reasoning, "competence" denotes the ability to reason, for example, to make inferences. People who make inferences in accordance with the principles of reasoning competence might be said to reason competently, and hence well. Thus, it is easy to think that people who satisfy the principles of reasoning competence are people who reason well – people whose inferences are correct. This is a mistake – an equivocation on "competent." People who construct inferences in accordance with the principles of inference competence are simply people who construct inferences, as opposed to drawing from hats or consulting their direct perceptions. The expression "people who construct their inferences in accordance with the principles of reasoning competence" is a pleonasm. It means simply "people who construct inferences." It is the same way with sentences. People who construct their sentences in accordance with the principles of linguistic competence are just people who construct sentences. Whether there is something to like or not like about their sentences is a further matter having to do with what sentences are for (to say nothing of considerations of, e.g., style). Equally, whether there is something to like or dislike in the inferences people make has something to do with what inferences are for (to say nothing of considerations of, e.g., elegance). In each case, there is a wide gap between the conditions that constitute the thing in question and norms under which we might like the thing or not. This gap has been obscured, or so I have been saying, through mismanagement of the concept of intuitions.

I am prepared, with Cohen, to hold to a strong analogy between linguistics and psychology. Correctly understood, our practice is where principles inhere for the making of sentences and inferences alike. For cases in which the theorist has an interest, and which are absent from practice, samples can be contrived for the untutored judgment of practitioners. What the practitioner testifies to is constrained by the question he is asked. If he answers bad questions he will give erroneous or misleading testimony. What he must never be asked by a descriptive scientist interested in constitutive conditions is "Would this ever be a member of any set of sentences you could see yourself making?" or "Would this other thing ever be in any set of inferences you could see yourself drawing?" Such questions obscure the constitutive-normative divide. In giving a negative answer, the practitioner could be saying that it is not a sentence (or an inference) he considers appropriate to make, given what sentences and inferences are for. In receiving the negative answer, the theorist misinfers as follows. Since it is not in any set of what the informant sees as sentences (or inferences) he would ever *make*, they are nonmembers of what he takes sets of sentences (or inferences) to *be*. So they are nonsentences (or they are noninferences). As we see, there is no concurrence between sentences (or inferences) that as the informant could see himself as ever making, and what the informant could ever see as a sentence (or inference). If the strong analogy between linguistics and psychology holds, there is in each case a gap between the descriptive and the normative. If so, Cohen's account fails. It will not have succeeded in showing that the norms, whether of linguistics or of psychology, have the same justification as the laws of descriptive theories. Whereupon Cohen's subscription to the method of reflective equilibrium produces a concept of normative justification no stronger than Goodman's this-is-the-best-we-can-do justification, a concept of justification that lies exposed to the objection that since it cannot objectify belief, it hardly deserves the name of justification. Cohen, like

Goodman, has a Peggy Lee problem, but with a crucial difference. The Peggy Lee problem is no problem for Goodman. But for Cohen it is a disaster.

One of my objectives is to judge the methodological fruitfulness of the concept of intuitions. Reflective equilibrium entered our discussion to assist the project of judging intuitions. There is much that is important and attractive about the method of reflective equilibrium and, as might be expected, it has its troublesome features, too. I shall not pursue these issues.[15] My aim, as I say, it to get to the bottom of intuitions.

Another casualty of the collapse of Cohen's position is the role of ideal models. All along we have conceded a role for ideal models in descriptive psychology. These are needed, we said, for the conversion of regularities into laws. But with the identity now lost between descriptive and normative theories, whether there is a real role for ideal models in normative psychology is something that will have to be renegotiated. Here, too, the introspective-behaviorist ambiguity of intuitions lurks nearby. It is reasonable to expect that the norms of reasoning will somehow be adumbrated by our intuitions, that is, by our normative intuitions. Until we have a good look at them, what cannot be clear is whether they motivate talk of ideal models or ideal reasoners. There is reason to tread warily. To this day, notwithstanding lots of indications to the contrary, theorists are still too ready to grant to the principles of logic and probability theory free membership in the club of reasoning norms. Disconformity is carelessly dismissed on grounds that the principles of logic and probability theory are true of ideal agents. This is seductive talk. Ask a practitioner, an actual agent, whether he thinks it a norm that in updating their beliefs, people should check their beliefs for consistency. It is hard to imagine the untutored giving anything but their assent to this. But they would be wrong. Given the size of the human brain and the number of its neurons, it would take a consistency checker all his life and then some, by a considerable stretch, to make even *one* such check. The check-for-consistency rule is free to be applied to the ideal reasoners of some theorist's fancy; it is not free to be a norm of actual agency. So we must distrust any careless identification of conditions on "ideal" agency and norms for real people.

The check-for-consistency rule is no norm. We were wrong in thinking so. What would be its successor? Presumably it would be a softened version, a version that eliminates the factor of intractability. Could we then say, with Cherniak (1986, p. 16), that the correct consistency norm is

Consistency: If inconsistencies arise in your belief set you should *sometimes* eliminate *some* of them.

Compare also with Harman's "nonmonotonicity" norm, which he calls

The Principle of Clutter Avoidance: Do not clutter your mind with needless irrelevancies (Harman, 1986, p. 12).

Something like *Consistency* and *Clutter Avoidance* is evident in the behavior of actual reasoners. In the behavioral sense of the word, *Consistency* has the backing of our intuitions. The testamentary perspective favors not *Consistency* but the full check-for-consistency rule. In the introspective sense, the full check-for-consistency rule has the backing of our intuitions. If we think that there is a serviceable concept of pretheoretical intuitions that incorporates both the behavioral and the introspective senses of the word, then the concept of intuitions would be no better off than Frege's intuitive concept of set. Inconsistency would dog them both.

Let us call the present conception of intuitions "intuitions in the BI sense," or "BI-intuitions." As we have seen, the very idea of BI-intuitions is at risk for inconsistency.[16] How, then, did it come to enjoy such a wide and distinguished provenance? What accounts for its abiding methodological appeal?

Human beings are fallible. Human practice is sometimes mistaken. *Pace* Cohen, human beings sometimes commit, and sometimes systematically commit, nonperformance errors. Perhaps these errors can be detected by way of our untutored avowals. If so, misbehavior stands out for its noncompliance with what we intuitively judge as correct. That is, I sometimes overrides B. Since human beings are fallible, they are sometimes mistaken in their I-judgments. What would show them to be wrong when they are? Perhaps it is the disconformity of these judgments with our behavior. So sometimes B overrides I.

As our cases show, there are occasions when our Bs and our Is share an error, for example, The Gambler's Fallacy. Sometimes our Bs and our Is are not helpful for – in fact bear only negatively on – discerning the norms of reasoning.

TUTELAGE

Cohen has a strategic, but mistaken, reason for insisting on a narrow interpretation of BI-intuitions.[17] In both respects they are untutored, that is, unaided by evidence or inference. Cohen prefers the untutored sense of BI-intuitions because he thinks that only these are relevant to the linguist's purpose. This matters for the theory of reasoning norms, since Cohen holds to a strong analogy between linguistics and psychology. I have said why it appears that Cohen's project fails. If this is right, there is no obvious strategic value that tells for the narrow interpretation. We would do well to leave methodological room for *considered* intuitions.

Considered intuitions are an improvement in linguistics. They enable the theorist to judge the following string as grammatical, notwithstanding that it fails the test of untutored intuitions.

The girl the cat the dog chased scratched fled (Stein, 1994, pp. 137–172).

Considered intuitions are not, however, much of an improvement in the theory of reasoning-norms. The Gambler's Fallacy is still embedded in gambling practice, and it is upheld by the considered intuitions of the laity, especially in Las Vegas (less so in Monte Carlo). This makes us see that "untutored" and "considered" are not natural opposites. What is wanted is the distinction between untutored and tutored. Enter the expert.[18] What is the tutor's position? How is he placed in this dynamic between beliefs and norms? He has access, or thinks he does, to the true principles of reasoning. Whether he does or does not is not the central question here. What matters is the nature of his doxastic relationship to those true principles. However we describe that relationship, it cannot be said to be one where the principles of correct reasoning are outputs of the mechanism of reflective equilibrium, with our BI-intuitions *as inputs*. Of course, the tutor will believe these principles and, if he is careful and lucky, he might also be able to act on them. He may even come to see them as obvious – as natural as breathing, so to speak. His confidence in them may induce him to create technical terms with which to express it; so there could be talk of "analyticity" or "conceptual truth." If this is indeed the tutor's leaning, he leans in the direction of Moorean analytic intuitions – an alarming tilt.

The tutor might say that the doxastic relation in which he stands to the true principles of reasoning is, no more and no less, one in which they are commended to him on the strength of considered intuitions. This is wrong, although it looks right. A person can consider a belief about X just by pondering it, without having had to take a course in X-hood, or to have been instructed in it. This is not the tutor's situation in the example under consideration. The tutor had to *learn logic*, however uncritically he may have done so. He thinks that this positions him to correct his own prior beliefs and those of his fellows, and he would be right in thinking so if the laws of logic and probability theory were norms of reasoning. It can be said that the tutor has reconsidered those prior intuitions, for what he now believes he believes on consideration of what logic and probability theory teach him. I will not grudge the word. It hardly matters, since, whether considered or unconsidered, the tutor's beliefs are not BI-intuitions. They are, as we might say, his *theoretical* beliefs. The confidence with which he marshalls his theoretical beliefs may lead him to say that they are intuitive. This could mean that if the tutor were to enter the long life of reason already stocked with the tutored beliefs he now has such confidence in, but without the aid of tutelage, they would be BI-intuitive for him.[19] But, of course, this is very much a counterfactual situation. In actuality, the intuitiveness of the theorist's theoretical beliefs is a matter of his easy confidence in them.

All of this puts considerable methodological weight on the tutor's theories. Suppose, for ease of exposition, that the theorems on the deducibility relation of logic were inference norms for human reasoners. How did the tutor (or *his* tutor, or ...) arrive at these norms? On the type of account that has dominated our discussion so far, he would have consulted his BI-intuitions about the concept of deducibility, or its converse, implication. But just as there was no good epistemic reason to privilege BI-intuitions when we were considering the case for untutored (although possibly considered) access to the norms of reasoning, there are not any reasons to privilege them in the theory of deducibility either. A good many theorems of first-order logic will fail the test of BI-intuitions, and notoriously so. The tutor will be unfazed by this disconformity. We should ask him why. If the tutor responds as before, he will cite his confidence in the disconforming theorems, and will take comfort in them as theoretical intuitions. Recall how the tutor's story began. We found BI-intuitions inadequate for grounding the norms of reasoning. We then admitted a new conception of intuition to bolster the fortunes of the BI-sort. These were tutored intuitions. In the example at hand, we imagined the tutor to have studied logic and probability theory, and we allowed the counterfactual assumption – "for ease of exposition" – that the laws of logic and probability theory are norms of reasoning. The laws of logic and probability theory are the tutor's theoretical beliefs. If the failure of BI-intuitions showed that the norms of reasoning are sometimes unintuitive, the same will be true of many laws of the tutor's theories of deducibility and probability. They, too, will be unintuitive. In the former case, theoretical intuitions were summoned to relieve our anxiety about BI-unintuitiveness. In the present case, the unintuitive nature of these laws of logic and probability theory may itself require explaining away. If we resort to the same measure as before, we will offer theoretical intuitions in support of the unintuitive cases. Since the unintuitive cases reside in the theories of deducibility and probability themselves, the relief offered by theoretical intuitions must be got, on pains of circularity, from some further theory. If this is indeed the tutor's position, it is a bad one. It starts him on a slippery slope to an unstoppable regress of redeeming theories. Better that we jettison the idea that unintuitive theorems require bailing out by intuitions of any kind.

This is not to say that the tutor need not be held accountable in the matter of his unintuitive theorems. Short of pleading an unending hierarchy of theoretical intuitions, the tutor has three options.

Ideal models. The theorems of deducibility or, if they are different, the true principles of reasoning, are stipulated in the model. They are analytically true in the model. Since they are analytically true, they are normatively canonical.

We encountered the ideal models approach in the first section of the present chapter, where we said why it did not work. Here it is again. It still does not work.

Analytic intuitions. The tutor digs in his heels. His unintuitive principles may not be self-evident, but they follow self-evidently from principles that are self-evident (Quine and Ullian, 1970, pp. 22–4). Self-evident principles of correct reasoning flow from what correct reasoning is, from what it is to be correct reasoning. They are the result of an analysis of the very idea of correct reasoning.

If this is not full-bore subscription to the methods of philosophical analysis in the manner of Moore, it is a near thing. Reasons for canceling the subscription were detailed in the preceding section. We need only add that self-evidence founders on the same shoal that holed the hull of analytic intuitions, the inability to discriminate between apparent self-evidence and the real thing.

Indispensability. Historically the indispensability argument was contrived for mathematics (Quine, 1990/1992, pp. 94–5). It is an argument to this broad effect:

a. Mathematics is indispensable for successful science.
b. Successful science is empirically adequate.
c. Therefore, mathematics is empirically adequate.

Care should be taken with (c), on pain of the fallacy of division. Better that we replace (c) with

(c′) Mathematics is at least as objectively justified by its indispensability to empirically adequate science, as empirical adequacy objectively justifies science itself.

A similar argument would apparently be available for logic, substituting "logic" for "mathematics" each time. Then, on the counterfactual assumption that the laws of logic are norms of reasoning, the unintuitiveness of the tutor's unintuitive norms is overriden by the objectivity conferred on them by their scientific indispensability.

It is worth noting that the indispensability argument preserves an attractive feature of Cohen's program. The laws of logic take on the justification of those empirically adequate theories to which they are indispensable. If the norms of rationality are the laws of logic, then they, too, have the same justification, however unintuitive they may be.

Of course, the trouble is that there is virtually nothing to be said for the view that the laws of logic are norms of reasoning. This should be evident to fanciers of the methodology of BI-intuitions. If BI-intuitions were even presumptively canonical for such norms, we should be obliged to note and to take seriously, *pace* Cohen, that our beliefs and practices are routinely inconsistent. Some theorists will think that privileging them epistemically as BI-intuitions would call for a paraconsistent logic. In Chapter 6, we explored the idea of a paraconsistent *logic* as a normative theory of doxastic inconsistency management, and found it wanting. Whatever an acceptable theory of such things turns out to be, it will not be the logic – whichever it may be – that is indispensable for science. It will not be classical logic; it will not be intuitionistic logic; not quantum logic; and not paraconsistent logic. What is more, whatever the correct theory of reasoning norms, *they* will not be required for science. It is one thing that science requires that its practitioners reason capably; it is another that the norms of reasoning are required to be among a scientific theory's *laws*. Thus, the norms of reasoning, whatever they may be, are not laws of logic, by the principle that anything identical to something indispensable to science is itself indispensable to science. It is true that there can be empirically adequate theories about what goes on when people think that "such and such" are norms of reasoning. These are theories about the normative beliefs of human agents. We might well discover, for example, that people have no clear idea of what justifies even their most cherished normative convictions. But the theory that records this fact, if it were a fact, would be a descriptive theory, not a normative one.

No one doubts that a theorist without pretheoretical beliefs has nothing to go on. Kant was right. Theories without intuitions are blind. Similarly, no one will doubt that a theorist's beliefs will constrain his theory, for no theorist aims to construct theories inconsistent with what he believes.[20] Something like the process of reflective equilibrium is at work here. Pretheoretical data suggest principles that, on consideration, require some pruning of the data, and so on. It is instructive to compare normative and descriptive theories. In their barest form descriptive theories are ones in which *data* adumbrate *laws*, whatever the details of whatever subtlety. In their barest form, normative theories are ones in which *intuitions* adumbrate *rules* or *norms*, whatever the details of whatever subtlety. There are lots of problems with this view of descriptive theories. It draws the complaint that there is no natural distinction between observation terms and theoretical terms, that observation is theory-laden, that holism discourages the unfiltered primacy of observation, and so on (Suppes, 1969; Hacking, 1983, Ch. 13). What is nowhere seriously in doubt, however, is that the data for a theory are independently and objectively true, or that they are as true as anything gets in our best science. The same applies to the theory's observational projections that figure centrally and indispensably in the matter of empirical adequacy. Holism has its attractions, of course. When faced with a counterexample, with what Quine calls "recalcitrant observations," the descriptive theorist has a lot of latitude. He may qualify a law, introduce a new theoretical construct, or enrich his stock of idealizations. Short of experimental error, there is something the descriptive theorist will not do even if he is a holist. He will not say that his theory simply overrides them.[21]

The normative theorist seeks to conform his theories to the descriptive paradigm. He too has "data" for his theory. They are his pretheoretical beliefs. Some of them will enjoy the standing of empirically justified descriptive beliefs. But these will exclude *normative* beliefs.[22] Like their descriptive counterparts, normative theories also have their projections, their outputs. Sometimes the theory will sanction a projection that the theorist dislikes, that is, which he disbelieves. When this happens, we say that the theory has counterintuitive consequences. Here, too, various accommodations are possible. He may avert the consequence by qualifying a norm; he may take cover in some presumed idealization, and so on. He can also do, and often does do, what no descriptive theorist would dream of doing: he can outright deny the recalcitrance of the recalcitrant consequence. If so, he will override this disbelief of the theory's counterintuitive consequence, and come to accept it as theoretically justified.

A counterexample to a descriptive theory is a counterobservational consequence that the theorist is unable to disarm in any of the ways we have taken note of. Without such a rescue, there is a strict identity between a counterobservational consequence and a counterexample. It is not that way with

normative theories. Counterintuitive consequences are the normative counterpart of the descriptivist's counterobservational consequences. Faced with
a counterintuitive consequence, the normative theorist has occasion to wonder whether he faces a counterexample. He, too, may try to save the theory
by making adjustments of a kind open to the descriptive theorist. When the
descriptive theorist attempts such a rescue and finds that it does not disarm
the counterobservational consequence, he has no option but to take it as a
counterexample. When the normative theorist finds that similar evasions fail
to disarm a counterintuitive consequence, he deems himself free to exercise
an option not open to his descriptivist colleagues. He can simply bite the bullet
and (with apologies for mixing) swallow the consequence. In this, he shows
himself committed to the view that counterintuitiveness is not, as such, counterexemplary. He binds himself to the idea that a normative theory need not
be required to conform its provisions to what the theorist believes. Thus, the
normative theorist finds himself in the same position as any abstract theorist,
the abstractness of whose theories denies them the constraints and comforts
of more or less direct empirical checks. That is, the normative theorist is like
the pure set theorists who, once launched into the transfinite, welcomes more
counterintuitiveness before lunch than a more earthbound enquirer will meet
in a lifetime.

I make this observation not to disapprove of it, but rather to make a point
about the theorist's beliefs. They are contingently required to get a theory
up and running. Beyond that they do not play much of a role. This can only
mean, as with the abstract theorist, that such beliefs are in no way canonical.
Observations are canonical and beliefs are not. We cannot have a descriptive
theory without some observations, and we cannot have a normative theory
without some beliefs. There the similarity ends. In conforming his view of
theory to the descriptive paradigm, however implicitly, the normative theorist
makes more than there is of the kinship between data and intuitions. He allows
himself to think that intuitions are canonical, and he forgets what, in many
cases, he will subsequently be drawn to, namely, refusing the counterexemplary
connotation of counterintuitive consequences.

This shows that normative theories are best reckoned on the model of formal or abstract theories rather than on the paradigm of physics or molecular
biology. Even so, we should not make too much of their affinity to such things as
set theory. The structure of normative theories is something like the reflective
equilibrium process taken in Aristotle's way. It is a process in which principles are adjusted to *endoxa*, and they to them, and in which further beliefs,
not necessarily endoxic, secure a purchase. The difference between Aristotle's
conception and that of the theorist of BI-intuitions is that belief is not epistemically privileged. *Endoxa* are beliefs everyone has – common knowledge,
as is said – or those endorsed by the wise – expert opinion, as is said. Their
high presumptive distribution and their expert provenance are important, for
we have to start somewhere, and we should wish to minimize the begging of

questions. These are nonepistemically privileged beliefs since any, in principle, lies open to refutation. But no amount of dialectical give-and-take can get us from belief to knowledge, that is, can *establish* those beliefs as objectively true.

Aristotle, too, made use of the concept of intuition, which is an unfortunate translation of his word *nous*. This is not the place for a lengthy exposition, but a brief glimpse may be helpful. Aristotle was heir to the central problem of Greek philosophy. By what means, if any, does a human being pass from belief to knowledge? Aristotle's answer is subtle and complex (and, as I believe, wrong). Let us consider the knowledge that a science makes possible. Aristotle thinks that the nomic core of a science, physics say, is the demonstrative closure of its first principles, which are self-justifying. First principles are grasped through the operation of *nous*. Even so, they are elusive. Most people will never grasp them. This is because most people lack the leisure and the interest required to track them down. First principles make themselves known to us through an arduous process of inductive enquiry and dialectical examination. Dialectic resembles reflective equilibrium. It is a process of testing our pretheoretic beliefs, our *endoxa*, by drawing out their consequences and by subjecting them to the intense scrutiny of putative counterexamples.

Some matters admit of knowledge, whereas others admit of what might be called "highly disciplined belief." If a matter admits of knowledge, then there is a science for it, and hence first principles of it. Dialectic is a way, though not a guarantee, of getting people to see first principles. Whatever else they are, first principles will be winners in the tough stakes of survival of the fittest. Part of what is involved in seeing something as a first principle is the experience, if eventually we come to it, of having no more ideas about how to put it under pressure. Dialectic helps to trigger *nous*, and being in a state of *nous* with respect to a first principle is having been made to see it, or intuit it, as necessary, prior, most intelligible, and indemonstrable. If first principles are unavailable for the matter at hand, dialectic serves more modest ends, namely, the ends of reflective equilibrium, or what we might call "reflective equilibrium in Aristotle's sense." This produces the best that we do do, namely, systems of belief that withstood the best we could throw at them. It is fair to ask what things are knowable. Aristotle is drawn to an autoepistemic argument. Let E be any enquiry that is seriously entrenched in our intellectual practice. Then if E possessed first principles we would know them by now. Let E be any normative theory likewise seriously entrenched. If E had first principles we would know them by now. E would have a standing and a stability similar to, say, biology. Notoriously, there is yet to be produced a single normative theory of which this is true. By the autoepistemic argument, knowledge is not possible for such matters.

We have been reviewing various methodological uses to which philosophers have put various conceptions of intuitions. For all their differences, they share a methodological presumption. The presumption is that the methodology of intuitions, in one or other variation if not all, transforms belief into knowledge, and is thus the epistemic *enabler* of normative theories. But, as I have tried to show, in no case is this presumption warranted. Goodman thinks that reflective equilibrium confers on any theory the same justification that it confers on logic. This, for him, is real justification; it is objectivity without transcendence. If there is a problem with most normative theories, it is not that when in reflective equilibrium they lack the justification that logic enjoys under classical sway. *Rather, it is that they are not in reflective equilibrium.*

The classical problem of epistemology presents the theorist with an extremely difficult challenge. The long history of epistemology is littered with the failure to negotiate the space between *endoxon* and *episteme* in the way set forth by the classical problem. It is an unwise epistemologist who has not prepared himself for the possibility that failure inheres in the problem itself. Intractable problems are grist for the mill of the Can-Do Principle. They are a standing inducement for the theorist to change the subject to do what he can do rather than to persist directly with what he cannot do, or has not been able to do yet. It is a legitimate inducement so long as the collateral work stands some chance of bearing fruitfully on the main task, something that is not always reliably discernible in advance. There are occasions, however, when the theorist cuts himself more slack than even the Can-Do Principle permits. When this happens he is in the ambit of the Make-Do Principle, which is a can-do corruption. I see all justifications rooted in the reflective equilibrium principle as providence of the Make-Do Principle. Here is how. Epistemology was launched by what we have been calling the classical problem. The classical epistemologist wanted to know what knowledge consists of – apart from its satisfaction of the condition of reputable belief. We may take it as given that the doctrine of reflective equilibrium is reasonably construed as output of a reputably intended application of the Can-Do Principle. It is reasonable to suppose that a theorist intent on solving the classical problem might well have thought that figuring out the empirical dynamics of reflective equilibrium could prove to be a fruitful contribution to the main project. But Can-Do loses its purchase when the theorist now proposes that reflective equilibrium solves the classical problem, as Goodman famously proposed in his all-the-justification-that-induction-can-get-or-needs remark. Can-Do now defers to Make-Do. In rough schematic form the Make-Do Principle is involved in response to a problem P when it is judged, however tacitly, (1) that P cannot be solved in the ways set by P itself; (2) that a problem bearing some resemblance to P can be solved by procedure M; and (3) that M therefore can be counted as an answer to P on grounds that none better is in evidence. In an especially loose form, the Make-Do Principle licenses an equivocation on "justification," a smudging of the difference between the meaning required by the classical problem, and the meaning with regard to which the reflective equilibrium dynamic *can* claim adequacy. This is not to overlook the fact that sometimes it must be acknowledged that when a problem is unresolvable – or perhaps illegitimate – the methods developed for its solution go into retirement. If they bear a recognized name – say, "epistemology" – then epistemology is out of business, and cannot without ambiguity be put to other purposes under the same name. By these lights, the reflective equilibrium approach cannot be epistemology except in a sense that cannot bear on the classical problem, and cannot provide a justification in any sense that that problem would see as relevant. Make-Do says in effect: "Let reflective equilibrium stand in for a solution to the classical problem." Saying so is taking Can-Do to illegitimate extremes.

VALUE INTUITIONS

The utterly central problem with the method of intuitions is its impotence when drawn into disagreements in the abstract sciences, normative disciplines included. Nothing is advanced in our disagreement by my forwarding intuitions that you reject. Tell any story we like about the privileged epistemological pedigree of intuitions, and it will count for precisely nothing for persons whose intuitions run contrary to one another. Even when some common ground reveals itself in the concurrence of some of our respective intuitions, the resultant dialectical accord flows not from their epistemological pedigree but rather from the fact that they are shared.

Imagine a dialectical community **dc** and a set of propositions $\{P_1, P_2, \ldots, P_n\}$ of which the following things are true. The P_i lack access to empirical checks and have attained the standing of intuitions, or derivations therefrom under intuitively impeccable transformations. As far as the members of **dc** are concerned, the P_i are analytically true, or conceptually true, or validated by reflective equilibrium, or some such thing. Thinking so is a domestic comfort only, and an unstable one. It takes but one troublemaker, one dissenter from a P_i, to making the ensuing disagreement irresolvable unless the epistemological presumptions is attacking to such propositions are lightened up on. "P_2 is analytically true!" cannot persuade someone who thinks that it is not even true. "You don't understand P_3-hood!" cuts no ice with anyone confident enough of his understanding of P_3-hood to charge P_3 with falsity. Shorn of their epistemological cachet, the P_i have precisely two points of salience in **dc**. They constitute a psychological commonality in **dc**; that is, they make it strictly true to say that **dc** *believes* P_i, $P_2, \ldots, P_n$. And they give members of **dc**, and of any like-minded community, non-question-begging purchase on one another. There is, as we saw, no analyticity by agreement. There is only agreement by agreement.

We may take it that an abstract theory is worth its salt only if it processes conflict resolution strategies for handling criticisms both within and without. A theory lacking such a strategy is best taken for playful, or some near thing. We said that the playful theorist need not be a lighthearted trifler, that his work can be playful and seriously pursued at the same time. The central fact about playful theories is that they have neither resources nor motive to defend themselves against certain lines of attack (e.g., that things are not usually like that).

Take such a theory and imagine that in some way its practitioners now hold its derivations to be intuitively correct derivations from intuitions. Such a theory is playful even if its corporate psychology has gone analytic. In going analytic, the theory has not increased its capacity for conflict resolution one iota.

A playful theory can be traumatized only in taking on a corporate analytic psychology. It is a transformation that makes definable certain accusations against it (e.g., that things are not really like that), and that motivates the answering of such objections without the means of doing so effectively.

I am a Humean about intuitions. It is unthinkable that we not have them and unimaginable as to how we might go about curing ourselves of them once they have struck. The Insistence of an Idea, the utter certainty that such and such a proposition is true nevertheless should and can be made disputationally nonnegotiable. Even if it is true, or analytically true, that it inheres in the very idea of sets that inconsistent sets exist, it will avail the dialethist nothing to cite the epistemological pedigree of his own utter certainty if he is squabbling with someone seized by classical intuitions, or even classical preferences. There is a general moral in this. Intuitions are fine in their place. They are grist for the theoretician's mill. They are his way of honoring his own utter certainties. In their epistemological state they do no good as dispute-resolvers, and in their de-epistemological state they do no disputational good short of acceptance by the other side. Such is the looming importance of the Heuristic Fallacy.

Notes

1. For an overview of Quine, the reader could consult Woods (1998).
2. See, for example, Goghossian (1996), and the lively discussion in subsequent issues.

Chapter 1

1. Kaufmann, 1906, p. 487. Kaufmann concludes: "*The results are not compatible with the fundamental assumption of Lorentz and Einstein*" (emphasis in the original). Lorentz admitted that "it seems very likely that we shall have to relinquish the idea altogether" (Lorentz, 1952, p. 213).
2. "The experimental evidence confirming the principle of relativity is actually overwhelming, in the sense that in no field has one ever discovered any dependence of the form of the laws of physics on the velocity of the reference frame" (Bohm, 1965, p. 106).
3. In van Fraassen's useful phrase (van Fraassen, 1980, p. 4).
4. "Abstract" conjures up images I want not to embed in my use of the word. The word is tangled with ambiguity, and that counts against our choosing it. But in its use here the choice is benign. It is lexical relief from the banality of "nonempirical." In particular, I need carry no brief for abstract objects in the manner, for example, of Zalta, interesting though this account surely is (Zalta, 1983).
5. Of course, in the philosophical aftermath of relativity theory, it came to be supposed by some writers that the empirical adequacy of a scientific theory that irreducibly imbibes a given geometry counts as empirical justification for said geometry. But this would not have settled the dispute between Frege and Hilbert; it only complicates our example in ways that we need not deal with until Chapter 3.
6. "The accepted wisdom is that mathematics lacks empirical content. This is not contradicted by the participation of mathematics in implying the [observation] categoricals, for we saw...that such participation does not confer empirical content" (Quine, 1995, p. 53). "A set of sentences that has *critical mass*, as we

327

Notes to Pages 10–26

may say – that is, that implies some synthetic observation categoricals – may be said to have those categoricals as its empirical content" (Quine, 1995, p. 48).

7. It should be emphasized, however, that the received triagic judgment of the Tarski paradox makes less of it than is deserved. See the Prologue and Chapter 7.

8. The references again are: Russell, 1937, pp. xv–xvi, xviii, and 2; van Heijenoort, 1967, 126–8; and Tarski, 1983, pp. 152–4. Here is Tarski on the moral of the Liar: "[t]he very possibility of a consistent use of the expression 'true sentence' which is in harmony with the laws of logic and the spirit of everyday language seems to be very questionable" (Tarski, 1983, pp. 164–5).

9. It is this latter consequence that is unemphasized in the received triage. But it is "there" all the same. If nothing is true, nothing is false either under the classical negation rules. If nothing is true or false, there are no bivalent sentences.

10. I owe this reference to Nekham to Stephen Read (Read, 1988, p. 31, n. 10, 11).

11. Locke, 1975, pp. 164–5. None of the other rules has escaped hostile scrutiny and rejection, although by far the largest consensus centers on the presumed inadequacy of Disjunctive Syllogism.

12. Here we entertain the possibility advanced by some philosophers of science that the law of causality fails in the microdomain, that is, that part of nature studied by quantum mechanics. But few doubt that the principle of causality holds in the *non*microdomain. I assume this qualification here.

13. Or reproduced it, since the proof may have originated with Alexander Nekham at the beginning of the thirteenth century.

14. Concerning which it is Johnstone's view that it is the only allowable way to transact a *philosophical* inquiry. See Johnstone (1978, p. 13). Feyerabend, too, sees the central importance of *ad hominem* case-making in the philosophy of science. See Feyerabend (1978).

15. This question is taken up in detail in Chapter 8.

16. *Sophistical Refutations* 167_b 8–9ff. Cf. *Prior Analytics* B27, 70^a 6–7 and *Rhetoric* Γ 13, 1414^a 31–37.

17. See *On Interpretation* 11, 21^a5ff. Cf. Δ5, 1015^b8, and *Posterior Analytics* A9, 76^a 13–15.

18. This is the Ross (1927), translation. Barens (1984), has it this way: "Now negative demonstration I distinguish from demonstration proper, because in a demonstration one might be thought to be assuming what is at issue, but if another person is responsible for the assumption we shall have negative-proof, not demonstration." Cf. *Sophistical Refutations* 6, 167^b 21–27; 1, 165^a 2–3; 15, 174^b 19–23; and *Rhetoric* B22, 1396^b22–27, 23, 1400^b26–29, Γ 17, 1418^b 1–4; and *Prior Analytics* B20, 66^b 11.

19. Cf. *Metaphysics* K5 1062^a30–31 and Γ4, 1006^a25–26 (Ross, 1927).

20. Cf. *Sophistical Refutations*, 22, 178^b17; 8, 170^a13, 17–18, 20; 177^b33–34; 183^a22, 24; and *Topics*, Φ11, 161^a21.

21. Cf. *Sophistical Refutations*, 20, 177^b31–34, and 22, 178^b 16–23.

22. Cf. *Sophistical Refutations*, 8, 170^a, 12–19.

23. Cf. *Sophistical Refutations*, 1.

24. *Metaphysics* K5.

25. Woods and Walton, 1988, Ch. 3. For a proof that the distinction is not disjoint, see Jacquette (1993).

26. An active research program in antiquity and the Middle Ages, its twentieth-century revival drives from the work of several writers, including: Hamblin, 1970; Woods and Walton, 1978; Woods and Walton, 1982, reprinted as Woods and Walton, 1989, Ch. 10 and 19; Mackenzie, 1979; Mackenzie, 1984; Woods and Walton, 1982, Ch. 6; Barth and Krabbe, 1992; Walton, 1984; Hintikka, 1976; Hintikka, 1987; and Carlsen, 1983. Influential pre-1970 work includes Belnap, (1963) and Harrah (1963).

   For "second-generation" developments see, for example, Walton and Krabbe, (1995); Girle (1993); Girle (1994); Girle (1996); and Belnap (1976); Gabbay and Woods, 2001a and 2001b.

27. There is a male barber in a village who shaves all and only those men who don't shave themselves. Who shaves the barber?

28. And its counterpart for the Liar, which we might dub **Koryé's Sorrow.**

Chapter 2

1. I follow Lewis and Langford's own cataloging notation, B_1, B_2, and so on.

2. For readers interested in a fuller discussion, see Woods (1975).

3. This is not undisputed. Some writers think that Quine mishandles probability sentences, dispositional predicates, and counterfactual conditionals. For these and other matters, Hookway (1988) affords an accessible and lucid *entré.*

   We can now see that the same expulsion policies apply to the epistemic modalities of the previous section. They, too, offend when quantifiers attempt to cross occurrences of the K-operator, as in "There is something compatible with everything Larry knows." Here, too, there is need to speak of epistemically possible worlds and the correlative need to maintain the identity of things across such worlds. More essentialism still.

4. For an excellent overview of relevance logics see Dunn (1986).

5. For example, that DS is invalid.

6. For example, that *ex falso* is true.

7. It was widely reported that the discovery required a substantial revision of the Standard Model, a huge overstatement in fact. Had the discovery been that *photons* have mass, that indeed would have overturned the Model.

8. Avron goes on to say: "To complete the picture, we add that in Routley-Meyer semantics $\wedge$, for itself, has a very simple and intuitive interpretation. The price is, however, that *implication* has a very complicated (and in our opinion) unintuitive semantics. Again – it is the attempt to have *together* both a relevant implication and an extensional conjunction which is incoherent and causes difficulties. Each of them is separately OK!" (Emphasis added in the second instance).

9. A decision problem is NP-hard just in case it can be converted in deterministic polynomial time into a problem that can be solved in nondeterministic polynomial time. PSPACE is the set of all decision problems recognizable by polynomial space-bounded deterministic Turing machine programs such that for any input the programs halt. A decision class K is PSPACE-*complete* if K ε PSPACE and, for all K*ε PSPACE, K* polynomially implies K. Thus, even if it were the case that P = NP, we could have it that P $\neq$ PSPACE (Garey and Johnson, 1979, p. 171). A problem is PSPACE-*hard* if some recognized PSPACE complexity

problem π is Turing reducible to it without the need to state that π itself is in PSPACE.

10. These and many other heuristics are accessibly diagnosed in Newell and Simon (1972).

11. A hint of how things might go. **IR** has much simpler decision problem than R, but it is less simple than that of classical logic. Thus, the full-use sense of relevance cannot be sustained by any prospect of optimal efficiency in proof-searches. Full-use relevance is a *cost*. On the other hand, logical structures such as proofs (or refutations) are also held to other desiderata. We want proofs, for example, to be epistemologically benign, which is what the proofs of **IR** are said to be. Abstractly considered, epistemic considerations might outweigh the inefficiency of proof-searches.

12. But not unknown to the empirical sciences either, as Graham Priest has reminded me in conversation.

Chapter 3

1. Provided that we are not here minded to interpret paraconsistently the inevitable vagueness of empirical predicates. See Priest and Routley (1989, pp. 389–90).

2. The term "cognitive dissonance" is borrowed from the writings of the cognitive psychologist Leon Festinger and his several colleagues. See Festinger (1957) and Veyne (1976). Cognitive dissonance is not in these writings primarily a matter of holding contradictory or paradoxical beliefs. In Festinger's employment of it, cognitive dissonance manifests itself when, for example, the recent purchaser of a product – a new washing machine, say – persists in reading advertisements praising the product purchased, as if the manufacturer's own propaganda could be rationally reassuring after the fact. Some critics of the Festinger school regret that the idea of cognitive dissonance did not sufficiently recognize paradoxical beliefs or preferences. In any event, the idea of cognitive dissonance has lately come to include such things as the having of inconsistent beliefs, and I shall so use it here.

3. "To recognize … 'inconsistency' [of beliefs or preferences] does not demolish [one's economic theory], since both the *feasibility* and the *necessity* of "consistency" also requires justification. … Whatever view one takes of, say, Agamemnon's dilemma, it can scarcely be solved by simply demanding that Agamemnon should lick his preference ordering in shape before he gets going" (Sen, 1987, p. 66).

4. This is a point I mean to make more of. See Chapter 8.

5. For more on this controversy, see [Cohen, 1981].

6. Contemporary AI theorists have amusing things to say, along the lines of apocalypse. Sherry Turkle quotes from Minsky (1987, p. 183), "A thinking child's mind … [needs no one] to tell it when some paradox engulfs and whirls it into a cyclone." Turkle writes: "Paradox, argues Minsky, is as dangerous as the primal scene [shades of Count Koryé] … Minsky adds: 'And what of sentences like *This statement is false*, which can throw the mind into a spin? I do not know anyone who recalls such incidents as frightening. But, then, Freud might say, this fact could be a hint that the area is subject to censorship.'" Turkle

likes the suggestion, "The idea of the cognitive unconscious supports this view: paradox and senselessness need to be repressed in the process of developing emergent intelligence, whether in machines or in people. Absurd results of reasoning are taboo, as threatening as sex. The censors work as hard to suppress them; they have no innocence" (Turkle, 1988). Cf. Sen (1987, p. 67), "Incompleteness or over completeness in overall judgements might well be a damned nuisance for decisions, but the *need* for a decision does not, on its own, *resolve* the conflict."

7. A lattice is a partially ordered set $\mathcal{L}$ such that every pair a, b of elements of $\mathcal{L}$ has a least upper bound (l.u.b), viz., the join $a \vee b$, and a greatest lower bound, (g.l.b), viz., the meet $a \wedge b$. A top element of $\mathcal{L}$, if there is one, is 1, and a bottom, if there is one, is 0. If for $a \in \mathcal{L}$ and their join $a \wedge b - 0$, *then* b is a complement of a. If b is a's sole complement in L, then it is the element $\neg b$. Boolean algebras can be conceived of as lattices $\mathcal{B}$ with 0, 1, and $0 \neq 1$, such that for $x, y, z \in \mathcal{B}, x \wedge (y \vee z) = (x \wedge y) \vee (x \wedge z)$, and such that for every element there is a compliment.

   Heyting algebras are lattices $\mathcal{H}$ with 0 such that for any $x, y \, \varepsilon \, \mathcal{H}$, the element $x \to y = \text{l.u.b.} \{a: a \wedge x \leq y\}$ exists.

   A lattice $\mathcal{L}$ is orthocomplemented if for every $x \in \mathcal{L}$ there is a unique element $\neg x$ such that $x \wedge \neg x = 1, x \wedge \neg x = 0$, and such that if $x \leq y$ then $\neg y \leq \neg x$. $\mathcal{L}$ is also orthomodular when $x \leq y$ implies $y = x \vee (y \wedge \neg x)$. Such lattices have interesting applications in quantum logic. (See, on this point, Gibbins [1985]. See also more generally De Vidi and Graham Solomon [1999].)

8. The Curry paradox licenses $\ulcorner \Phi \to (\Phi \to \psi). \to (\Phi \to \psi) \urcorner$, and is discussed in greater detail in Chapter 7.

9. Cf. Donald Davidson: "...I think that we are justified in carrying on without having disinfected this particular source of conceptual anxiety" (Davidson, 1967, p. 314).

10. All papers by KGH are reprinted in Martin (1984).

11. We note in passing, Austin's mistranslation of "the method of *inference*" ("Schlussweise") as "argument" (Frege, 1950).

12. An interesting and elaborate account of commitments, different from the approach I take here, can be found in Walton and Krabbe (1995).

13. The notion of modal independence is discussed by, for example, von Wright (1957), Smiley (1958), and Geach (1972). Initially introduced as a claim about the entailment relation, I have adapted it here to the commitment relation. Under this adaptation, commitment is a *pragmatic restriction* of the semantic relation of implication.

14. I borrow the term "pragma-dialectical" from the work of the Dutch argumentation theorists, Frans H. van Eemeren and Rob Grootendorst. See, for example, van Eemeren and Grootendorst (1984). In their usage the term denotes a particular theory of argument, their own in fact. In my usage it simply denotes anything that could be taken as a rule of n-person argumentation.

15. Here is a difference. In Brown's logic, ambiguation can take the sting out of even $\{(\Phi \wedge \neg \Phi)\}$, whereas for Schotch and Jennings, this is an inconsistent set whose sting stays in.

16. For example, as we saw in our discussion of Aristotle on refutations, in Chapter 1, a would-be refuter may use as a premise in the syllogism against his opponent nothing but what his opponent has conceded in answering the refuter's questions.

This is not a condition on syllogisms, but rather on syllogisms-in-use, as we might say. It is a pragma-dialectical rule, not a logical one.

17. A detailed discussion may be found in Woods (2001, Ch 4).

18. Saying so is problematic for **DL3** in other respects. We saw in Chapter 1, **DL3** cannot resolve disputes about whether something, for example, Adjunction, *is* a principle of logic. But if it is a principle of *logic*, **DL3** requires that it not be challenged or retracted.

19. The idea that inconsistent databases are not only not rare occurrences but indeed are often quite harmless is nicely developed in work that has come to my attention too late to incorporate into the present chapter. See Gabbay and Hunter (1991) and Gabbay and Hunter (1993). See also Gabbay (2000, Ch. 10).

Chapter 4

1. "Introduction to Part II," in Boolos (1998).

2. For historical accuracy, it should be remarked in passing that Frege could not have thought that the Russell Paradox destroyed Frege's analysis of sets. Frege distrusted the concept of sets independently of the Paradox, and certainly had no theory of them. Frege acknowledged *courses of values of concepts*, and he thought the Paradox as damaging to his courses of values as to Russell's propositional functions. That Frege subscribed to the naive theory of sets is a confusion of commentators who see in Frege's courses of value what Russell saw in intuitive sets. Even so, I shall conform my remarks here to the traditional view, in which Frege himself would later in his life seem to have acquiesced, that there is an early Fregean set theory and that the Paradox wrecked it.

3. Although I shall say below why I think we must take Russell as meaning not *believed*, but *assented to*.

4. Again, "'Intuition' is in general not to be trusted.... [It] is bankrupt, for it wound up in contradiction.... The logician has had to resort to mythmaking" (Quine, 1966/1995, p. 27). So "when we pursue general set theory, we must grapple with the paradoxes, whether by von Neumann's method of non-elements, or by Russell's hierarchy of types, or by some other probably equally artificial device" (Quine, 1951a, p. 138).

5. Which sanctions, among other things, this inference: Since it is possible that there are intelligent Martians, there exist things of which it is true that they might be Martians.

6. This is precisely what they do in fact do according to the ambiguation strategist. Which is another reason to distrust ambiguation as a *general* pacifier of philosophic discord.

7. There are apparent exceptions. If I doubted that Harry is pusillanimous while granting that he is fainthearted, you could direct me to a good dictionary and get me to see that "Harry is fainthearted" is true only if "Harry is pusillanimous" is true. But this is a case of your getting me to see the truth of "Harry is pusillanimous" by getting me to see what 'pusillanimous' means. It is not a case of getting me to see that "Harry is pusillanimous" is a truth true by virtue of meaning.

8. The demographic and other conditions on attributing beliefs, and disagreements about beliefs, to communities are extremely elusive. The concept of group

or community is itself a theoretically recalcitrant one, as Gilbert (1993) ably attests.

9. See here Barth (1974) and Carlson and Pelletier (1995). Since neither plural quantification nor genericity knows yet the comforts of settled theory, here, too, I shall take the idea of a *community's* belief that Φ as primitive.

10. To this day there are people, some of them quite sensible, for whom Cantor's Paradise cannot be a matter of belief. The notion of the transfinite, of actual but uncompletable infinites, is "too bad to be true," or counterintuitive enough to preclude positive belief. But if the transfinite goes, so do significant chunks of modern mathematics to which the transfinite-doubter gives his untroubled assent. This is an arrangement in which the charge of counterintuitiveness against the transfinite is acquitted, and it leaves in its wake not belief but assent.

11. The definitions also extend to dialectic communities in the obvious way.

12. Hume, 1739/1969, p. 119: "When two numbers are so combined, as that the one has always an unite answering to every unite of the other, we pronounce them equal.... "

13. Not to be confused with the "Paradox of Analysis," which, later, would so greatly exercise Moore, and that was anticipated by Husserl.

14. In this I concur with Gratton-Guinness (1978): "[J]une (1901) appears to be the most likely date" (p. 135). Judging from Hilbert's letter to Frege, it is apparent that Zermelo derived the paradox that bears Russell's name in 1899. It would also appear that Cantor was aware of "Russellian" difficulties: "Cantor's 'definitions' [from 1899] only allow as sets those collections which are *wholes* and this does not at all imply that any collection can be a set. Nothing like the comprehension principle of so-called 'naive set theory' follows from Cantor's statements. If 'naive set theory' is characterized as set theory based on the comprehension principle, then this goes back, not to Cantor, but to Russell [1903]" (Hallett, 1984, p. 38).

15. A change that was mediated by the publication in mid-course of the book on Leibniz (Russell, 1903/1937), concerning which see Hylton (1990).

16. Aristotle, *Metaphysics* Γ3, 1005^b 19–13: "The firmest of all first principles is that it is impossible for the same thing to belong and not to belong to the same thing at the same time in the same respect." However, "*we must presuppose, in face of dialectical objections, any further qualification which might be added*" (emphasis added).

17. cf. Zermelo, 1930, p. 47: "And so the set theoretic 'antinomies' properly understood instead of leading to a contradiction and a mutilation of mathematical science, lead rather to an *unsurveyable* unfolding and enrichment of that science." Emphasis added.

18. "Both logic and psychology, if they are to exist at all, must remain each in principle independent. The undistinguished use of both at once must, even where instructive, remain in principle confusion. And the subordination of one to the other, whenever seriously attempted, will never, I think, fail to make manifest in its result the absurdity of its leading idea" (Bradley,1883/1922, p. 613).

19. Thus Peirce, for example, gave to his theory of truth a markedly idealistic cast in a series of six papers published in 1897 and 1888 in *The Popular Science Monthly.*

20. Except, of course, for those parts of reality that are in their own right mental: minds and their contents.

21. Russell, 1903/1937. Note that Russell says that the presence of a given class is "insured", not "ensured" or "assured." Russell was no illiterate; his choice of words is significant. In a *mathematical* definition of a class, the defining condition on the price for its "presence," they are the premiums which the theorist must pay.
22. Russell, 1903/1937, p. 112 It is interesting that Frege's Rule V for "sets" is sometimes called the Abstraction Axiom.
23. As I said in the Prologue, wary readers should read all of page 9 of Quine (1992).
24. I agree with Hintikka that there is reason to think that set theories lack intended models. See (1996, p. 170).
25. We come back to these issues in Chapter 6.

Chapter 5

1. Invocation of the same of Routley does not indicate historical priority, although it is not far off that mark. "Routley" is my choice rather than its bearer's subsequent arrangement to favor "Sylvan," since it is the former name, not the latter, which names a coauthor of the – to date – single most important work in that field, *Paraconsistent Logic*.
2. This point needs handling with care. Although Quine does sometimes say that sentences closer to the periphery are more vulnerable to revision, he does not say that this is why this is so. Vulnerability is rather a matter of the degree of disruptiveness to science attendant upon a sentence's fall. Sentences of logic are least vulnerable since their departure would be most disruptive, in the spirit of the maxim of minimum mutilation.
3. Cf. Benardete, 1989, p. 124: "...[w]hat we want to know is precisely what it is about sets as such, sets *qua* sets, that precludes there being a set of all sets that are not members of themselves. Insight being above all what we seek, Zermelo would seem to be the least promising source to consult; for he was positively ostentatious in insisting that this was no more than a home remedy, suited to the needs of the working mathematician on the most pedestrian level."
4. Cf: "Widely regarded today as the iterative (or cumulative) conception of a set, the ZF approach is still more illuminatingly characterized as an ordinal conception, drawing on the distinction between ordinal and cardinal numbers. There are sets on the first level..., sets on the second level, and third level, and so on, each set being understood to emerge at a particular level that may be assigned and ordinal numbers as its index.... Sets are thus seen to *supervene* on their members, as they are not merely constituted by, just as there is nothing for it to supervene upon; and yet if one takes the cumulative or even the ordinal conception of a set in the standard way..., the empty set is securely lodged on level 1" (Bernardete, 1989, pp. 126–7).
5. The slightness lies in the narrowness of the changes to DS itself. But **PR** is a relevant logic, and we know that in such logics narrow adjustments to DS are widely consequential. The same is true here. (See Chapter 2.)
6. It is worthwhile noticing that the method of analytical intuitions does not require, but is nevertheless compatible with the cost-benefit perspective and with holism. If one is analytically minded about concepts then the Russell paradox reveals *either* that there is no such thing as the very idea of set, *or* that the very idea of set is such that some sets are actually inconsistent.

7. See also Chapter 7.

8. We return to this point in the section to follow.

9. For respects in which Gupta's and Herzberger's accounts differ, see Belnap (1982). Belnap points out similarities between Kripke's account and that of Martin and Woodruff (1979).

10. The fixed point method was originated by Gilmore, who sees an anticipation of it in a Persistence Lemma of Robert Lyndon in 1959 (Gilmore, 1967/1974).

11. In Gilmore's original paper, the method of fixed points was used to model the Abstraction Axiom in a partial theory of sets. Here too, inconsistency defers to incompleteness. Neither "RεR" nor "R & R" receives an interpretation of which it is true.

12. Here there is a difference between Gupta and Herzberger and Kripke. Paradoxical sentences are non-truth-valued on Kripke's account.

13. Not all grounded sentences have finite rank. In fact, as Gupta has said (personal correspondence, July 28, 1987), "some unproblematic statements may 'wobble' for a long time (through transfinitely many stages) before acquiring a stable truth value."

14. The appearance, possibly in 2000, of Brady's two volumes on *Universal Logic* is awaited with much interest.

15. Another extensional system is that of Batens (1980).

16. Parsons (1990) considers a tie-breaking measure that, if adequate, would still leave gappers and glutters at parity, since it would reject them both. His tie-breaker he calls *agnostaletheism* and in it he proposes a "neutral value," N, with which to evaluate problematic sentences neutrally as between the "seeing" of gaps and the "seeing" of gluts. But, at this stage, agnostaletheism is more a jest than a solution.

17. A partial valuation of a set of sentences assigns T to some sentences, F to others, and no truth value to those left over. Assume a partial valuation PV for some set Σ. Consider all consistent classical extensions of PV to a complete valuation. Consider, in other words, all consistent extensions of PV in which every truth valueless sentence under PV is given T or F by classical valuation rules. Then *truth in the supervaluation* over PV is truth in all classical extensions of it. If it should happen that sentences take different truth values in *different* classical extensions, then it is truth-valueless in the supervaluation. Suppose that Φ has no value under PV. Then all classical extensions Φ are either T or $\ulcorner\neg\Phi\urcorner$ is T; hence $\ulcorner\Phi \vee \neg\Phi\urcorner$ collects T in the supervaluation. So Excluded Middle stands, but Bivalence falters, since in the supervaluation there can be sentences without truth values. (See van Fraassen and Lambert, 1972.)

18. Priest deals with the Barber differently. Both the Russell set and the Liar arise because of a regress of truth conditions that, just so, are non-well-founded. The Barber, on the other hand, owes nothing to a regress of truth conditions. See Priest (1987 p. 186, f. 14).

19. For an illustrative attempt, see Walton (1988).

20. A development of some of these ideas can be found in Woods (1995).

21. I am, of course, leaving to one side the battery of cost-benefit arguments that a dialethist can bring to bear, and that we sampled earlier in the present chapter.

Chapter 6

1. In some approaches to fiction and fictionalism alike, the target is not the semantic property of truth but the epistemic (and pragmatic) property of warranted assertability. The defense of fictionalism is easily objected to reflect this difference.
2. Semanticists are sometimes rather bloodless about the artist's creative authority. But here is Cézanne to his son in a letter of 1906: "What is needed is to merge the two parallel texts, seen in nature and experienced nature, nature out there (he points to the green and blue) and nature in here ... (he strikes his forehead), in order to last, in order to live, a half-human, half-divine life, the life of art ... – the life of God" (Cézanne, 1980, p. 29).
3. Considered but not quite accepted by Pavel (1986, Ch. 3). See also Doležel (1979) and Currie (1990).
4. But, a word of warning, on the make-believe approach fictional sentences are not true.
5. The example is John Heintz's via Ray Bradbury's "A Sound of Thunder." See Heintz (1979, pp. 92–3).
6. The maximal account of a text in the story it tells expressly and the further parts of it that a competent reader will understand as unexpressed.
7. Wherewith the need for existence-neutral quantification, concerning which see Priest (1987, p. 191 ff), Routley (1966), and Woods (1971).
8. I am not convinced that it is so, in fact. I have tried to motivate the distinction in Woods (1971) and more especially in the last chapter of Woods (1978).
9. Meyer and Routley, 1975. "Dialetical" here suggests dialetical materialism, true contradictions and all, rather than its more modern sense of n-person arguments.
10. This is the trouble with the fictive-operator 'O' of the *Logic of the Fiction* (Routely, 1979, pp. 11–12).
11. For a somewhat different approach to the factor of make-believe, see also Currie (1990).
12. Hyphenation-blindness generally comes in degrees. From the stipulatively bold "true-in-S" (e.g., "true-in-intuitionistic logic"), there is a further stage, which we might call adverbilization (e.g., "intuitionistically true"), thence to adjectivisms ("an intuitionistic truth"), and, finally, in the absence of rival adjectivisms, we have *substantiation* (e.g., "a truth").

Chapter 7

1. In Russell (1899, p. 410–12), Russell distinguishes between a philosophical and a mathematical approach to concepts. The mathematician's function is to assert relationships between concepts. The philosopher's function is to determine what the concepts mean. Nicholas Griffin calls attention to this passage and suggests that the distinction at hand undermines Russell's remarks on the purpose of the philosophy of mathematics in the Preface to *Principles of Mathematics* (Griffin, 1991, p. 272). I am unable to see that this is so.

   Frege did not learn this philosophical analysis at the feet of Moore or from his early correspondence with Russell. It was up and running in 1902 in a sufficiently robust condition to make his response to Russell immediate and decisive. We have lost the concept of set; knowledge of sets is impossible. For some years, I was prepared to conjecture that Frege's analytical proclivities were a Kantian inheritance; that Frege had come to think of courses of values as (Kantian) *categories*, as conditions of the very possibility of arithmetical knowledge. I owe to Griffin the comfort of knowing that this is how Russell in his early analytic days viewed mathematical concepts (Griffin, 1977, p. 273). Prior to the Paradox, Russell thought that the basic concepts of mathematics are categories. In the case of geometry, basic concepts are categories of intuition, fixing the spatio-temporal conditions for experience. Both here and in pure logic, categories are fundamental and irreducible. These are Russell's views in "On the Axioms of Geometry," which is a self-described neo-Kantian enterprise. For arithmetic, quantity is a category. Pure categories cannot be defined, and they possess "an unanalyzable and intuitively apprehended meaning" (Griffin, 1977, p. 163). Cf. Russell (1903/1937, p. xv).

The indefinable concepts listed in Russell (1899) are addition, number, order, equality, less, and greater. And *manifold*.

2. At each level, we might wish to speak of the sets lodged there as proper subsets of English. For this to be strictly true, there would have to be a set of all English sentences, hence a set Σ of all true English sentences. But consider:

$$\text{(i) (i)} \notin \Sigma$$

If (i) is true, it is and is not a member of Σ. If not true, it is and is not a member of Σ. This may incline to suppose that Σ does not exist which is precisely the moral of the Knower's Paradox. See Kaplan and Montague (1960).

3. For some recent notable work, apart from that cited in Chapter 5, see: Barwise and Etchemendy, 1984; Burge, 1984; Feferman, 1984; Gaifman, 1988; Gupta, 1984; Gupta and Belnap, 1993; Herzberger, 1970; McGee, 1990; Parsons, 1984; and Simmons, 1993.

4. See Curry, 1941 and Curry, 1942. My presentation derives from that of Boolos and Jeffrey (1989, p. 186). Cf. Barwise and Etchemendy (1984). As Boolos and Jeffrey point out, the proof mimics that of Löb's theorem, which asserts that $\ulcorner Bew\,(\Phi) \rightarrow \Phi \urcorner$ is a theorem of L only if Φ is a theorem of L ('Bew' denotes the property of being a logical truth): Löb's own proof makes use of Gödel's self-referential lemma: For any formula $\ulcorner \Phi\,(x, v_1, \ldots, v_n) \urcorner$ of L, there is a presentable formula $\ulcorner \psi\,(v_1, \ldots, v_n) \urcorner$ such that it is provable in Robinson arithmetic that

$$\forall v_1, \ldots, \forall v_n(\psi(v_1, \ldots, v_n) \leftrightarrow \ulcorner \psi \urcorner, v_1, \ldots v_n)).$$

It could be said that Löb's theorem stands to the Curry Paradox as Tarski's indefinability theorem stands to the Liar. See M. H. Löb, "Solution of a Problem of Leon Henken," *Journal of Symbolic Logic*, 20 (1955), 115–18. A lemma-free version of the proof, in which Gödel's second incompleteness theorem is assumed, was discovered by Kripke. See further McGee (1990, p. 47).

5. And, in saying earlier that it does not matter whether LLT is true, it is clear that this must mean that it does not matter for the question whether there is a *primary datum* that motivates dialethic logic.

6. Those familiar with the approach to the Liar taken by hyperset theorists such as Barwise and Moss will see an affinity with the approach developed in this

chapter. Suppose that λ is a Liar sentence ⌜¬**True**ₕ **(this)**⌝. Assume that there are contexts K in which **this** actually does refer to λ, **h** actually does refer to K ('**h**' means 'here'), and, Excluded Middle obtains. Barwise and Moss point out that most "solutions" (sneer-quotes are theirs) to the Liar require that one of these three assumptions be abandoned. Tarski's approach gives up on the first assumption. Gappers give up on the third assumption. Barwise and Moss propose to drop the second assumption. "Proposals which see the problem in some sort of indexicality or context sensitivity can be seen as giving up (2) [= this second assumption]. The idea, roughly, is that the act of asserting the Liar about the whole world results in some sort of pragmatic shift in context, a shift that is overlooked in the reasoning that seems to lead to paradox. Put crudely, here (as referred to by **h**) before the claim and here after the claim are slightly different." Shades of *Reconciliation* (Barwise and Moss, 1996, p. 188).

7. Also bearing on the place of the Liar in IF logic is the collapse of Convention T, of which we had a brief glimpse pages ago. I shall not take the time to expatiate on this interesting development.

8. It is known that Paul duBois-Reymond made use of diagonal arguments two decades before Cantor's innovation. There is, however, a difference between the two types of diagonal reasoning that justifies Cantor's paternity. In Cantor's diagonalization, but not duBois-Reymond's, the concept of "countervalue" plays a central role. See, for example, duBois-Reymond (1873) and Grattan-Guinness (1978).

9. For details of the proof, see Kalish, Montague, and Mar (1980).

10. Richard, 1967, p. 143. See also Russell (1967), in which Richard's Paradox is solved "by remarking that 'all definitions' is an illegitimate notion."

11. Hintikka, 1996, p. 176. Indeed, "the apparent dependence of Tarski-type truth definitions on set theory is in my view one of the most disconcerting features of the current scene in logic and the foundations of mathematics. I am sorely tempted to call it 'Tarski's curse'" (Hintikka, 1996, p. 16).

Chapter 8

1. Lakatos, 1980, p. 42.

2. This is a crude picture, of course, with most of the subtlety left out. But it will do for my purposes here. Those wishing to regain some of the subtlety would be rewarded by consulting Suppes (1960) and Suppes (1962). See also Starmans's notion of a *conceptual model* for a normative theory (Starmans, 1996, pp. 22–6).

3. An interesting problem emerges in so supposing, although I shall not deal with it here. It is the problem of fitting together actual performance and ideal performance by way of *describable* approximation relations.

4. Cohen, 1986, p. 201. The example *is* implausible, and is for illustrative purposes only. In actual theoretical practice, L would fail because of its incompatibility with pretheoretical data. Any theorist of argument "already knows what everyone else already knows"; that question-begging is not a good thing.

5. The extent to which the social sciences are normative is still a matter of lively controversy. Keynes in a letter to Harrod want[s] "to emphasize strongly the point about economics being a moral science. I mentioned before that it deals with introspection and with values. I might have added that it deals with motives,

expectations and psychological uncertainties" (Keynes, 1973, p. 300). Against this, many economists see their enterprise as "value free." Against *this*, is the classic riposte of Myrdal (1970). There is little point in allowing these contentions to distract us. Some social scientists who see their work as value-free evidently incorporate into their theories *normative concepts*, such as rationality, and often they let their ideal models authorize prescriptions for rational conduct. Even so, they see their work as nonnormative in the further sense that, for example, one can describe a mechanism for reducing unemployment without in any sense endorsing or recommending it (but see again Myrdal [1970]). For our purposes, it is enough that the discussion admit instances of social scientific theories that are *normative* in first sense but not the second. In the case of economics, there is an attractive classification in Suppes (1960, p. 606).

	Individual Decisions	*Group Decisions*
Normative theory	Classical economics	Game theory
	Statistical decision theory	Welfare economics
	Moral philosophy	Political theory
	Experimental decision studies	Social Psychology
Descriptive theory	Learning theory	Political science
	Survey studies of voting behavior	

6. Exceptions are theorists such as Keynes, for whom economics is a branch of logic. Another exception is the Austrian school of economics, for which the fundamental postulates of economics are synthetic a priori. See von Mises (1949) and von Mises (1981).

7. With Socrates, for example, who repeatedly made the following *autoepistemic* argument: If there were knowledge of such things, I more than anyone would have it. But I have no such knowledge. Therefore, there is no knowledge of such things. Furthermore, since theories are a bridge from belief to knowledge, there are no theories about these matters. Indeed, we just have our beliefs and the dialectical mechanisms that drive and reshape them. Wisdom is knowing this.

8. It is worth emphasizing that the otioriness of the ideal model methodology is compatible with asserting the normative soundness of theories, such as pragma-dialectics, it seeks to serve. My complaint here is not about pragma-dialective norms, but, rather, about the pragma-dialectical penchant for ideal models.

9. Readers having no acquaintance with such theories may wish to consult, for example, French (1986).

10. After all quantum theory accurately describes the color of gases, the heat potential of solids, and the nature of the chemical bond. It explains the periods in the periodic table of the elements, "it can fathom the electrical properties of conductors, semiconductors, and insulators; it gives a theoretical foundation for lasers and masers, superconductors and superfluids; it adumbrates neutron diffraction and electron imaging; it fixes numbers of properties of the elementary particles and of radioactivity; and so on.

The difficulty had by economists in hitting empirical targets was noticed by Mill (1833/1967), in what is still one of the best essays on the methodology of economics yet to be written. A hundred and two years later, Keynes was noticing the same thing but not apologizing for it. Mill and Keynes had the same diagnosis.

To get smooth and powerful economic laws, the economist must engage in a hefty abstraction from the raw world. Beyond a certain point abstraction is distortion. One of the more interesting empirical failures of neoclassical economics was the discovery (Lester, 1946 and 1947) that for-profit companies do *not* seek to maximize profits.

11. Short of psychological theories about the pd_i themselves.

12. True, intuitionistic negation fails the Double Negation Law, but there is no reason to think that our subject is operating intuitionistically in the Fred-inference.

13. "Is That All There Is?," a song by J. Lieber and M. Stoller, popularized by the vocalist Peggy Lee.

14. The distinction is drawn expressly in [Rawls, 1974/1975, p. 8]. Cohen rejects wide equilibrium in 1981 (p. 320).

15. Consider, for example, Catherine Elgin (1996).

16. In the case of the Gambler's fallacy, there is concurrence between B and I, an error each time. Conjunctivitis also shows the concurrence of B and I, also a mistake each time. The check-for-consistency principle fails the B-test and passes the I-test, rightly in the first instance and mistakenly in the second. Cherniak's more subdued principle *Consistency* passes the B-test and (probably) fails the I-test, rightly in the first instance and mistakenly in the second. *Modus ponens* fails the B-test and passes the I-test, rightly in the first instance and mistakenly in the second. And so on.

17. Stein, 1994. Cf. a rather unconvincing reply by Cohen [1994].

18. On expert reflective equilibrium, see Stich and Nisbett (1980). For a recantation, see Stich (1990).

19. Cf. the situation in set theory. For at least a generation, post-Paradox set theorists allowed that their beliefs about sets were not underwritten by BI-intuitions. Recall Quine's strong anti-intuitionism in mathematics and his acceptance of artificial devices. Today things appear to be different. Here again is Ullian: "Doing set theory nowadays all but means working in ZF or some extension of it. The subject has become a sophisticated and beautiful part of mathematics. For many it is underlain by what is regarded as a *highly intuitive concept*, the iterative notion of set . . . [D. M.] Martin calls it 'the standard concept of set' . . . " (Ullian, 1986, p. 584, emphasis added).

20. Not to overlook his subscription to a benign form of Moore's Paradox, also in that form a version of the Preface Paradox. What the theorist aims for is a theory that confirms what he thinks he already knows. But if he is at all *au fait* with the dynamics of theory construction he will recognize that as the theory matures it may give him principled occasion to reject some of what he thought he knew at the beginning.

21. Prior to the 1930s, it was staple of neoclassical economics that an extended depression was not possible. Keynes sought empirical reconciliation (to the extent that such, for him, was possible and desirable in economics) in an alternative theory that long since has borne his name. Keynesian economics, for all its attractions, did not put neoclassicism into early retirement. Neoclassical economics has had enormous staying power. What it did to accommodate the problem of lengthy depressions was to split itself into macroeconomics and microeconomics (which is probably inconsistent with *Keynesian* macroeconomics!).

22. Harman tells this story in detail (1977).

Bibliography

Ackermann, W., "Beggrüdung einer strengere Implikation," *Journal of Symbolic Logic*, 21 (1956), 113–28.

Ainslie, G., "A Behavioral Economic Approach to the Defense Mechanism: Freud's Energy Theory Revisited," *Social Science Information*, 21, 6 (1982), 735–79.

Alchourrón, C. E., Gärdenfors, P., and Makinson, D., "On the Logic of Theory Change: Partial Meet Functions for Contraction and Revision.," *Journal of Symbolic Logic*, 50 (1985), 510–30.

Anderson, Alan Ross, "Mathematics and the 'Language Game'," *Review of Metaphysics*, 11 (1958), 446–58.

Anderson, Alan Ross, and Belnap, Nuel D. Jr., *Entailment: The Logic of Relevance and Necessity*, Vol. 1, Princeton, N.J.: Princeton University Press, 1975.

Anderson, Alan Ross, and Belnap, Nuel D. Jr., "A Simple Treatment of Truth Functions," *The Journal of Symbolic Logic*, 24 (1959), 301–12.

Anderson, Alan Ross, and Belnap, Nuel D. Jr., *Entailment: The Logic of Relevance and Necessity*, Vol. 2, Princeton, N.J.: Princeton University Press, 1995.

Apostoli, Peter, and Brown, Bryson, "A Solution to the Completeness Problem for Weakly Aggregative Modal Logic," *The Journal of Symbolic Logic*, 60–3 (1995), 832–42.

Aristotle, *Sophistical Refutations* and *Prior Analytics, Rhetoric, On Interpretation, Posterior Analytics* and *Metaphysics*, in Jonathan Barnes (ed.), *The Complete Works of Aristotle*, The Revised Oxford Translation, Vols. 1 and 2., Princeton, N.J.: Princeton University Press, 1984.

Arrow, Kenneth J., *Social Choice and Individual Values*, New York: John Wiley, 1951.

Avron, Arnon, "Whither Relevance Logic?," *Journal of Philosophical Logic*, 21 (1992), 243–81.

Bar-Hillel, Y. (ed.), *Logic, Methodology and Philosophy of Science*, Amsterdam: North-Holland, 1965.

Barnes, Jonathan (ed.), *The Complete Works of Aristotle*, The revised Oxford translation Vol. 1 and 2., Princeton, N.J.: Princeton University Press, 1984.

Barth, E. M., *The Logic of Articles in Traditional Philosophy*, Dordrecht and Boston: Reidel, 1974.

341

Barth, E. M., and Krabbe, E. C. W. (eds.), *Logic and Political Culture*, Amsterdam: North-Holland, 1990.

Barth, E. M., and Krabbe, E. C. W., *From Axiom to Dialogue: A philosophical study of logic and argumentation.*, Berlin: de Gruyter, 1992.

Bartlett, F. C., *Remembering*, Cambridge, Mass.: Cambridge University Press, 1932.

Bartley, W. W., III (ed.), *The Retreat to Commitment*, revised and enlarged, LaSalle, Ill.: Open Court, 1984.

Barwise, John (ed.), *Handbook of Mathematical Logic*, Amsterdam: North-Holland, 1977.

Barwise, John, and Etchemendy, John, *The Liar*, Oxford: Oxford University Press, 1984.

Barwise, John, and Moss, Lawrence, *Vicious Circles*, Stanford, Calif.: CSLI Publications 1996.

Batens, D., "Extensional Paraconsistent Logics," *Logique et Analyse*, 90–91 (1980), 195–234.

Becherand, W., and Essler, W. K. (eds.), *Konzepte der Dialektik*, Frankfurt/Main: Klostermann, 1981.

Bell, John S., "On the Einstein-Podolsky-Rosen Paradox," *Physics*, 1 (1964), 195.

Bell, J. L., "Category Theory and the Foundation of Mathematics," *British Journal of the Philosophy of Science*, 32 (1981), 349–58.

Belnap, Nuel D., Jr., *An Analysis of Questions: Preliminary Report*, Santa Monica, Calif.: System Development Corporation, 1963.

Belnap, Nuel D., Jr., "Gupta's Rule of Revision Theory of Truth," *Journal of Philosophical Logic*, 11 (1982), 103–16.

Belnap, Nuel D., Jr., and Steel, Thomas B., *The Logic of Questions and Answers*, New Haven, Conn.: Yale University Press, 1976.

Benacerraf, Paul, "Mathematical Truth," *Journal of Philosophy*, LXX (1973), 661–79.

Benardete, José A., *Metaphysics*, Oxford: Oxford University Press, 1989.

Beth, Evert W., "Towards An Up-to-date Philosophy of Natural Sciences," *Methods* 1 (1949), 178–84.

Birkhoff, G., "Lattices in Applied Mathematics," *American Mathematical Society*, Proceedings of Symposia in Pure Mathematics, 2 (1961), 155–84.

Black, Max, "A Critical Review of [Quine's] Mathematical Logic," *Mind*, 52 (1943), 264–75.

Bläu, Ulrich, *Die dreiwertige Logik der Sprache: Ihre Syntax, Semantik und Anwendung in der Sprachanalyse*, Berlin: deGruyer, 1978.

Bohm, David, *Special Theory of Relativity*, Amsterdam: W. A. Benjamins, 1965.

Boolos, George S., "On Second Order Logic," *Journal of Philosophy*, 72, 16 (1975), 509–27.

Boolos, George, "Saving Frege From Contradiction," in Richard Jeffrey (ed.), *Logic, Logic and Logic*, Cambridge, Mass.: Harvard University Press, 1998, 171–82.

Boolos, George, and Jeffrey, Richard, *Computability and Logic*, 3rd edition, Cambridge: Cambridge University Press, 1989.

Brady, R. T., "Depth Relevance of Some Paraconsistent Logics," *Studia Logica*, 43 (1984), 63–74.

Brady, R. T., "The Non-triviality of Dialectical Set Theory," in Priest, Routley, and Norman (eds.), *Paraconsistent Logic: Essays on the Inconsistent.*, Münich: Philosophia Verlag, 1989, 437–71.

Brady, R. T., *Universal Logic*, unpublished.

Bradley, F. H., *The Principles of Logic*, two volumes, Oxford: Oxford University Press, 1883; corrected impression 1922.

Brown, Bryson, "Simple Natural Deduction for Weakly Aggregative Paraconsistent Logics," in Dov Gabbay (ed.), *Frontiers of Paraconsistent Logic*, London: Research Studies Press, 2001.

Brown, Bryson, "Approximate Truth: A Paraconsistent Account," in Joke Meheus (ed.), *Inconsistency in Science*, vol. 2 in the series *Origins: Studies in the Sources of Scientific Creativity*, Amsterdam: Kluwer Academic, 2002.

Brown, Bryson, and Schotch, Peter, "Logic and Aggregation," *The Journal of Philosophical Logic*, 28 (1999), 265–87.

Burali-Forti, "Una Questione Sui Numeri Transfiniti," *Rendiconti del Circolo Mathematico di Palermo*, 11 (1897), translated as "A Question on Transfinite Numbers," Jean van Heijenoort (trans.), in van Heijenoort (ed.), *From Frege to Gödel*, Cambridge, Mass.: Harvard University Press, 1967, 104–12.

Burge, Tyler, "Semantic Paradox," *Journal of Philosophy*, 76 (1979), 169–98, reprinted in Robert L. Martin (ed.), *Recent Essays on Truth and The Liar Paradox*, Oxford: Oxford University Press, 1984, 83–117.

Burgess, John P., "Introduction to Part II," in Richard Jeffrey (ed.), *Logic, Logic and Logic*, Cambridge, Mass.: Harvard University Press, 1998, 135–42.

Cantor, George, "Beiträge zer Begründung der transfiniten Mengenlehre," parts one and two, *Mathematischen Annalen*, 46 (1895), 481–512 and 49 (1897), 207–46; translated as *Contributions to the Founding of the Theory of Transfinite Numbers*, by P. E. B. Jourdain (trans.), Chicago: Open Court, 1915; reprinted New York: Dover, 1952, 85–201.

Cantor, George, "Ueber eine elementare Frage der Mannigfaltigkeitslehre," *Jahresbericht der Deutschen Mathematiker-Vereinigung* 1 (1890–91), 75–8, reprinted in Ernst Zermelo (ed.), *Gesammelte Abhandlungen mathematischen und philosophischen Inhalts*, Berlin: Springer, 1932, 278–81.

Cargile, James, *Paradoxes*, Cambridge: Cambridge University Press, 1979.

Carlsen, Laurie, *Dialogue Games: An Approach to Discourse Analysis.*, Dordrecht: Reidel, 1983.

Carlsen, Laurie, and ter Meulen, Alice, "Informational Independence in Intensional Contexts" in E. Saarinen et al. (eds.), *Essays In Honor of Jaakko Hintikka*, Dordrecht: Reidel, 1979, 61–72.

Carlson, Gregory N., and Pelletier, Francis Jeffry (eds.), *The Generic Book*, Chicago: Chicago University Press, 1995.

Carnap, Rudolf, *The Logical Syntax of Language*, London: Routledge & Kegan Paul, 1937.

Cartwright, N., *How the Laws of Physics Lie*, Oxford: Clarendon Press, 1983.

Cartwright, Richard, "Implications and Entailments," in Richard Cartwright, *Philosophical Essays,* Cambridge, Mass.: MIT Press, 1987, 237–58.

Cartwright, Richard, *Philosophical Essays*, Cambridge, Mass.: MIT Press, 1987.

Cartwright, Richard L., "Propositions," R. J. Butler (ed.), *Analytical Philosophy*, 1st Series, Oxford: Basil Blackwell, 1962, 81–103.

Cartwright, Richard L., "Propositions Again," *Nôus* 2 (1968), 229–46.

Castañeda, Hector-Neri, "Fiction and Reality: Their Fundamental Connections," in John Woods and Thomas Pavel (eds.), *Formal Semantics and Literary Theory*, a special issue of *Poetics* 8 (1979), 31–62.

Cézanne, Paul, in W. Hess (ed.), *Über die Kunst: Gespräche mit Gasquet*, Mittenwald: Mäander, 1980.

Chellas, Brian F., *Modal Logic: An Introduction.*, Cambridge: Cambridge University Press, 1980.

Cherniak, Christopher, *Minimal Rationality*, Cambridge, Mass.: MIT Press, 1986.

Church, Alonzo, "The Weak Theory of Implication," in Albert Menne, Alexander Wilhelmy, and Helmut Angstil (eds.), *Untersuchungen zum Logikkalkül und zur Logik der Einzelwissenschaften*, Freiburg: Kontrolliertes Denken, 1951, 22–37.

Churchland, Paul M., "Eliminative Materialism," *Journal of Philosophy*, 78 (1981), 67–90.

Cohen, L. Jonathan, "On the Psychology of Prediction: Whose Is the Fallacy?," *Cognition* 7 (1979), 385–407.

Cohen, L. Jonathan, "Can Human Irrationality Be Experimentally Demonstrated?," *The Behavior and Brain Sciences* 4 (1981), 317–70.

Cohen, L. Jonathan, *The Dialogue of Reason: An Analysis of Analytical Philosophy*, Oxford: Clarendon, 1986.

Cohen, L. Jonathan, *An Introduction to the Philosophy of Induction and Probability*, Oxford: The Clarendon Press, 1989.

Cohen, L. Jonathan, "Reply to Stein," *Synthese* 99 (1994), 173–6.

Collins, A., and Quillam, M., "Retrieval Time From Semantic Memory," *Journal of Verbal Learning and Verbal Behavior*, 8 (1969), 240–9.

Craig, William, "On Axiomatizability Within a System," *Journal of Symbolic Logic* 18 (1953), 1, 30–2.

Craig, William, "Three Uses of the Herbrand-Gentzen Theorem," *Journal of Symbolic Logic* 22 (1957), 269–85.

Currie, Gregory, *The Nature of Fiction*, Cambridge: Cambridge University Press, 1990.

Curry, Haskell B., "The Paradox of Kleene and Rosser," *Transactions of the American Mathematical Society*, 50 (1941), 454–516.

Curry, Haskell B., "The Inconsistency of Certain Formal Logics," *Journal of Symbolic Logic*, 7 (1942), 115–117.

da Costa, Newton C. A., "On the Theory of Inconsistent Formal Systems," *Notre Dame Journal of Formal Logic*, XV (1974), 497–510.

Dalla Chiara, M. L., "The Relevance of Quantum Logic," *Proceedings of the International Congress for Logic, Methodology and the Philosophy of Science*, Salzburg: North-Holland, forthcoming.

Davidson, Donald, "Theories of Meaning and Learnable Languages," in Y. Bar-Hillel (ed.), *Logic, Methodology, and Philosophy of Science*, Amsterdam: North Holland, 1965, 383–94.

Davidson, Donald, "Truth and Meaning," *Synthese*, 17 (1967), 304–23.

Davidson, Donald, "Psychology as Philosophy," in J. Glover (ed.), *The Philosophy of Mind*, Oxford: Oxford University Press, 1976, 101–10.

DeVidi, David, and Solomon, Graham, "On Confusions About Bivalence and Excluded Middle," *Dialogue: Canadian Philosophical Review*, xxxviii (1999), 785–99.

Diaz, M. Richard, *Topics in the Logic of Relevance*, Munich: Philosophia Verlag, 1981.

Doležel, Lubomír, "Extensional and Intentional Narrative Worlds," John Woods and Thomas Pavel (eds.), *Formal Semantics and Literary Theory*, a special issue of *Poetics*, 8 (1979), 193–211.

duBois-Reymond, Paul, "Eine neue Theorie der Convergenz und Divergenz von Reihen mit poitiven Gliedern," *Journal für die reine und angewandte Mathematik* 76 (1873), 61–91.

Dummet, Michael, *Frege: Philosophy of Language*, London: Duckworth, 1973.

Dunn, J. Michael, "Relevant Logic and Entailment," in D. Gabbay and F. Guenthner (eds.), *Handbook of Philosophical Logic*, Vol. 3: Alternatives to Classical Logic, *Synthese Library*, Vol. 166, Dordrecht and Boston: Kluwer Academic Publishers, 986, 117–224.

Dunn, J. Michael, "Intuitive Semantics for First Degree Entailments and 'Coupled Trees,'" *Philosophical Studies*, 29 (1976), 149–68.

Eagleton, Terry, *The Illusions of Postmodernism*, Oxford: Blackwell, 1996.

Einstein, Albert, "Ist die Traegheit eines Koerpers von seinem Energieinhalt abhaengig?," *Annalen der Physik* 18 (1905), 639–41.

Elgin, Catherine Z., *Considered Judgement*, Princeton, N.J.: Princeton University Press, 1996.

Elster, Jon, *Sour Grapes: Studies in the Subversion of Rationality*, Cambridge: Cambridge University Press, 1985.

Emerson, Ralph Waldo, "Self Reliance," in *Essays*, London: Aldine Press, 1967.

Feferman, Solomon, "Toward Useful Type-Free Theories, I," *Journal of Symbolic Logic*, 49 (1984), 75–111, reprinted in Robert L. Martin (ed.), *Truth and the Liar Paradox*, Oxford: Clarendon Press, 1984, 237–87.

Festinger, Leon, *A Theory of Cognitive Dissonance*, Stanford, Calif.: Stanford University Press, 1957.

Feyerabend, Paul, *Against Method: Outline of an Anarchistic Theory of Knowledge*, London: Verso, 1978.

Fine, Kit, "Incompleteness for Quantified Relevance Logics," in Jean Norman and Richard Sylvan (eds.), *Directions in Relevant Logic*, Dordrecht: Kluwer Academic, 1989, 205–25.

Flew, Anthony (ed.), *Essays in Conceptual Analysis*, London: Macmillan, 1960.

Forrest, Peter, *Quantum Metaphysics*, Oxford: Blackwell, 1988.

Frege, Gottlob, 1893–1903 partial translation as *The Basic Laws of Arithmetic: Exposition of the System* (Montgomery Furth, ed. and trans.), Berkeley and Los Angeles: University of California Press, 1964.

Frege, Gottlob, *Grundgeseze der Arithmetik*, Jena: Pohle, 1903; partial translation as *The Basic Laws of Arithmetic: Exposition of the System*, Montgomery Furth (trans. and ed.), Berkeley and Los Angeles: University of California Press, 1964.

Frege, Gottlob, Letter to Russell, in van Heijenoort (ed.), *From Frege to Gödel*, Cambridge, Mass.: Harvard University Press, 1967, 127–8.

Frege, Gottlob, *Foundations of Arithmetic*, J. L. Austin (trans.), Oxford: Basil Blackwell, 1950.

French, Simon, *Decision Theory: An Introduction to the Mathematics of Rationality*, Chichester, U.K.: Ellis Horwood, 1986.

Friedman, Harvey, and Meyer, Robert K., "Can We Implement Relevant Arithmetic?," *Technical Report TF-ARP-12/88*, Automated Reasoning Project, Research School of Social Science, Australian National University, 1988.

Frye, Northrop, *The Great Code*, Toronto: Academic Press, 1981.

Fuhrmann, A., *Relevant Logics, Modal Logics and Theory Change*, Ph.D. Thesis, Australian National University, 1988.

Gabbay, Dov M., "Dynamics of Practical Reasoning: A Position Paper," in *Advances in Modal Logic 2*, K. Segerberg, M. Zkharayashev, M. de Rijke, and H. Wansing, (eds.), CSLI Publications, Cambridge: Cambridge University Press, 2001, 179–224.

Gabbay, Dov M., and Guenthner, F. (eds.), *Handbook of Philosophical Logic*, Vol. 3, Dordrecht and Boston: Reidel, 1984.

Gabbay, Dov M., and Hunter, Anthony, "Making Inconsistency Respectable, Part I," in P. Jorrand and J. Kelemen (eds.), *Fundamentals of Artificial Intelligence Research (FAIR '91)*, LNAI, Vol. 535, Springer-Verlag, 1991.

Gabbay, Dov M., and Hunter, Anthony, "Making Inconsistency Respectable, Part II," in M. Clarke, R. Kruse, and S. Seraffin (eds.), *Symbolic and Quantitative Approaches to Reasoning and Uncertainty: European Conference ECSQARU '93*, Granada, Spain, LNCS, Vol. 747, Berlin: Springer-Verlag, 1993.

Gabbay, Dov M., and Woods, John, "Non-cooperation in Dialogue Logic: Getting Beyond the Goody Two-Shoes Model of Argument," *Synthese*, forthcoming.

Gabbay, Dov M., and Woods, John, "More on Non-cooperation in Dialogue Logic," *Logic Journal of the IGPL*, 9 (2001), 321–39.

Gabriel, Gottfried, "Fiction: A Semantic Approach," *Poetics: Journal of Empirical Research on Literature, the Media and the Arts*, 8.2 (1979), 245–65.

Gaifman, Haim, "Operational Pointer Semantics: Solution to Self-referential Puzzlers I," M. Vardi (ed.), *Proceedings of the Second Conference on Theoretical Aspects of Reasoning About Knowledge*, Los Altos, Calif.: Morgan Kaufman, 1988, 43–60.

Gäredenfors, Peter, *Knowledge in Flux: Modeling the Dynamics of Epistemic States*, Cambridge, Mass.: MIT Press, 1987.

Garey, Michael R., and Johnson, David S., *Computers and Intractability*, San Francisco: W. H. Freeman, 1979.

Peter, Geach, "Entailment," *Proceedings of the Aristotelian Society Supplementary* Vol. 32 (1958), 143–72; reprinted in Peter Geach, *Logic Matters*, Oxford: Blackwell, 1972, 174–86.

Gellner, Ernest, *Words and Things*, Harmondsworth, U.K.: Penguin, 1968.

Gibbins, Peter, "A User-Friendly Quantum Logic," *Logique et Analyse*, 112 (1985), 353–62.

Gilles, Donald, *Philosophy of Science in the Twentieth Century: Four Central Themes*, Oxford: Blackwell, 1993.

Gilbert, Margaret, *On Social Facts*, Princeton, N.J.: Princeton University Press, 1993.

Gilmore, P. C., "The Consistency of Partial Set Theory Without Extensionality," in Dana Scott (ed.), *Symposium in Pure Mathematics*, University of California, Los Angeles, 1967; Axiomatic Set Theory, *Proceedings of Symposia in Pure Mathematics*, XIII Pt. 2, American Mathematical Society, 1974.

Girle, Roderic A., "Dialogue and Entrenchment," *Proceedings of the Sixth Florida Artificial Intelligence Research Symposium*, Fort Lauderdale, 1993, 105–89.

Girle, Roderic A., "Knowledge: Organized and Disorganized," *Proceedings of the Seventh Florida Artificial Intellegence Research Symposium*, Pensacola Beach, Fla., 1994, 198–203.

Girle, Roderic A., "Commands in Dialogue Logic," *Proceedings of the International Conference on Formal and Applied Practical Reasoning*, Bonn, 1996, 246–60.

Girle, Roderic A., "Belief Sets and Commitment Stores," in Hans V. Hansen, Christopher W. Tindale, and Athena V. Colman (eds.), *Argumentation and Rhetoric*, St. Catharines, Ont.: OSSA, 1998.

Glover, J. (ed.), *The Philosophy of Mind*, Oxford: Oxford University Press, 1976.

Gochet, Paul, *Ascent to Truth*, Münich: Philosophia Verlag, 1986.

Gödel, Kurt, "What Is Cantor's Continuum Problem?," *American Mathematical Monthly*, 54 (1947), 515–25.

Goghossian, Dave, "What the Sokal Hoax Ought to Teach Us," *Times Literary Supplement December* 13 (1996), 14–15.

Goodman, Nelson, *Fact, Fiction and Forecast*, 2nd ed., New York: Bobbs-Merrill, 1965.

Grattan-Guiness, Ivor, "How Bertrand Russell Discovered His Paradox," *Historia Mathematica*, 5 (1978), 127–37.

Griffin, Nicholas, *Russell's Idealist Apprenticeship*, Oxford: Clarendon Press, 1991.

Gupta, Anil, "Truth and Paradox," in Robert L. Martin (ed.) *Recent Essays on Truth and the Liar Paradox*, Oxford: Clarendon Press, 1984, 175–236.

Gupta, Anil, and Belnap, Nuel D., Jr., *The Revision of Theory of Truth*, Cambridge, Mass.: MIT Press, 1993.

Haack, Susan, *Deviant Logic*, Cambridge: Cambridge University Press, 1974.

Hacking, Ian, "What Is Logic?," *The Journal of Philosophy*, LXXVI (1979), 285–319.

Hacking, Ian, *Representing and Intervening*, Cambridge: Cambridge University Press, 1983.

Hahn, Lewis Edwin (ed.), *The Philosophy of W. V. Quine*, second rev. ed., LaSalle, Ill.: Open Court, 1998.

Hahn, Lewis Edwin, and Schilpp, Paul Arthur (eds.), *The Philosophy of W. V. Quine*, LaSalle, Ill.: Open Court, 1986.

Hallett, Michael, *Cantorian Set Theory and Limitation of Size*, Oxford: Clarendon Press, 1984.

Hamblin, C. L., *Fallacies*, London: Methuen, 1970.

Hansen, Hans V., and Pinto, Robert C. (eds.), *Fallacies: Classical and Contemporary Readings*, University Park: Pennsylvania State University, 1995.

Harman, Gilbert, "Is Modal Logic Logic?," *Philosophia* 2 (1972), 75–84.

Harman, Gilbert, *The Nature of Morality*, Oxford: Oxford University Press, 1977.

Harman, Gilbert, *Change in View*, Cambridge, Mass.: MIT Press, 1986.

Harrah, David, *Communication: A Logical Model*, Cambridge, Mass.: MIT Press, 1963.

Hart, W. D., "Access and Inference," in W. D. Hart (ed.), *Philosophy of Mathematics*, Oxford: Oxford University Press, 1996, 52–62.

Hausman, Daniel M., "Introduction" in Daniel M. Hausman (ed.), *The Philosophy of Economics: An Anthology*, Cambridge: Cambridge University Press, 1984, 1–50.

Heintz, John, "Reference and Inference in Fiction," John Woods and Thomas Pavel (eds.), *Formal Semantics and Literary Theory*, a special issue of *Poetics* 8 (1979), 85–99.

Henkin, Leon, "Some Remarks on Infinitely Long Formulas," in *Infinite Models*, New York: Pergamon, 1961, 167–83.

Henson, C. Ward, "Reviews of Jensen, R. B. reply to Quine and Quine to Jenson," *The Journal of Symbolic Logic*, 40 (1975), 241–2.

Hermes, Hans, Kambartel, Friedrich, and Kaulbach, Friedrich (eds.), *Frege: Posthumous Writings*, Chicago: University of Chicago Press, 1979.

Herzberger, Hans G., "Paradoxes of Grounding in Semantics," *Journal of Philosophy* 67 (1970), 145–67.

Herzberger, Hans G., "Notes on Naive Semantics," *Journal of Philosophical Logic* 11 (1982), 62–102, reprinted in Robert L. Martin (ed.), *Recent Essays on Truth and The Liar Paradox*, Oxford: Clarendon Press, 1984, 134–74.

Hilbert, David, *Grundlagen der Geometrie*, Leipzig: Teubner; translated as *The Foundations of Geometry*, E. J. Townsend (trans.), Chicago: Open Court, 1902; 10th edition, with a supplement by Paul Bernays translated under the previous English title, La Salle, Ill.: Open Court, 1971.

Hinman, P., Kim, J., and Stich, S., "Logical Truth Revisited," *The Journal of Philosophy*, LXV (1968), 495–500.

Hintikka, Jaakko, *Knowledge and Belief*, Ithaca, N.Y.: Cornell University Press, 1962.

Hintikka, Jaakko, "A Counter-example to Tarski-type Truth definitions as Applied to Natural Languages," *Philosophia*, 5 (1975), 207–12.

Hintikka, Jaakko, *The Semantics of Questions and the Questions of Semantics* [= *Acta Philosophica Fernica* 28 (1976)].

Hintikka, Jaakko, "The Role of Logic in Argumentation," *The Monist*, 72 (1989), 3–24.

Hintikka, Jaakko, "Information-Seeking Dialogues," in W. Becherand and W. K. Essler (eds.), *Konzepte der Dialektik*, Frankfurt/Main: Klostermann, 1981, 212–31.

Hintikka, Jaakko, *The Principles of Mathematics Revisited*, Cambridge: Cambridge University Press, 1996.

Holland, J. H., Holyoak, K. J., Nisbett, R. E., and Thagard, P. R., *Induction: Processes of Inference, Learning and Discovery*, Cambridge, Mass.: MIT Press, 1987.

Hooker, C. A. (ed.), *The Logico-Algebraic Approach to Quantum Mechanics, Vol. 1: Historical Evolution*, Dordrecht: Reidel, 1979.

Hookway, Christopher, *Quine*, Oxford: Polity Press, 1988.

Howe, M., *Introduction to Human Memory*, New York: Harper & Row, 1970.

Howell, Robert, "Fictional Objects: How They Are and How They Aren't," in John Woods and Thomas Pavel (eds.), *Formal Semantics and Literary Theory*, a special issue of *Poetics* 8 (1979), 50–72.

Howell, Robert, "Review of [Parsons's] 'Non-existent Objects,'" *Journal of Philosophy*, 80 (1983), 163–73.

Hughes, G. E., and Cresswell, M. J., *An Introduction to Modal Logic*, London: Methuen, 1968.

Hume, David, *A Treatise of Human Nature*, Ernest C. Mossner (ed.), Harmondsworth, U.K.: Penguin Books, 1969.

Hylton, Peter, *Russell, Idealism and the Emergence of Analytic Philosophy*, Oxford: Clarendon Press, 1990.

Iseminger, Gary, "Relatedness Logic and Entailment," *The Journal of Non-Classical Logic* 3 (1986), 5–23.

Jacquette, Dale, "Logical Dimensions of Question-Begging Argument," *American Philosophical Quarterly* 30 (1993), 317–27.

Jáskowski, Leon, "Propositional Calculus for Contradictory Systems," *Studia Logica* XXVI (1969), 143–57.

Jech, Thomas J., "About the Axiom of Choice," in Jon Barwise (ed.), *Handbook of Mathematical Logic*, Amsterdam: North-Holland, 1977, 345–70.

Johnson-Liard, P. H., and Byrne, R. M. J., *Deduction*, Hillsdale, N.J.: Erlbaum, 1991.

Johnson-Laird, P. N., and Wason, P. C. (eds.), *Thinking: Readings in Cognitive Science*, Cambridge: Cambridge University Press, 1977.

Johnstone, Henry W., Jr., *Validity and Rhetoric in Philosophical Argument*, University Park, Penn.: The Dialogue Press of Man and World, 1978.

Johnston, David, and Jennings, R. E., "Paradox Tolerant Logics" *Logique et Analyse* 26 (1983), 291–308.

Kalish, Donald, Montague, Richard, and Mar, Gary, *Logic: Techniques of Formal Reasoning*, 2nd ed., San Diego, Calif.: Harcourt Brace Jovanovich, 1980.

Kanigel, Robert, *The Man Who Knew Infinity: A Life of the Genius Ramanujan*, New York: Scribner, 1991.

Kant, Imanuel, *Inquiry Concerning the Distinctness of Principles of Natural Theology and Morality, and Logic*, Indianapolis, Ind.: Bobbs-Merrill, 1764/1800/1974.

Kant, Imanuel, Hartman, Robert S., and Schwartz, Wolfgang (trans. with introduction) *Logic*, Indianapolis, Ind.: Bobbs-Merrill, 1974.

Kaplan, David, "Bob and Carl and Ted and Alice," in Jaakko Hintikka et al. (eds.), *Approaches to Natural Languages*, Dordrecht: Reidel, 1973, 490–518.

Kaplan, David, and Montague, Richard, "A Paradox Regained," *Notre Dame Journal of Formal Logic* 1 (1960), 79–90; reprinted in Richard Montague, *Formal Philosophy*, New Haven, Conn.: Yale University Press, 1974, 271–85.

Kaufmann, W., "Uber die Konstitution des Elektrons," *Annalen der Physik*, 19 (1906), 487.

Keynes, John Maynard, "Letter to R. F. Harrod, 4 July 1938," in Moggridge (ed.), *Collected Writings of John Maynard Keynes*, Vol. 14, *The General Theory and After: Part II: Defense and Development*, London: Macmillan, 1973.

A., Klatzley, *Human Memory: Structures and Processes*, San Francisco: W. H. Freeman, 1975.

Kline, Morris, *Mathematics: The Loss of Certainty*, New York: Oxford University Press, 1980.

Koryé, A., "The Liar," *Philosophy and Phenomenological Research*, 6 (1946), 344–62.

Kripke, Saul, "Outline of a Theory of Truth," *Journal of Philosophy*, 72 (1975), 690–716 reprinted in Robert L. Martin (ed.), *The Paradox of the Liar*, Oxford: Clarendon Press, 1984, 134–74.

Kripke, Saul A., *Naming and Necessity*, Cambridge, Mass.: Harvard University Press, 1980.

Lakatos, Imre, *Mathematics, Science and Epistemology*, Cambridge and New York: Cambridge University Press, 1980.

Lambert, Karel, "Logic and Microphysics," in Karel Lambert (ed.), *The Logical Way of Doing Things*, New Haven, Conn.: Yale University Press, 1969, 93–117.

Lester, R., "Shortcomings of Marginal Analysis for Wage-Employment Problems," *American Economic Review*, 36 (1946), 63–82.

Lester, R., "Marginalism, Minimum Wages and Labour Markets," *American Economic Review*, 37 (1947), 135–48.

Lewis, C. I., "Implication and the Algebra of Logic," *Mind* 21 (1912), 522–531. Collected Papers, pp. 351–359.

Lewis, C. I., and Langford, C. H., *Symbolic Logic*, New York: Dover, 1932.

Lewis, David, "Truth in Fiction," *American Philosophical Quarterly*, 15 (1978), 37–46, reprinted with postscripts in David Lewis (ed.), *Philosophical Papers*, Vol. 1, Oxford: Oxford University Press, 1983, 261–75.

Lewis, David K., *Convention*, Cambridge, Mass.: Harvard University Press, 1969.

Lieber, J., and Stoller, M., "Is that all there is?," recorded on Capital Records by Peggy Lee in 1969.

Lindsay, P., and Norman, D., *Human Information Processing*, New York: Academic Press, 1977.

Lindstrom, Per, "On Extensions of Elementary Logic," *Theoria*, XXXV (1969), 1–11.

Löb, M. H., "Solution of a Problem of Leon Henken," *Journal of Symbolic Logic*, 20 (1955), 115–18.

Locke, John, *Essay On Human Understanding*, Peter H. Niddith (ed. with introduction), Oxford: Clarendon Press, 1975.

Lorentz, Hendrik Antoon, *The Theory of Electrons and Its Application to the Phenomena of Light and Radiant Heat*, 3rd ed., New York: Dover, 1952.

Mackenzie, J. D., "Question-Begging in Non-Cumulative Systems," *Journal of Philosophical Logic*, 8 (1979), 117–33.

Mackenzie, J. D., "Begging the Question in Dialogue," *Australian Journal of Philosophy*, 62 (1984), 174–81.

Martin, Donald, "Review of Jean van Heijenoort (ed.), *From Frege to Gödel*," *Journal of Philosophy*, 67 (1970), 113.

Martin, Robert L. (ed.), *Recent Essays On Truth and the Liar Paradox*, Oxford: Clarendon Press, 1984.

Martin, Robert, and Woodruff, Peter, "On Representing 'True-in-L' in L," *Philosophia* 5 (1975), 213–17, reprinted in Robert L. Martin (ed.), *The Paradox of the Liar*, Oxford: Clarendon Press, 1984, 47–51.

McGee, Vann, *Truth, Vagueness and Paradox*, Indianapolis, Ind.: Hackett, 1990.

McKinsey, J. C. C., Sugar, A. C., and Suppes, Patrick, "Axiomatic Foundations of Classical Particle Mechanics," *Journal of Rational Mechanics and Analysis*, 2 (1953), 253–72.

Menne, Albert, Wilhelmy, Alexander, and Angstil, Helmut (eds.), *Untersuchungen zum Logikkalkül und zur Logik der Einzelwissenschaften*, Freiburg: Kontrolliertes Denken, 1951

Mermin, N. B., "Quantum Mysteries for Everyone," *Journal of Philosophy*, 78 (1981), 397–408.

Meyer, R. K., "Entailment," *The Journal of Philosophy*, 68 (1971), 808–18.

Meyer, Robert, and Routley, Richard, "Dialectical Logic and Consistency of the World," *Studies in Soviet Thought*, 16 (1975), 1–25.

Mill, John Stuart, "On the Definition of Political Economy and the Method of Investigation Proper to It," in John Robson (ed.), *Collected Works of John Stuart Mill, Vol. IV, Essays on Economics and Society 1824–1845*, Toronto: Toronto University Press, 1967, 309–39.

Mill, John Stuart, *A System of Logic Ratiocinative and Inductive*, in John Robson (ed.) *Collected Works of John Stuart Mill, Vol. VIII Book II of Reasoning and Book III of Induction*, Toronto: University of Toronto Press, 1974, 157–640.

Millikan, Ruth Garrett, *Language, Thought, and Other Biological Categories: New Foundations for Realism*, Cambridge, Mass.: MIT Press, 1984.

Minsky, Marvin, *The Society of the Mind*, New York: Simon and Schuster, 1987.

Moore, G. E., "The Nature of Judgement," *Mind* 7 (1898), 176–93.

Moore, G. E., "Critical Notice on Russell's *An Essay on the Foundations of Geometry*," *Mind* 8 (1899), 397–405.

Moore, G. E., *Principia Ethica*, Cambridge, Mass.: Cambridge University Press, 1903.

Moore, G. E., "The Refutation of Idealism," *Mind* 12 (1903), 433–53; reprinted in G. E. Moore, *Philosophical Studies*, London: Routledge and Kegan Paul, 1922, 1–30.

Moore, G. E., *Philosophical Studies*, London: Routledge and Kegan Paul, 1922.

Moore, G. E., *Philosophical Papers*, London: George Allen and Unwin, 1959.

Moore, G. E., "Russell's Theory of Descriptions," in G. E. Moore, *Philosophical Papers*, London: George Allen and Unwin, 1959, 151–95.

Mortensen, Chris, *Inconsistent Mathematics*, Dordrecht and Boston: Kluwer, 1995.

Mortensen, Chris, "Paraconsistency and C_1," in Graham Priest, Richard Routely, and Jean Norman (eds.), *Paraconsistent Logic: Essays on the Inconsistent*, Münich: Philosophia Verlag, 1989, 289–305.

Myrdal, Gunnar, *Objectivity in Social Research*, London: Duckworth, 1970.

Nagel, Ernest, Suppes, Patrick, and Tarski, Alfred (eds.), *Logic, Methodology and Philosophy of Science*, Stanford, Calif.: Stanford University Press, 1962.

Nekham, Alexander, *De Naturis Rerum*, T. Wright (ed.), London: Longman, 1863.

Newell, Allen, and Simon, Herbert A., *Human Problem Solving*, Englewood Cliffs, N.J.: Prentice Hall, 1972.

Nisbett, R., and Ross, L., *Human Inference: Strategies and Shortcomings of Social Judgement*, Englewood Cliffs, N.J.: Prentice Hall, 1980.

Norman, Jean, and Sylvan, Richard, *Directions in Relevant Logic*, Dordrecht: Kluwer Academic, 1989.

Nozick, Robert, "Experience, Theory and Language," in Lewis Edwin Hahn and Paul Arthur Schilpp (eds.), *The Philosophy of Quine*, LaSalle, Ill.: Open Court, 1986, 339–63.

Parsons, Charles, "The Liar Paradox," *Journal of Philosophical Logic*, 3 (1974), 381–412, reprinted in Robert L. Martin (ed.), *Truth and The Liar Paradox*, Oxford: Clarendon Press, 1984, 9–45.

Parsons, Terence, *Non-existent Objects*, New Haven, Conn.: Yale University Press, 1980.

Parsons, Terence, "True Contradictions," *Canadian Journal of Philosophy*, 20 (1990), 335–82.

Pavel, Thomas, *Fictional Worlds*, Cambridge, Mass.: Harvard University Press, 1986.

Peacocke, Christopher, "What Is a Logical Constant?", *The Journal of Philosophy*, LXXIII (1976), 221–40.

Plantinga, Alvin, *The Nature of Necessity*, Oxford: Oxford University Press, 1974.

Plantinga, Alvin, *Warrant and Proper Function*, Oxford: Oxford University Press, 1993.

Plato, *Meno*, in E. Hamilton and H. Cairns (eds.), *Plato: The Collected Dialogues.*, Princeton, N.J.: Princeton University Press, 1980, 353–84.

Post, John F., "A Gödelian Theorem for Theories of Rationality," in G. Radnitzky and W. W. Bartley, III (eds.), *Evolutionary Epistemology, Rationality, and the Sociology of Knowledge*, LaSalle, Ill.: Open Court, 1987.

Priest, Graham, "The Logic of Paradox," *Journal of Philosophical Logic*, 8 (1979), 219–41.

Priest, Graham, *In Contradiction: A Study of the Transconsistent*, Dordrecht: Kluwer Academic, 1987.

Priest, Graham, *Beyond the Limits of Thought*, Cambridge, Mass.: Cambridge University Press, 1995.

Priest, Graham, and Routley, Richard, "Applications of Paraconsistent Logic," in Graham Priest, Richard Routley, and Jean Norman (eds.) *Paraconsistent Logic: Essays on the Inconsistent*, München: Philosophia Verlag, 1989, 367–93.

Priest, Graham, and Routley, Richard, "Systems of Paraconsistent Logic" in Graham Priest, Richard Routley, and Jean Norman (eds.), *Paraconsistent Logic: Essays on the Inconsistent*, Münich: Philosophia Verlag, 1989, 151–86.

Priest, Graham, Routley, Richard, and Norman, Jean, *Paraconsistent Logic: Essays on the Inconsistent*, Münich: Philosophia Verlag, 1989.

Putnam, Hilary, " The Logic of Quantum Mechanics," in *Mathematics, Matter and Method: Philosophical Papers, Vol. I*, Cambridge: Cambridge University Press, 1975, 174–97.

Quine, W. V., "Truth by Convention," in O. H. Lee (ed.), *Philosophical Essays for A. N. Whitehead*, New York: Longmans, 1936, 90–124.

Quine, W. V., "Notes on Existence and Necessity," *Journal of Philosophy*, 40 (1943), 113–27.

Quine, W. V., "The Problem of Interpreting Modal Logic," *Journal of Symbolic Logic*, 12 (1947), 43–8.

Quine, W. V., "Review of Geach, Subject and Predicate," *Journal of Symbolic Logic*, 16 (1951), 138.

Quine, W. V., *Mathematical Logic*, rev. ed., Cambridge, Mass.: Harvard University Press, 1951.

Quine, W. V., *From a Logical Point of View*, Cambridge, Mass.: Harvard University Press, 1953.

Quine, W. V., "New Foundations for Mathematical Logic," reprinted with additions in *From a Logical Point of View*, Cambridge, Mass.: Harvard University Press, 1953, 80–101.

Quine, W. V., *Word and Object*, Cambridge, Mass.: MIT Press, 1960.

Quine, W. V., *The Ways of Paradox and Other Essays*, rev. and enlarged ed., Cambridge, Mass.: Harvard University Press, 1976.

Quine, W. V., *Selected Logic Papers*, New York: Random House, 1966; enlarged edition, Cambridge, Mass.: Harvard University Press, 1995.

Quine, W. V., *Ontological Relativity and Other Essays*, New York: Columbia University Press, 1969.

Quine, W. V., *Set Theory and Its Logic*, rev. ed., Cambridge, Mass.: Harvard University Press, 1969.

Quine, W. V., *Philosophy of Logic*, Englewood Cliffs, N.J.: Prentice Hall, 1970; second ed., Cambridge, Mass.: Harvard University Press, 1986.

Quine, W. V., *Roots of Reference: The Paul Carus Lectures*, LaSalle, Ill.: Open Court 1973.

Quine, W. V., "Reply to Manley Thompson," in Lewis Edwin Hahn and Paul Arthur Schilpp (eds.), *The Philosophy of W. V. Quine*, LaSalle, Ill.: Open Court 1986, 564–8.

Quine, W. V., "Empirically Equivalent Systems of the World," *Erkenntnis* 9 (1975), 313–28.

Quine, W. V., "Reply to Robert Nozick," in Lewis Edwin Hahn and Paul Aurthur Schilpp (eds.), *The Philosophy of W. V. Quine*, LaSalle, Ill.: Open Court, 1986, 364–8.

Quine, W. V., *Theories and Things*, Cambridge, Mass.: Harvard University Press, 1981.

Quine, W. V., "Things and Their Place in Theories," *Theories and Things*, Cambridge, Mass.: Harvard University Press, 1981, 1–23.

Quine, W. V., "Ontology and Ideology Revisited," *Journal of Philosophy* 80 (1983), 499–501.

Quine, W. V., *Quiddities*, Cambridge, Mass.: Harvard University Press, 1987.

Quine, W. V., *The Pursuit of Truth*, revised edition, Cambridge, Mass.: Harvard University Press, 1990/1992.

Quine, W. V., "Structure and Nature," *The Journal of Philosophy*, LXXXIX (1992), 5–9.

Quine, W. V., *From Stimulus to Science*, Cambridge, Mass.: Harvard University Press, 1995.

Quine, W. V., and Ullian, J. W., *The Web of Belief*, New York: Random House, 1970.

Raiffa, Howard, *Decision Analysis*, Reading, Mass.: Addison-Wesley, 1968.

Ramsey, Frank P., *The Foundations of Mathematics and Other Logical Essays*, London: Routledge and Kegan Paul, 1931.

Rawls, John, *A Theory of Justice*, Cambridge, Mass.: Harvard University Press, 1971.

Rawls, John, "The Independence of Moral Theory," *Proceedings and Addresses of the American Philosophical Association* 48 (1974–5), 5–22.

Read, Stephen, *Relevant Logic: A Philosophical Examination of Inference*, Oxford: Blackwell, 1988.

Rescher, N. (ed.), *Studies in the Philosophy of Mind*, Oxford: Basil Blackwell, 1972.

Rescher, Nicholas, and Brandon, Robert, *The Logic of Inconsistency: A Study in Non-standard Possible-world Semantics and Ontology*, Oxford: Basil Blackwell, 1980.

Richard, Jules, "The Principles of Mathematics and the Problem of Sets," in Jean van Heijenoort (ed.), *From Frege to Gödel*, Cambridge, Mass.: Harvard University Press 1967, 143–4.

Rips, Lance J., *The Psychology of Proof: Deductive Reasoning in Human Thinking*, Cambridge, Mass.: MIT Press, 1994.

Ross, W. D., *Aristotle*, Oxford: Clarendon Press, 1953.

Rosser, Barkley, "The Burali-Forti Paradox," *Journal of Symbolic Logic*, 7 (1942), 1–17.

Routley, Richard, "Some Things Do Not Exist," *Notre Dame Journal of Formal Logic*, VII (1966), 251–76.

Routley, Richard, "The Semantic Structure of Fictional Discourse," in John Woods and Thomas Pavel (eds.), *Formal Semantics and Literary Theory*, a special issue of *Poetics*, 8 (1979), 3–30.

Routley, Richard, and Routley, Val, "The Semantics of First Degree Entailment," *Noûs*, 6 (1972), 335–59.

Russell, Bertrand, "On The Axioms of Geometry," in Nicholas Griffin and Albert C. Lewis (eds.), *The Collected Papers of Bertrand Russell Volume II, Philosophical Papers 1896–99*, London: G. Allen and Unwin, 1983, 390–415.

Russell, Bertrand, *A Critical Exposition of the Philosophy of Leibniz*, Cambridge: Cambridge University Press, 1900; republished London: George Allen and Unwin, 1937.

Russell, Bertrand, *The Principles of Mathematics*, 2nd ed., London: Allen and Unwin, 1937.

Russell, Bertrand, "On Denoting," *Mind* N.S. XIV (1905), 479–93.

Russell, Bertrand, *An Essay on the Foundations of Geometry*, New York: Dover, 1956.

Russell, Bertrand, *My Philosophical Development*, London: Allen and Unwin, 1959.

Russell, Bertrand, *Our Knowledge of the External World: As a Field for Scientific Method in Philosophy*, London: Allen and Unwin, 1961.

Russell, Bertrand, "Mathematical Logic as Based on the Theory of Types," in Jean van Heijenoort (ed.), *From Frege to Gödel*, Cambridge, Mass.: Harvard University Press, 1967, 152–82.

Ryle, Gilbert, "Are There Propositions?," in *Collected Papers 2: Collected Essays 1929–1968*, London: Hutchinson and Co., 1971, 12–38. Reprinted from the *Proceedings of the Aristotelian Society*, vol. xxx, 1930.

Ryle, Gilbert, *Philosophical Arguments*, Oxford: Clarendon Press, 1945.

Ryle, Gilbert, "Ordinary Language," *The Philosophical Review* LXII (1953), 167–86.

Ryle, Gilbert, "Proofs in Philosophy," in Gilbert Ryle (ed.) *Collected Papers 2: Collected Essays 1929–1968*, London: Hutchinson, 1972, 319–25.

Ryle, Gilbert, "Use, Usage and Meaning," *Proceedings of the Aristotelian Society*, supplementary vol. XXXV (1961), 223–42.

Sainsbury, R. M., *Paradoxes*, Cambridge: Cambridge University Press, 1988.

Savage, Leonard J., *The Foundations of Statistics*, New York: John Wiley, 1954.

Schotch, P. K., "Paraconsistent Logic: The View from the Right," *Philosophy of Science Association* 2 (1993), 421–9.

Schotch, P. K., and Jennings, R. E., "On Detonating," in Graham Priest, Richard Routley, and Jean Norman (eds.), *Paraconsistent Logic: Essays on the Inconsistent*, Munich: Philosophia, 1989, 306–27.

Sen, Amartya, "Social Choice Theory," in Kenneth J. Arrow and M. D. Intriligator (eds.), *Handbook of Mathematical Economics*, Vol. III, Los Altos, Calif.: Morgan Kaufmann, 1986, 1070–181.

Sen, Amartya, *On Ethics and Economics*, Oxford: Blackwell, 1987.

Sher, Gila, *The Bounds of Logic*, Cambridge, Mass.: MIT Press, 1991.

Shoenfield, Joseph R., *Mathematical Logic*, Reading, Mass.: Allison-Wesley, 1967.

Simmons, Keith, *Universality of the Liar*, Cambridge: Cambridge University Press, 1993.

Skolem, Th., "Untersuchungen über die Axiome des Klassenlalkuss," in Th. Skolem (ed.), *Selected Works in Logic*, Oslo: Universitetsforlaget 1970, 67–101.

Slater, B. H., "Paraconsistent Logics?," *Journal of Philosophical Logic*, 24 (1995), 451–4.

Smiley, Timothy J., "Entailment and Deducibility," *Proceedings of the Aristotelian Society Supplementary* Vol. 32 (1958), 123–42.

Soames, Scott, *Understanding Truth*, Oxford: Oxford University Press, 1999.

Sorenson, Roy A., *Blindspots*, Oxford: Clarendon Press, 1988.

Sperber, Dan, and Wilson, Deirdre, *Relevance*, Oxford: Blackwell, 1986.

Starmans, Richard, *Logic, Argumentation and Common Sense*, Tilburg: Tilburg University Print, 1996.

Stein, Edward, "Rationality and Reflective Equilibrium," *Synthese*, 99 (1994), 137–72.

Stich, Stephen, *The Fragmentation of Reason*, Cambridge, Mass.: MIT Press, 1990.

Stich, Stephen, and Nisbett, Richard, "Justification and Psychology of Human Reasoning," *Philosophy of Science*, 47 (1980), 188–202.

Suppes, Patrick, *Introduction to Logic*, New York: Van Nostrand Reinhold, 1957.

Suppes, Patrick, "Models of Data," in Ernest Nagel, Patrick Suppes, and Alfred Tarski (eds.), *Logic, Methodology and Philosophy of Science*, Stanford, Calif.: Stanford University Press, 1962, 252–61.

Suppes, Patrick, "A Comparison of the Meaning and Uses of Models in Mathematics and the Empirical Sciences," *Synthese* 12 (1960), 287–301.

Suppes, Patrick, "The Philosophical Relevance of Decision Theory," *The Journal of Philosophy*, 58 (1960), 606–14.

Suppes, Patrick, "Logics Appropriate to Empirical Theories," in C. A. Hooker (ed.), *The Logico-Algebraic Approach to Quantum Mechanics Vol I: Historical Evaluation*, Dordrecht: Reidel, 1979, 329–40.

Suppes, Patrick, *Studies in the Methodology and Foundations of Science*, New York: Humanities Press, 1969.

Tarski, Alfred, "On the Concept of Logical Consequence," in Alfred Tarski, *Logic, Semantics, Metamathemathics: Papers from 1923 to 1938*, J. H. Woodger (trans.), Oxford: Oxford University Press: 2nd ed., John Corcoran (ed.), Indianapolis, Ind.: Hackett, 1983, 409–20.

Tarski, Alfred, *Logic, Semantics, Metamathematics: Papers from 1923 to 1938*, J. H. Woodger (trans.), Oxford: Oxford University Press: 2nd ed., John Corcoran (ed.), Indianapolis, Ind.: Hackett, 1983.

Tarski, Alfred, "The Concept of Truth in Formalized Languages," in Alfred Tarski, *Logic, Semantics, Metamathematics: Papers from 1923 to 1938*, J. H. Woodger (trans.), Oxford: Oxford University Press: 2nd ed., John Corcoran (ed.), Indianapolis, Ind.: Hackett, 1983, 152–278.

Tennant, Neil, "Natural Deduction and Gentzen Sequent Systems for Intuitionistic Relevant Logic," *Journal of Symbolic Logic*, 52 (1987), 665–80.

Tennant, Neil, *Autologic*, Edinburgh: Edinburgh University Press, 1993.

ter Meulen, Alice G. B. (ed.), *Studies in Model Theoretic Semantics*, Dordrecht: Foris, 1983.

Turkle, Sherry, "Artificial Intelligence and Psychoanalysis," in Graubard (ed.), *The Artificial Intelligence Debate: False Starts, Real Foundations*, Cambridge, Mass.: MIT Press, 1988, 260–1.

Ullian, Joseph S., "Quine and the Field of Mathematical Logic," Lewis E. Hahn and Paul A. Schilpp (eds.), *The Philosophy of W. V. Quine*, LaSalle, Ill.: Open Court, 1986, 569–89.

Urmson, J. O., *Philosophical Analysis: Its Development Between the Two World Wars*, Oxford: Clarendon Press, 1956.

Urquhart, Alasdair, "The Undecidability of Entailment and Relevant Implication," *Journal of Symbolic Logic* 49 (1984), 1059–73.

van Benthem, Johan, "Five Easy Pieces," in Johan van Benthem and Alice G. B. ter Meulen (eds.), *Studies in Model Theoretic Semantics*, Dordrecht: Foris, 1983, 1–17.

van Benthem, Johan, "What Is Dialectical Logic?," *Erkenntnis* 14 (1979), 333–47.

van Benthem, Johan, and ter Meulen, Alice G. B. (eds.), *Generalized Quantifiers in Natural Language*, Dordrecht: Foris, 1985.

van Eemeren Frans H., et al., *Fundamentals of Argumentation Theory: A Handbook of Historical Backgrounds and Contemporary Developments*, Mahwah, N.J.: Lawrence Erlbaum Associates, 1996.

van Eemeren, Frans H., and Grootendorst, Rob, *Speech Acts in Argumentation Discussions*, Dordrecht: Foris, 1984.

van Fraassen, Bas C., "Meaning Relations Among Predicates," *Noûs* 1 (1967), 161–79.

van Fraassen, Bas C., *Formal Semantics and Logic*, New York: Macmillan, 1971.

van Fraassen, Bas C., "The Labyrinth of Quantum Logics," in C. A. Hooker (ed.), *The Logico-Algebraic Approach to Quantum Mechanics Vol. I: Historical Evolution*, Dordrecht: Reidel, 1975, 577–607.

van Fraassen, Bas C., *The Scientific Image*, Oxford: Clarendon Press, 1980.

van Heijenoort, Jean (ed.), *From Frege to Gödel: A Source Book of Mathematical Logic, 1879–1931*, Cambridge, Mass.: Harvard University Press, 1967.

Van Inwagen, Peter, "Creatures of Fiction," *American Philosophical Quarterly*, 14 (1997), 299–308.

Veyne, Paul, *Le Pain et le Cirque*, Paris: Seuil, 1976.

von Mises, Ludwig, *Human Action: A Treatise on Economics*, New Haven, Conn.: Yale University Press, 1949.

von Mises, Ludwig, *Epistemological Problems of Economics*, G. Riseman (trans.), New York: New York University Press, 1981.

von Neumann, John, *Mathematical Foundations of Quantum Mechanics*, Princeton, N.J.: Princeton University Press, 1953.

von Wright, G. H., "The Concept of Entailment," *Logical Studies*, London: Routledge and Kegan Paul, 1957, 166–91.

Walton, D. N., *Logical Dialogue – Games and Fallacies*, London: University Press of America, 1984.

Walton, D. N., "Burden of Proof," *Argumentation*, 2 (1988), 233–54.

Walton, D. N., and Krabbe, E. C. W., *Commitment in Dialogue*, Albany: State University of New York, 1995.

Walton, Douglas, *A Pragmatic Theory of Fallacy*, Tuscaloosa: University of Alabama Press, 1995.

Walton, Kendall, *Mimesis as Make-Believe: On the Foundations of the Representational Arts*, Cambridge, Mass.: Harvard University Press, 1990.

Wang, Hao, "Quine's Logical Ideas in Historical Perspective," in Lewis E. Hahn and Paul A. Schilpp (eds.), *The Philosophy of W. V. Quine*, LaSalle, Ill.: Open Court, 1986, 623–43.

Whately, Richard, *Elements of Logic*, 1st edition, Book 3, Section 13, New York: Harper and Row, 1826/1853.

Wilson, N. L., "Substances Without Substrata," *Review of Metaphysics*, 12 (1959), 521–39.

Wittgenstein, Ludwig, *Tractatus Logico-Philosophicus*, London: Routledge and Keagan Paul, 1958.

Wittgenstein, Ludwig, *Remarks on the Foundations of Mathematics*, Oxford: Blackwell, 1964.

Wittgenstein, Ludwig, in C. Diamond (ed.), *Wittgenstein's Lectures on the Foundations of Mathematics, Cambridge, 1939*, Hassocks, U.K.: Harvester Press 1976.

Woods, John, "The Contradiction-Exterminator," *Analysis*, XXV (1965), 49–53.

Woods, John, "Fictionality and the Logic of Relations," *The Southern Journal of Philosophy*, 7 (1969), 51–64.

Woods, John, "Essentialism, Self-Identity, and Quantifying In," in Milton K. Muntitz (ed.), *Identity and Individuation*, New York: New York University Press, 1971, 165–98.

Woods, John, *The Logic of Fiction: A Philosophical Sounding of Deviant Logic*, The Hague and Paris: Mouton, 1974.

Woods, John, "Identity and Modality," *Philosophia* V (1975), 69–120.

Woods, John, "Critical Notice of Deviant Logic," *Canadian Journal of Philsophy* VII (1977), 651–66.

Woods, John, *Engineered Death: Abortion, Suicide, Euthanasia, Senecide*, Ottawa, Ont.: University of Ottawa Press, 1978.

Woods, John, "Ideals of Rationality in Dialogic," *Argumentation*, 2 (1988), 419–24.

Woods, John, "The Relevance of Relevant Logic," in J. Norman and R. Sylvan (eds.), *Directions in Relevant Logics*, Dordrecht: Kluwer Academic Publisher, 1989, 77–86.

Woods, John, "And So Indeed Are Perfect Cheat," *Argumentation*, 9 (1995), 654–68.

Woods, John, "Semantic Intuitions" in Johan van Bentham, Frans H. van Eemeren, Rob Grootendorst, and Frank Veltman (eds.), *Logic and Argumentation*, Amsterdam: North-Holland, 1996, 179–208.

Woods, John, "A Captious Nicety of Argument: The Philosophy of W. V. Quine," in Lewis E. Hahn (ed.), *The Philosophy of W. V. Quine*, Library of Living Philosophers, 2nd rev. ed., LaSalle, Ill.: Open Court Press, 1998, 687–725.

Woods, John, *Aristotle's Earlier Logic*, Oxford: Hermes Science Publications, 2001.

Woods, John, and Pavel, Thomas (eds.), *Formal Semantics and Literary Theory*, a special issue of *Poetics* 8 (1979).

Woods, John, and Peacock, Kent, "What Physics Does to Logic: The Current State of Quantum Logic Is Incompatible With," forthcoming.

Woods, John, and Walton, Douglas, "Arresting Circles in Formal Dialogues," *Journal of Philosophical Logic* 7 (1978), 73–90; reprinted in John Woods and Douglas Walton, *Fallacies: Selected Papers 1972–1982*, Berlin: Foris/deGruyter, 1988, 143–159.

Woods, John, and Walton, Douglas, "Question-Begging and Cumulativeness in Dialectical Games," *Nôus* 16 (1982), 585–606; reprinted in John Woods and Douglas Walton, *Fallacies: Selected Papers 1972–1982*, Berlin: Foris/deGruyter 1988, 253–72.

Woods, John, and Walton, Douglas, *Argument: The Logic of Fallacies.*, Toronto: McGraw-Hill Ryerson, 1982.

Woods, John, and Walton, Douglas, *Fallacies: Selected Papers 1972–1982*, Berlin: Foris/de Gruyter, 1988.

Wright, Crispin, *Frege's Conception of Numbers as Objects*, Aberdeen: Aberdeen University Press, 1983.

Zalta, Edward N., *Abstract Objects: An Introduction to Axiomatic Metaphysics*, Dordrecht: Reidel, 1983.

Zermelo, Ernst, "Beweis, das jede Menge wohlgeoronet werden kann," *Mathematischen Annalen* 59 (1904), 514–16; translated as "Proof That Every Set Can Be Well-ordered," S. Bauer-Mengelburge (trans.), in Jean van Heijenvoort (ed.), *From Frege to Gödel: A Source Book in Mathematical Logic, 1879–1931*, Cambridge, Mass.: Harvard University Press, 1967, 139–41.

Zermelo, Ernst, "Uber Grenzzahlen und Megenbereiche: neue Untersuchungen über die Grundlagen der Mengenlehre," *Fundamentae Mathematicae*, 16 (1930), 29–47.

Zucker, J., "The Adequacy Problem for Classical Logic," *Journal of Philosophical Logic*, 7, 4 (1978), 517–35.

Index